WASHINGTON IN TIME

75	1900	1925	1950	1975	2000
	1876 Olmsted, Sr. Plan U. S. Capitol grounds 1890 Rock Creek Park authorized by Congress 1893 Permanent System of Highways Act	1901-02 Senate Park / McMillan Commission Plan (park system, The Mall, Federal Triangle) 1910 Commissionof Fine Arts 1913 Rock Creek and Potomac Parkway authorized by Congress 1920 Zoning Regulations	1926 Public Buildings Act (authorized Federal Triangle) 1928 Potomac River Parks Plan 1928-43 National Capital Park and Planning Commission: Park system (The Mall, Federal Triangle, Rock Creek Parkway) 1950-74 Urban Renewal Project Plans	1958 Current DC Zoning Regulations 1961 National Capital Planning Commission Plan: Year 2000 Policy Plan 1966 NCRPC Plan: Regional Development Guide, 2000 1972 Pennsylvania Ave. Commission 1972-95 NCPC Plans: Comprehensive Plan for National Capital (Federal and District Elements) 1974-75 DC Self-Government and Governmental Reorganization Act (Home Rule, ANCs formed)	1974-96 PADC Plans: Pennsylvania Ave Development Corporation 1985 Krier Plan: Master Plan for the Bicentennial: Year 2000 1986-94 DC Ward Plans 1991 Miller Plan: 'Visions of Washington' Composite Plan of Urban Interventions NCPC Federal Sector Plans for Monumental Core 1998 NCPC Plan: Legacy Framework
	Bartholdi Fountain (1876-1877) Peace-Naval Monument (1877) Tidal Basin (1882-1897)	Union Station (1908) Lincoln Memorial (1911-1922) Ulysses S. Grant Memorial (1921)	DC World War I Memorial (1931) Supreme Court Building (1935) Jefferson Memorial (1943)	Iwo Jima Memorial (1954) Theodore Roosevelt Memorial (1960) Kennedy Memorial Grave (1966) John F. Kennedy Center for Performing Arts (1971) ...ary McLeod Bethune Statue (1974)	...terans Memorial (1982) ...Gateway (1986) ...ial (1990) ... Enforcement Officers' ...1) ...terans Memorial (1995) ...in Delano Roosevelt Memorial (1997) African-American Civil War Memorial (1997-1998) Martin Luther King Memorial (proposed-2000)
	1878 Congress creates DC "Organic Act," a municipal corporation form of government 1887 The Catholic University of America established 1889 American University established 1898 Spanish-American War 1900 *Population 278,718*	1901 Senate Park Commission established 1910 Height Limitation Act established 1912 Cherry trees for Tidal Basin donated by Japan 1917 US enters WWI 1918 WWI ends 1920 Congress establishes Zoning Commision in DC	1926 National Capital Park and Planning Commission established Public Buildings Act leads to construction of many federal buildings 1930 Capper-Cranston Act (DC regional parks and parkway program) Shipstead-Luce Act (regional highway systems plan) 1934 DC Harbor Regulations 1941 US enters WWII 1945 WWII ends DC Redevelopment Act creates Redevelopment Land Agency 1950 Korean War begins *Population 800,000*	1953 Korean War ends 1958 Japan donates a Lantern and Pagoda commemorating Commodore Perry's voyage to open Japan (1853-54) 1961 23rd Amendment to Constitution ratified, giving Washington residents right to vote in presidential elections 1963 President Kennedy assassinated; Martin Luther King Jr. leads "March on Washington for Jobs and Freedom", delivers "I Have a Dream" speech at Lincoln Memorial 1964 Capital Beltway completed 1968 Martin Luther King assassinated resulting in riots 1970 Anti-Vietnam War demonstrators on The Mall reach record numbers 1974 President Nixon resigns in wake of Watergate scandal	1976 US celebrates Bicentennial The Metro, Washington's subway system, begins operation 1979 Farmers demonstrate for farm relief on The Mall 1986 Commemorative Works Act 1993 Bill proposing DC Statehood defeated in Congress 1995 Pennsylvania Ave. closed to motorists to protect White House 2000 *Population 650,000; area population 3,400,000*

...1

...JR, 1881-1885

...EVELAND, 1885-1889

...ENJAMIN HARRISON, 1889-1893

GROVER CLEVELAND, 1893-1897

WILLIAM MCKINLEY, 1897-1901

THEODORE ROOSEVELT, 1901-1909

WILLIAM HOWARD TAFT, 1909-1913

WOODROW WILSON, 1913-1921

WARREN GAMALIEL HARDING, 1921-1923

CALVIN COOLIDGE, 1923-1929

HERBERT CLARK HOOVER, 1929-1933

FRANKLIN DELANO ROOSEVELT, 1933-1945

HARRY S. TRUMAN, 1945-1953

DWIGHT DAVID EISENHOWER 1953-1961

JOHN FITZGERALD KENNEDY, 1961-1963

LYNDON BAINES JOHNSON, 1963-1969

RICHARD MILHOUS NIXON...

GERALD RUDOLP...

JAMES EA... CARTER, JR., 1977-1981

RONALD WILSON REAGAN, 1981-1989

GEORGE HERBERT WALKER BUSH, 1989-1993

WILLIAM JEFFERSON CLINTON, 1993-2001

GEORGE WALKER BUSH, 2001-

WASHINGTON IN MAPS

1606–2000

IRIS MILLER

First published in the United States of America in 2002 by
RIZZOLI INTERNATIONAL PUBLICATIONS, INC.
300 Park Avenue South, New York, NY 10010

ISBN: 0-8478-2477-0
Library of Congress Control Number: 2002102946

Cover: The A. Sachse View and Vignettes (p. 106)
Page 3: Plan de la Ville de Washington (p. 51)
Page 4, left: E. Sachse View of the Capitol (p. 80)
Page 4, right: The Colton Atlas (p. 82)
Page 5, left: Topographical Map of the District of Columbia (p. 86)
Page 5, right: E. Sachse View of the Capitol (p. 92)
Page 6, left: The Peters Co. Map for International Christian Endeavor (p. 102)
Page 6, right: Bird's-eye view of the National Capital (p. 108)
Page 7, right: Metro System Map (p. 153)

Designed by Bonfilio Design, New York

Printed and bound in China

2002 2003 2004 2005 2006 / 10 9 8 7 6 5 4 3 2 1

CONTENTS

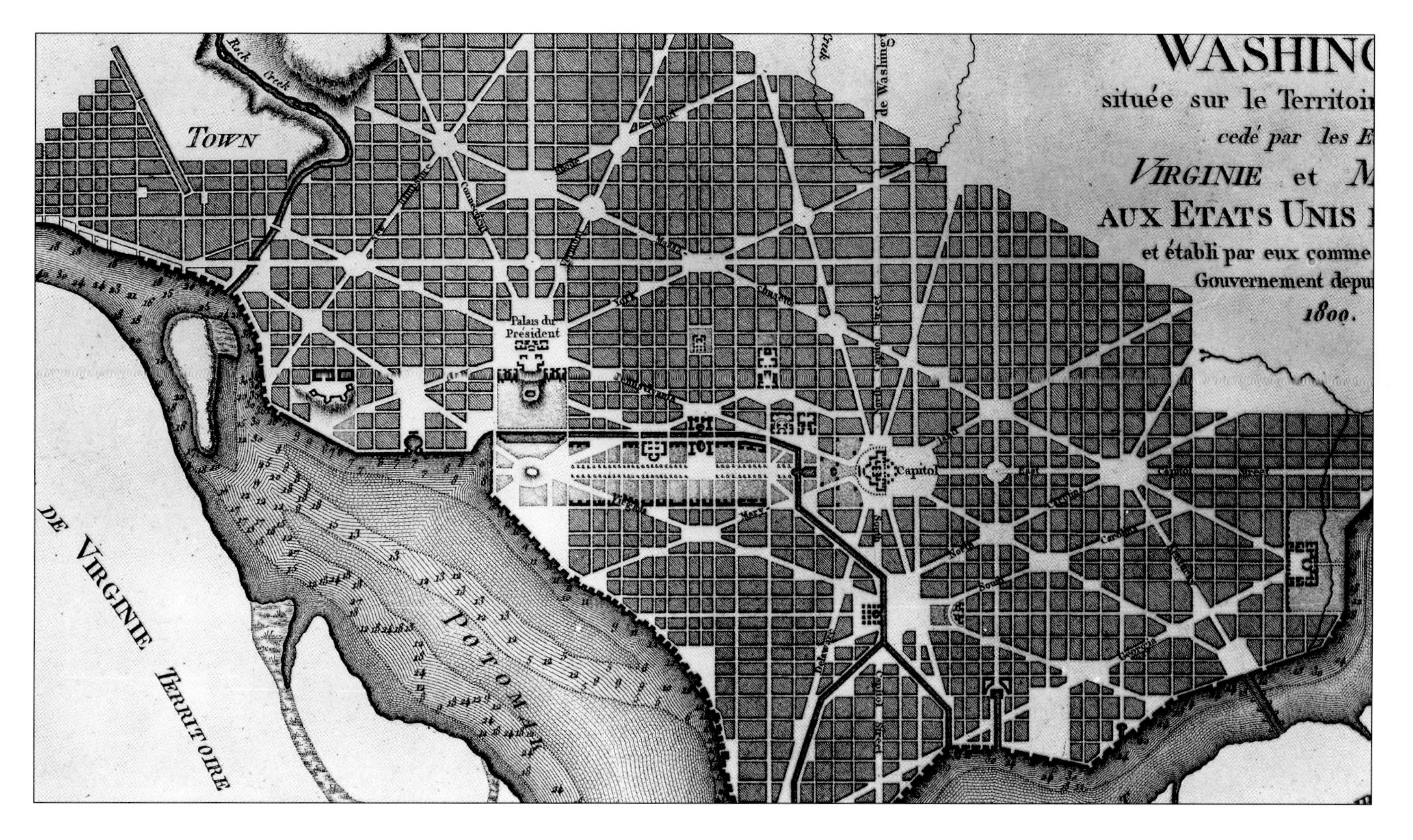
WASHING
située sur le Territoi
cedé par les E
VIRGINIE et M
AUX ETATS UNIS
et établi par eux comme
Gouvernement depu
1800.
TOWN
Palais du Président
Capitol
DE VIRGINIE
TERRITOIRE
POTOMAK

DESIGNING WASHINGTON, DISTRICT OF COLUMBIA—SPIRIT OF DEMOCRACY

REWORKING AND PUBLICIZING THE NEW CITY PLAN

URBAN COMPOSITION AT THE MICRO-SCALE

INTERNAL URBAN IMPROVEMENTS

TOPOGRAPHICAL
Map
DISTRICT COLUMBIA
IN THE YEARS 1856'57'58&59
A. BOSCHKE.

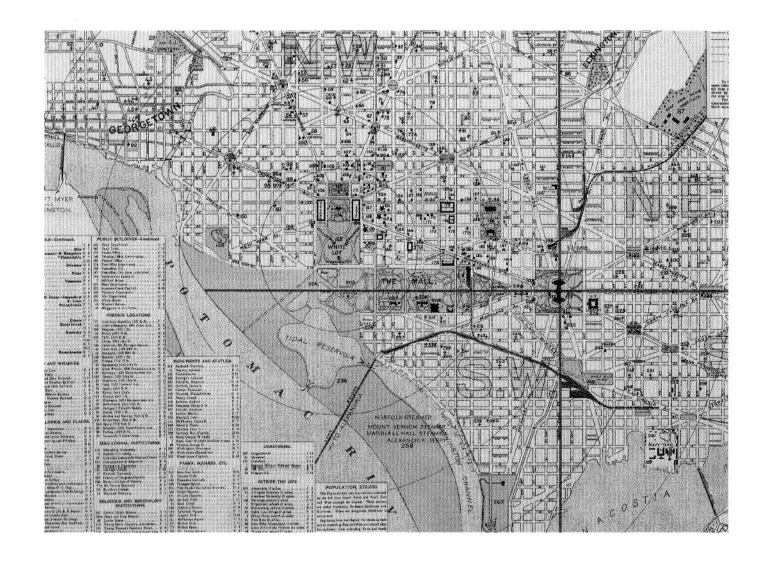

VIEWING THE CITY IN 3-D

REVISITING L'ENFANT'S URBAN AESTHETIC—CITY BEAUTIFUL MOVEMENT

PROMOTING PARKS

SPATIAL LAYERS IN TIME

PLACE AND SPACE—GOVERNMENT INTERVENTIONS

STRADDLING TECHNOLOGY AND URBAN DESIGN

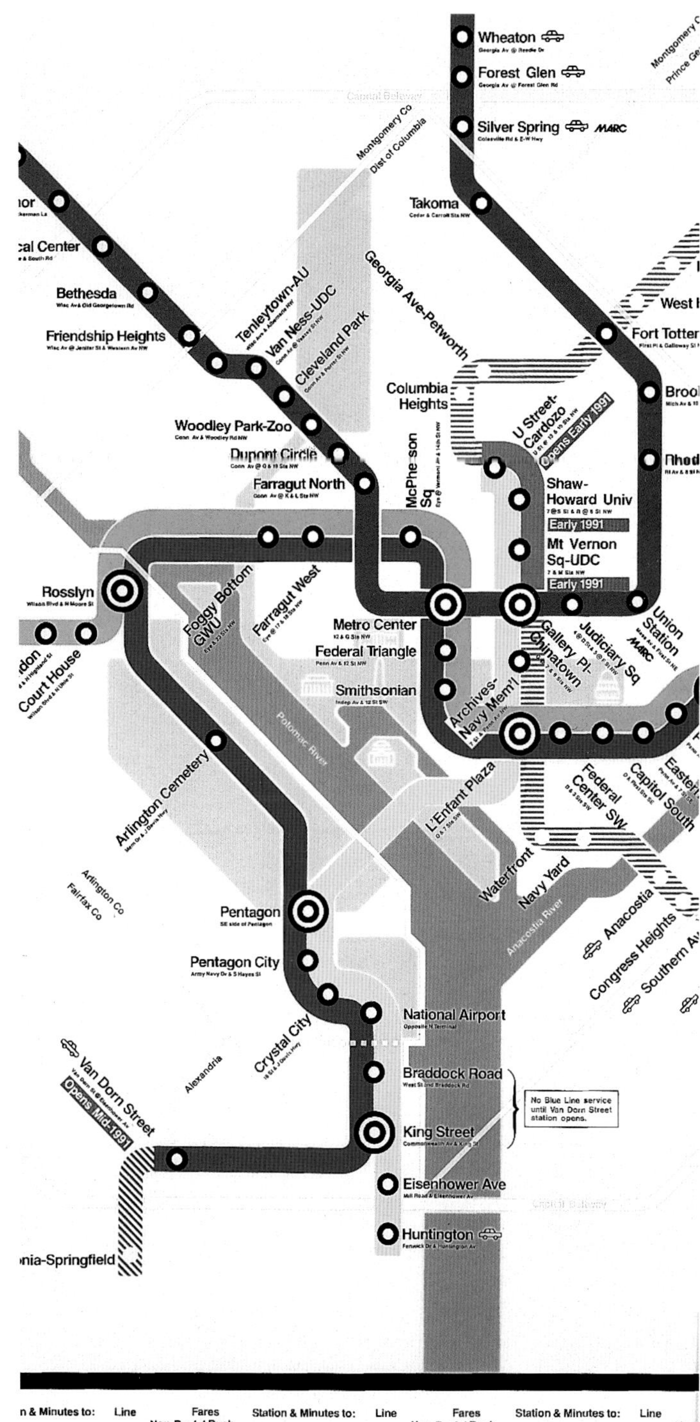

Station & Minutes to:	Line	Fares Non-Rush	Rush
Fort Totten			
Franconia-Springfield			
Friendship Heights			
Gallery Pl-Chinatown			
Georgia Ave-Petworth			
Glenmont			
Greenbelt			
Grosvenor			
Huntington			
Judiciary Square			
King Street			
Landover			
L'Enfant Plaza			
McPherson Square			
Medical Center			
Metro Center			

Station & Minutes to:	Line
Mt Vernon Sq-UDC	
National Airport	
Navy Yard	
Naylor Road	
New Carrollton	
Pentagon	
Pentagon City	
Potomac Ave	
Prince George's Plaza	
Rhode Island Ave	
Rockville	
Rosslyn	
Shady Grove	
Shaw-Howard Univ	
Silver Spring	
Smithsonian	

DEDICATION

To those who value urbanism, who are awed by the splendor of nature, and take pleasure in the multiple layers of interest implicit in maps of cities and countryside—most especially, to my dearest family, friends, and mentors who have encouraged these endeavors . . . in particular, to Larry, mom and dad . . . .

ACKNOWLEDGMENTS

How did I become interested in maps? Since early childhood, my family's recreation included exploration of city and countryside in our automobile. We marveled at the bridges crossing Pittsburgh's three rivers, the building and extension of the Pennsylvania Turnpike, and the natural and historic sites of our country. Maps were our companions in these adventures by car or by foot. When our children were young, my husband Larry and I delighted in similar family outings, and I was invariably the navigator.

My appreciation for maps grew with my admiration for their artistry and historic content, as well as for the information they conveyed. Deciphering maps has not always been easy, though I welcomed the challenge. I encountered another kind of challenge in the 1980s when I was researching Washington history, lecturing and preparing exhibitions on the city. Combing numerous archives, libraries, government offices, and private sources in Washington and elsewhere, it was often difficult to find or obtain the information I was seeking. I shared my frustration with colleagues about the circuitous routes and detective work required to locate maps. This book of collected Washington maps represents the germination of an idea planted about fifteen years ago when Steven Hurtt and several other architect friends responded to my frustration by suggesting that I write a book on the topic. Allan Greenberg sent me a booklet about early maps of New Haven, Connecticut with a note, "Do it."

The intervening years were rich with related activity and encouragement from many sources. Always, Rizzoli editor David Morton was an advocate for the map I designed, and later for this book. Special appreciation goes to sources from which I received grants to support the photography and research for the book. The Catholic University of America (CUA) provided a major Research Grant-in-Aid; its School of Architecture and Planning, and Gregory Hunt, its Dean, and Professor Stanley Hallet (former Dean) supported my grant effort and inclusion of a Leon Krier map. Of immense importance to this project was a grant from the U.S. Department of Interior-Historic Preservation Fund/D.C. State Historic Preservation Grant-in-Aid (Steven J. Raiche, Donald McCleary, Stanley Onye, and David Maloney) for photography, research, and the inclusion of several splendid maps. Sincere gratitude goes to the School of Architecture, University of Maryland, University of Maryland Foundation; and The Washington Map Society and its president Thomas F. Sander. These organizations have provided magnanimous stipends toward the completion of this work. New York's Museum of Modern Art and its Architecture Curator, Terence Riley, generously donated two visionary Krier maps.

Primary sources of maps and images have come from two collections, to which I am deeply indebted. Albert H. Small, Washington philanthropist and collector of Washingtoniana, graciously contributed a substantial number of images and frequently informed me about the latest map publications. Without his extraordinary generosity, this book would not have its far-reaching scope. The Library of Congress's Geography and Map Division, the other major source of images, generously provided workspace and great amounts of staff time. In both cases I must acknowledge their unsurpassed generosity, inspiration, expertise and guidance, without which this book would not have been possible. Indispensable to this effort was Geography and Map Division's Ronald E. Grim, Specialist in Cartographic History who untiringly oversaw photography, sought answers and solutions with a smile, and persevered as a reader and advisor. John Hebert, chief, Ralph Ehrenberg, former chief, Richard Stephenson, and James Flatness—cartographic specialists on Washington—each provided much information and thought-provoking discussions. The writings of Messrs. Ehrenberg and Stephenson have added greatly to the dialogue on Washington maps. Kathryn Engstrom and Charlotte Houtz also facilitated my research efforts.

To those who collaborated on aspects of the text, inventories and photography, or contributed maps and images from their collections, I am immeasurably grateful. Without the help of four current and former CUA students, this book may not have come together. Lisa Pallo collaborated in writing several texts and inventories. Kim Johnson offered cogent advice and creative technical and organizational assistance. Rich Blanch and George Mohasseb lent their computer graphic skills toward construction of the Time Line. Austin Kiplinger, Timothy Davis, Herbert Franklin, Charlene Drew Jarvis, Lucinda Janke, Robert L. Miller, and Whayne Quin contributed their expertise and insights in the texts they wrote for this book. Walter Smalling, an incredible photographer, with his assistant Forrest MacCormack photographed a third of the maps, and Chrome, Inc. reproduced numerous images. Map-makers Joseph Passonneau, Dhiru Thadani, and SPOT Satellite Corporation's Clark Nelson and Patricia Ravelo graciously provided their maps and satellite images, respectively.

Consultants and readers offered invaluable editorial and content expertise: George Oberlander (formerly of National Capital Planning Commission), David Maloney (Historic Preservation Review Board), Barbara Saffir (editor and journalist), Lucinda Janke (Kiplinger Collection), Sibley Jennings (Architect, Macon, Georgia), Dean Steven Hurtt, and Professor Roger Lewis (School of Architecture, University of Maryland). Curators of collections and many government agency specialists allowed me to use their remarkable rare maps and assisted with my detective work: Sue Kohler (U.S. Commission of Fine Arts), Fred Bauman (Manuscript Division, Library of Congress), Sam Davidson and Marilyn Ibach (Prints and Photographs Division, Library of Congress), Joanne Neuhaus (formerly Pennsylvania Avenue Development Corporation), Paul Gosselin (Government Services Administration/Pennsylvania Avenue Development Corporation Archives), Patty Thomas (Albert H. Small Collection), Nicole Wells (New York Historical Society), Barbara Wolanin (Curator for Architect of the Capitol), Ben Myers, Jim Corbus, Eddie Ashby (Architect of the Capitol, Records Management Branch), Richard Smith (National Archives), John Fondersmith (D.C. Office of Planning).

Particularly important, have been the publications of historians who have devoted many research hours and have shared their knowledge of Washington and its maps with eager readers. Among those, are the writings of John Reps and Paul H. Caemmerer, and Senate Report No. 166, 1902 by the Senate Park Commission.

The French connection has been a special part of my quest to understand the nuances and inspiration for L'Enfant's Plan. Since receiving a grant in 1985 from the Government of France to research this topic in France, many have assisted with my inquiries. Sophie Join-Lambert (Musee des Beaux Arts, Tours), and her colleague, Annie Gilet, helped with research and acquisition of material relating to L'Enfant's father and colleagues. Others include Isabelle Lemaistre (Musee du Louvre, Paris), Annie Jacques (Musee des Beaux Arts, Paris), Dominique Malicet and Lasare Paupert (Embassy of France, Washington, D.C.), and Catherine Michaeloff, Paris.

For help in the early years of my research on L'Enfant, I must not overlook the many who guided me through an exciting world of French urbanism and landscape: Monique Mosser, Jean Louis Cohen, Daniel Rabreau, Pierre Rosenberg, Jacques Foucart, Anton Picon, Bruno Fortier, Jacques Allegret, Isabelle Allegret, Pierre Noel and Francoise Drain, Charles d'Yturbe, Isabelle de Laroulliere, Phillippe Grunchec, Claude Bilgorag, Yolande and Laurens d'Albis, Bruna Rizzi Bond, Roselyne de la Ferte Senectere, Pierre Antoine Gatier, Didier Reppelin, Marie Gallup, Laurent and Claire Gaillard, Antoine Bouchayer, Mary Mallet, Michele le Menestrel Ullrich, Marie Sol de la Tour d'Auvergne, Robert Coustet, Yves Buffet and the late Helen Buffet, Philippe Martial, Lini Janesson, Paul Spreiregen, Laurence Coffin, Beatriz Coffin, Robin Middleton, Henry Millon, and the late Herbert Stein-Schneider.

Further I especially wish to acknowledge and thank editor David Morton, his assistant Douglas Curran, and their colleagues at Rizzoli who made this publication possible. It has been a privilege to work with them. To the many others who have helped, I extend my sincerest gratitude.

And certainly, my admiration for the work of the cartographers, designers, and printers of these maps knows no bounds.

FOREWORD

BY AUSTIN H. KIPLINGER

For some, maps are a fascination. For others, they are an impenetrable jungle. But one thing stands clear to all: maps have been the starting point for exploration since time immemorial and are an invaluable record of discovery. In the case of Washington, they are also the tangible form of a vision that took shape in the minds of some farsighted founders of the American Republic. George Washington, of course, was preeminent among them. To be noted is Pierre Charles L'Enfant, the author of the geographic plan for the then-newest capital in the world. And on top of its original skeleton, Washington is still evolving.

What Iris Miller has done in her fascinating book is show, by means of clear historical perspective, how Washington, D.C., came into being as a place—from the seventeenth century to the twenty-first, from ground level to monumental heights.

Mapmaking, which started out with the use of rods and chains and sextants, has expanded into the realm of high technology. Now, in less than an hour of flight time, an airplane can photograph every square inch of an American state. And, in one pass, a satellite can record the topographic character of the world. These records are the daily grist of road-builders, golf course designers, construction engineers, traffic regulators, forest preservationists, as well as sailors, soldiers, pilots and military people the world over. They are the raw material for national and international decisions. Wars are fought with their guidance—and over the boundaries they delineate. Peace treaties are never signed without them.

Maps are also social histories. Within them are contained the elements of life for people at their own home sites: where they live, work and play.

As you read this book, or luxuriate in the wonders of its lines and illustrations, you will be transported through more than three hundred years of human history. Like all history, that of Washington is an unfinished story—so, as you turn the following pages, let your imagination run, and allow yourself to speculate how things may be in another three hundred years. I am certain you will enjoy the experience as much as I have.

INTRODUCTION—THE FRENCH INSPIRATION

Seventeenth- to Eighteenth-Century Cultural Milieu

L'Enfant in France and America

The shaping of a city is an historic act—as indeed it was with Washington, District of Columbia. Born of a profound utopian notion rooted in eighteenth-century ideals, the nation and its federal city were envisioned as hallmarks of equality, justice, and liberty.

The innovative shaping of a wilderness and farmlands into a coherent city plan, assimilating spatial features of common American urban fabric with those models from the great European cities, was the implicit challenge that President George Washington put forth in 1791 to Pierre Charles L'Enfant—the inspired designer on whose seminal design Washington's foundations were laid.[1]

The new republic would offer a realm of philosophical and social pilgrimage for all people: a remedy for the anxiety and dogma of European daily life, for the disparity and excesses between the nobility and common people. The national capital would be an icon of an exemplary city, an ordered place in which the physical environment would influence individual deeds. Classicism, as the architectural lexicon, would provide a setting for formal ritual. Maps, as a universal language of city-making, would launch a new narrative to ascribe meaning and amass collective memory.

Washington's city plans unveil a pattern of a city in flux, accommodating growth, evolving through time—a geographical record of population and place-making. Maps, like photographs, are non-objective documents revealing the biases and talents of the cartographer and researcher. The station point from which the image is drawn, the graphic technique and items included or excluded, and sociopolitical implications all expose the designer's predilections.

More importantly, maps affirm that urban design is both an art form, and a response to site, societal, economic, and contextual conditions. Form itself is neutral. "Reading" urban form, without the cultural context, cannot accurately convey political intention or social structure. In truth, similar plans have been advanced to promote and justify both autocratic and democratic regimes. It is the designer, or others, who imbue the plan with substance.

The images and text of this book parallel the courageous story of the formation of the national capital of the United States of America. Rarely can a city boast such a sampling of maps that fully exposes its process of formation and growth, defining its boundaries and geometries. The reader may become immersed in the history. On the other hand, the exquisite illustrations speak for themselves.

The introduction explores the milieu of L'Enfant—artistic and urban precedents culled from his French past and post-colonial American experiences that inspired a unique design in a new place for a new order. As opposed to his American counterparts and successors, L'Enfant's plan (see p. 34) reflects poetic acumen. Not until the twentieth-century Senate Park (McMillan) Commission proposals (1901-1902) was this urban landscape imagery revisited (see p. 116). In spite of a myriad of pressures, for 200 years this powerful design has not unraveled.

It is necessary to understand L'Enfant's milieu—a vibrant atmosphere typified by a constant flow of ideas and artistic training across disciplines—in order to grasp the unusual circumstances of this extraordinary design. Design of the public realm had become high art. L'Enfant would anticipate the potential flowering of the city, infusing his plan with elements of his European heritage.

FEDERAL PLAN AND IDEOLOGY

Washington is an elaborately developed and controlled composition. The Plan—itself the physical symbol of the grandeur of this country and its people, its social goals and noble democratic ideal—was conceived as a setting for human civilization and achievement. It is the embodiment of man asserting his identity. As a late-eighteenth-century "new town" plan, it is an artful and articulate model for urban culture and civic function. Its iconography—city as garden—personifies rational order and aesthetic harmony.

This plan, a metaphor for a just humanity, was singular in its intent to express the glory and valor foreseen for the nation. The federal city of this new democracy was an experiment upon which the eyes of the world were focused.

In its search for an ideology, each generation sustains an attitude about the past that tempers its vision of the future. During the eighteenth century, a keen interest in history was a prevailing influence upon architecture, urbanism, and equally upon other arts. Idealizations of earlier eras were depicted in a multiplicity of paintings, engravings, and a vast number of publications. The design and redesign of cities, especially in France, evoked powerful symbols of classical imagery—symbols derived from the glorious Golden Age of antiquity and, in part, a result of the rediscovery of Pompeii, Herculaneum, and ancient Rome. The revelation of these city planning and spatial settings stimulated great enthusiasm, resulting in numerous folios and replicas.

FIGURE 1.

FIGURE 2.

FIGURE 3.

FIGURE 4.

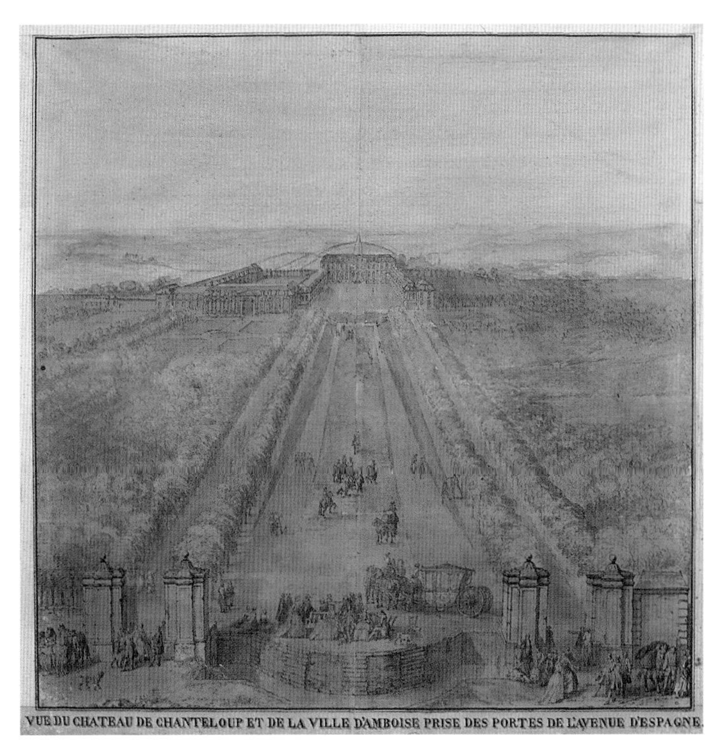

FIGURE 5.

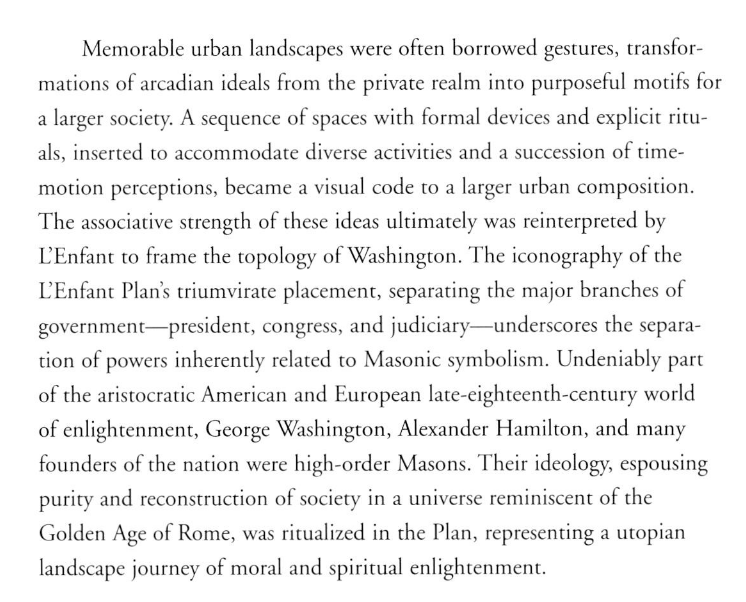

Memorable urban landscapes were often borrowed gestures, transformations of arcadian ideals from the private realm into purposeful motifs for a larger society. A sequence of spaces with formal devices and explicit rituals, inserted to accommodate diverse activities and a succession of time-motion perceptions, became a visual code to a larger urban composition. The associative strength of these ideas ultimately was reinterpreted by L'Enfant to frame the topology of Washington. The iconography of the L'Enfant Plan's triumvirate placement, separating the major branches of government—president, congress, and judiciary—underscores the separation of powers inherently related to Masonic symbolism. Undeniably part of the aristocratic American and European late-eighteenth-century world of enlightenment, George Washington, Alexander Hamilton, and many founders of the nation were high-order Masons. Their ideology, espousing purity and reconstruction of society in a universe reminiscent of the Golden Age of Rome, was ritualized in the Plan, representing a utopian landscape journey of moral and spiritual enlightenment.

FROM FATHER TO SON

A sensibility of eighteenth-century historicism influenced Pierre Charles L'Enfant's formative years. Born in 1754, a period marked by cataclysmic change, he grew up at Manufacture des Gobelins, the national tapestry factory, in Paris. Here his artist father, Pierre L'Enfant, inherited a *lodgement,* an artist's studio apartment, from his distinguished mentor Charles Parrocel. The father, Pierre, painter *ordinaire du roi* and a designer of cartoon motifs for tapestries, was admitted in 1745 as a member and professor at *Academie Royale de Peintre et Sculpture* (an institution supported by the King since its inception in 1648). His works were in demand—with paintings destined for salons, private collections, and for King Louis XV, and *Ecole Militaire Superieure de Genie at Versailles,* where they can be seen to this day.

Known as a genre painter of landscape and military battlefield scenes, Pierre L'Enfant documented troop maneuvers in the 1740s—many of which he composed as splendid large oil paintings, including Lawfeld (Fig. 1)[2] and Fontenoy (Fig. 2)[3]. Charming, but now faded watercolor and ink drawings, such as campaigns of Fribourg (Fig. 3)[4], Strasbourg, and Lille, are in the collection of Musée de l'Armée. Like landscape and other genre paintings, the combination of elements in these combat scenes included foreground elements, perspective alignments, depth, volume, light, focal terminus, emphasis upon sky as a figural object, vegetation, and classical imagery of structures and figures. Layered picture planes incorporated a semblance of both verticality and horizontality, assiduously emulating 100 years of the battlefield genre tradition.

Integrated into the warfare panoramas were distant spires and town walls fortified by the engineer Sebastien Le Prestre Vauban. City composition was decisively articulated, exposing formal urban relationships and view sheds, along with troop positions. Typical of military artists, the elder L'Enfant followed the French Royal Army beside the military engineers and cartographers, learning their craft. These skills, he seems to have conveyed to his son.

In addition to battlefield paintings, the father painted landscapes juxtaposing town and garden. Among those completed in conjunction with his close colleague at the Gobelins, Charles Cozette, were paintings for the "Duc de Choiseul. Pierre L'Enfant's two views of Chateau de Chanteloup and the town of Amboise reinforce the unity of earth and sky (Figs. 4 and 5)[5, 6]. A provocative relationship between the private world of the noble families and the public realm—the contrast between sacred and profane—comes alive in these paintings. The garden microcosm constituted the ideal

FIGURE 6.

from which to ascertain aspects of urban placement in civic design.

Consequent to the intrigue of Louis XV's court at Versailles, the Duc de Choiseul returned to his Chateau de Chanteloup, where he entertained the noblesse with fantastic receptions, theatrical performances, and hunting parties. Cozette's immense painting of 1770, now in Musée des Beaux-Arts at Tours, combined both a regional map and a landscape military scene (Fig.6)[7]. The latter depicted artillery troops gathered along a road, with additional infantry encamped nearby, as two officers stood under a tree examining the terrain with their instruments. Chanteloup forest, situated in proximity to Amboise and Tours, exploited the character of the site. The map was compiled by the same Col. Jean-Baptiste Berthier who assembled a vicinity map of Paris and Versailles (see Fig. 8) and oversaw installation of Pierre L'Enfant's paintings at Versailles, and whose renowned son, Alexander, also fought for American Independence along with Pierre Charles L'Enfant.

The plan of Chanteloup and its forest typology deserve serious study in reference to Pierre Charles L'Enfant's Washington Plan. Previously, his design has been compared foremost to Versailles and Paris. However, a look at Chanteloup offers fresh insight into his remarkable plan. To a degree, we might speculate on whether Pierre L'Enfant, or his son, assisted on Cozette's huge painting. In contrast to their other collaborations, only Cozette's signature appears, but the fact that they were both working for Choiseul suggests some possibility. Nevertheless, there is no doubt that L'Enfant and his son Pierre Charles would have been quite familiar with this work and the forest of Chanteloup.

As opposed to most hunting forests, here a rigid grid pattern of pathways was simultaneously integrated with radial tree-lined allees, creating a multiplicity of transparent uninterrupted views leading to *ronds points* (space at the confluence of routes), and artifices. Extending from a grand axis and hemicycle just beyond the formal garden, an irregular distribution of "avenues" responded to terrain and natural features, offering a cogent example for urban design, as described by the influential l'abbé Marc-Antoine Laugier, in *Essai sur l'architecture,* 1753, and *Observations sur l'architecture,* 1765. As models for ritualized urban walks—a prominent social feature of urban life—hunting forests impacted nearly all redesigned eighteenth-century French towns.

Little is known about Pierre Charles L'Enfant's youth because of the destruction of records during the French Revolution. Certainly we can imagine that he followed the art curriculum available to children of resident artists at the Gobelins. This embraced a range of artistic techniques and interdisciplinary skills based upon the Academie's required courses, where he later studied. No doubt he was well acquainted with oeuvres of his father's famous colleagues (noted below), whose work depicted urban landscape and architecture. In 1771, his name appears on the list of students at the Academie Royale de Peinture et Sculpture, "Pierre Charles L'Enfant, Un eleve de son pere" [a student of his father].

Beside his matriculation records as a student of painting, only one of his art works, "Deux Lutteurs" [two wrestlers], remains (Fig. 7)[8]. This little-known drawing, for which he received a prize, has never been exhibited or published. French records show that L'Enfant attended neither the

FIGURE 7.

FIGURE 8.

architecture academy, nor military, nor engineering schools. Yet the vibrant atmosphere of experimentation and intellectual fervor of his milieu, typified by a constant flow of ideas across disciplines, enabled artist Pierre Charles L'Enfant to apply his ample skills to urban scale and spatial organization in the design of Washington, D.C.

REINTERPRETATION OF FRENCH URBAN LANDSCAPE

The Baroque garden and hunting forest as a quintessential archetype informed the city planner. New maps of France by three generations of Cassini family cartographers indicated the breadth of these impressive gardens and forests, especially surrounding Paris. Andre LeNotre's landscape metaphor joining town and garden was applied routinely. Radial *patte d'oie* (crows feet road pattern) imposed upon orthagonal streets, emanating from a symbolic space, offered an urban typology for expedient circulation, enchanting entry sequences and vistas, and possibilities for defining neighborhoods. The city as artifact merged a sense of reason with a memorable sense of space. At once orderly and intriguing with sudden bursts of the unexpected, Laugier advocated the imposition of symmetry interspersed with clever buffers of chaos at intervals to effect a magical unity.

Throughout France, as national unity was being achieved, older eighteenth-century cities were enlarged and connected to one another by a network of roads. Civil engineers from Ecole des Ponts et Chausees in Paris modeled their thesis examinations and real projects after the prolific playgrounds of the elite, such as Versailles and Marly-le-Roi (formerly within Versailles' forest), St. Germain-en-Laye, St. Cloud, and the Tuilleries[9] in central Paris (Fig. 8). Synthesizing their mathematical skills and their aesthetic training, their interventions altered the French urban landscape.

Concurrently, architects and landscape designers were commissioned to fashion the urban scene. Innovative mediation between the old town and new, on a grand or small scale, emphasized an identity with evolving ideas about republican space. Typically, these Grandes Projects Urbaines (great urban projects) reflected the period's intellectual rigor and cultural heritage.

Celebratory public spaces, named on behalf of nobility, gave towns an air of grandeur. Announcing entry or arrival at honorific sites, vivid and varied park-like compositions fostered delight and edification. These royal squares and hemicycles paired national-political goals with economic necessity. Pierre Patte's 1765 composite map of Paris competitions, for Place Louis XV, was a telling document. Reinforcing landscape interventions for spatial adornment, solutions contemplated fantasy with traces of reality. Jacques-Ange Gabriel's treatment of Place de la Concorde vies with equally exquisite set pieces across France—Reims, Dijon, Nantes, and others.

Like these urban landscape schemes, young L'Enfant's Washington plan recognized the significance of episodic interplay of linear movement and spatial volume, vista, and narrative in a continuous urban theme. Executed on the scale of an entire city, his brilliant design exhibited obvious connections with these compositions.

When L'Enfant returned to Paris from America in 1783, he found many changes. The newly built Pantheon, and its urban setting, had been designed by architect Jacques-Germain Soufflot. Soufflot was director of the nearby

FIGURE 9.

Manufacture des Gobelins, L'Enfant's boyhood home, where many artists lived and worked. Located in a dense older section, numerous structures had been cleared to make way for the Pantheon's huge plaza and a wide avenue terminating at Luxembourg Palace (Fig.8)[9]. In 1773, Soufflot prepared an urban intervention for Lyon—a loose street grid with a series of distinctive public squares, ingeniously linked in a splendid manner.

Claude-Nicholas Ledoux's 1775-1779 plan for the saltworks, Saline de Chaux at Arc-et-Senans, while less far-reaching than L'Enfant's Washington plan, was based upon a vision of an ideal city. At Aix-en-Provence, architect Ledoux applied an inventive interpretation of a landscape metaphor of ordered geometry to the redesign of the old city and the extension of the new town. The hierarchy of streets, boulevards, and abundant squares with enticing fountains are among the most pleasing in France.

Chateau Tromphet, a fortification depicted in paintings by Claude-Joseph Vernet circa 1755, was demolished to make way for a heroic plaza near the Bordeaux waterfront. This 1787 proposal highlights one of many transformations in the city's urban landscape—streets fan out to engage the old city in an ordered radial composition (Fig.9)[10], reminiscent of the politically symbolic 1610 project for Place de France by Henri IV in Paris.

The city of Nancy evolved from an enclosed fortified town within the 1583 south walls and 1610-1620 north walls. The walls were removed, and the exceptional 1750-1755 plan by Emmanuel Heré was inserted to unite Place Carriére and Place Stanislas and the Pepiniére garden. A screen of facades emanated from the palace along Place Carriére, thus directing a vista through an ancient arched portal to an immense enclosed square with honorific statue to Stanislas Leszczynski (father-in-law of King Louis XV). The ceremonial thoroughfare of Cour Leopold, linking Place Carnot to Porte des Villes, was a late-eighteenth-century design. Orthagonal street morphology adhered to the undulating terrain, essentially an artistic European response (Fig.10)[11], in contrast to the American surveyor's customary technique exercising excessive regularity over the land.

In France, L'Enfant grew up in the midst of a public debate over the urgency for reform in the urban environment. Concerns about the well-being of the populace focused upon eighteenth-century urban law, and

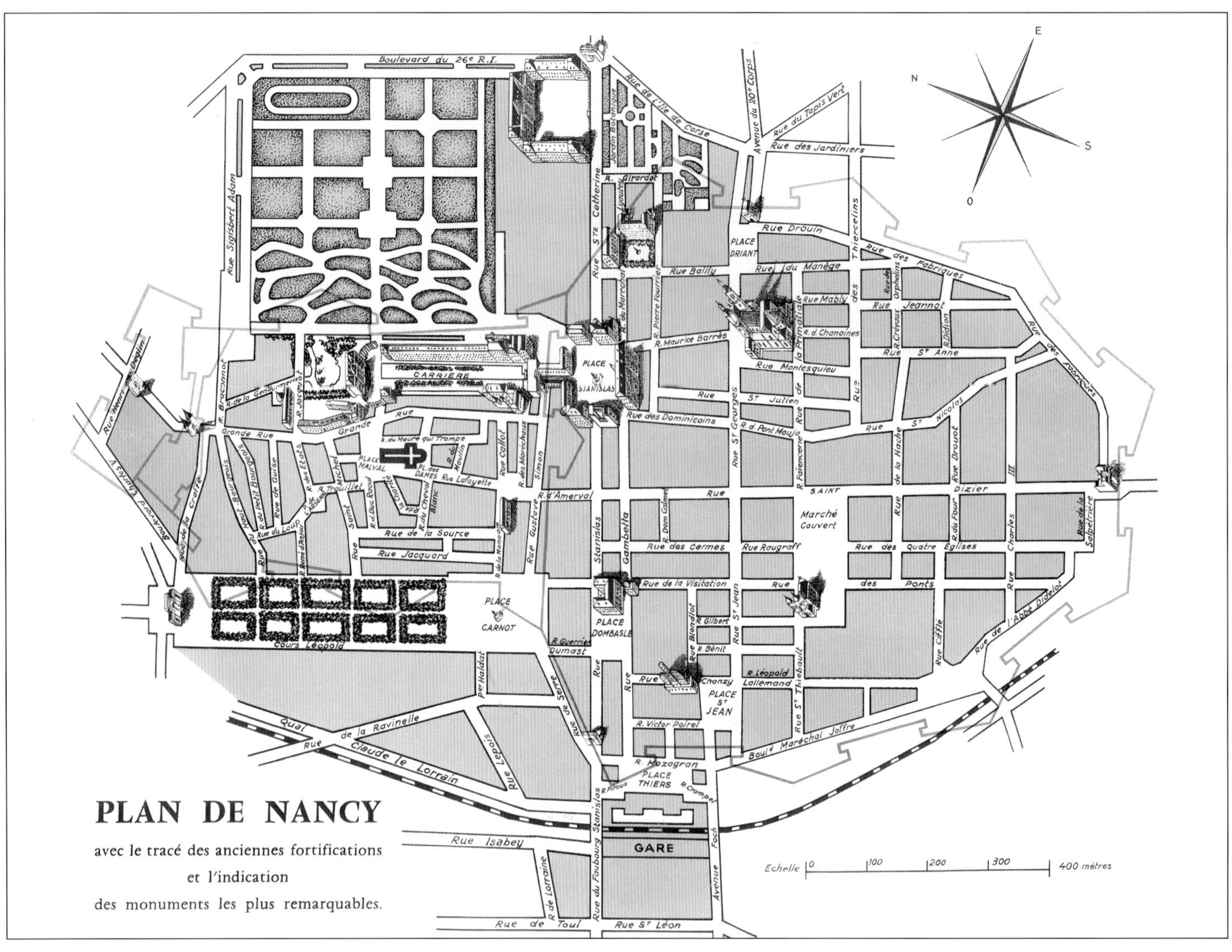

FIGURE 10.

considerations of health and corollary architectural issues. Radically changing opinions about death, coupled with unsanitary burial procedures, led to the design of new gardenesque cemeteries on the outskirts of town, which became popular places for outings.

Similarly, laws pertaining to heights of buildings to width of street ratios were well-established and strictly maintained as a means to control the spread of disease. Assurance of sunlight and pure air for society, a critical aspect of transforming a city, required broad streets and sizable squares. In his plan for Washington, L'Enfant attains a spatial quality to accommodate both dramatic visual effect and rational adaptations of similar regulations, receiving the endorsements of George Washington and Thomas Jefferson.

The milieu's creative spirit and political turmoil was expressed in the works of an elite cadre of genre painters, engravers, and tapestry artists trained in the academies and promoted in the salons through patronage by nobility of educated taste. Jean-Baptiste Oudry, for example, portrayed landscape patterns of hunting forests with tree-lined allees crossing at intersections. Nostalgic imagery of ruins and classical fragments exerted a forceful influence. The drama of symbolic *fabriques*, elements in pastoral and urban landscape played out as vignettes or stage sets, evoked new social ideals. Purity of reductivist geometries was balanced against the transience of nature. Extended volumetric space as background was often contrasted by action figures as foreground, a device to imply layered picture planes. Stylistic strategies generated by artists such as Hubert Robert, Honore Fragonard, Joseph-Marie Vien, Jean-Baptiste Greuze, and Jacques-Louis David resembled those by architects and urban designers.

TRANSFORMATION AND INNOVATION OF A TIMELESS IDEA

In America, L'Enfant, a young man with only a formal art education, rose to the rank of major by the end of the Revolutionary War—despite his lack of military, engineering, or architecture schooling. He was known as an artist to George Washington for his portraits of officers and later for his designs for the Society of Cincinnatus. In 1783, following the War, he visited France at an opportune moment, to find an outpouring of urban proj-

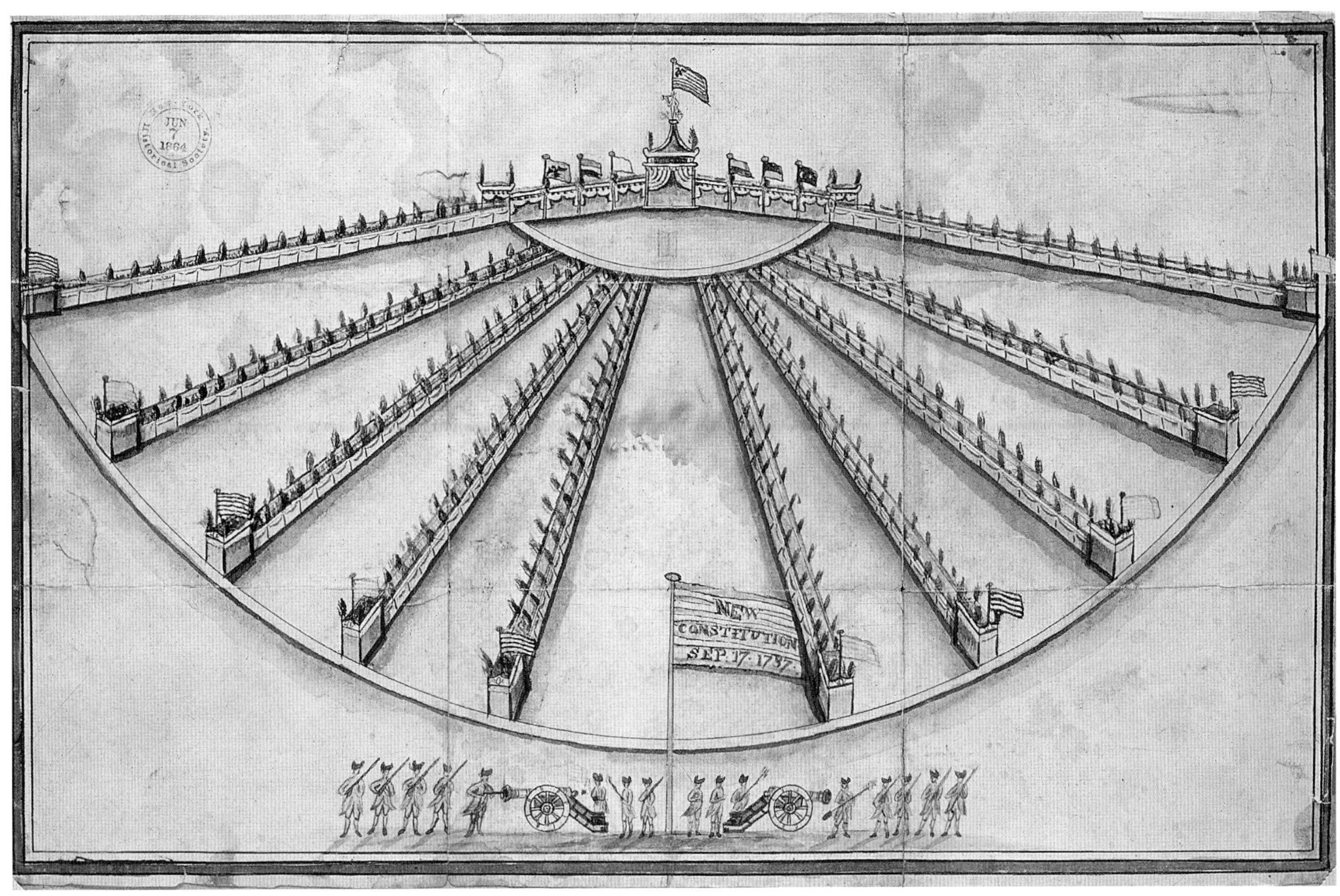

FIGURE 11.

ects. Subsequently, he established himself as an architect, working in New York and Philadelphia. The redesign of New York's old city hall by L'Enfant can still be seen on Wall Street. His Federal Banquet Pavilion plan (Fig.11)[12], erected on July 23, 1788 near Broadway and Broome Streets to serve marchers in a Federal procession supporting the new Constitution, employed a French urban landscape technique—tables radiating from a hemicycle reminiscent of proposals for Place de France (1610) and Place de Bordeaux (1784).

Thus known to President Washington, L'Enfant was poised to become the designer of the new federal city, a city blazing the way for the transformation in human existence through the association of ideas, a city with limited collective memory. In history, such an opportunity has been rare. Why was L'Enfant, a Frenchman, selected for this honor? What iconography was appropriate to portray the first true democracy? Answers are explored throughout the text.

L'Enfant's plan reinforces the iconography of democratic ideals and republican space, the foundation of the new nation. Aesthetically, it commands a spatial quality to accommodate both a collage of dramatic visual effects and rational adaptations to societal conditions, thereby integrating design and symbol.

While Washington is unique in many ways, the tradition from which it was created and its profound influence remain enduring models, providing a basis for generating civic design and vitality in urban life. Such nineteenth-century cities as Indianapolis and Detroit take their place among cities whose resurgence in the early twentieth century ensued from urban interventions of Neoclassical spatial order. Decidedly, the influence of Ecole des Beaux Arts and World's Columbian Exposition in Chicago, 1893, spurred new impulses for urban design. Amplification upon L'Enfant's plan, set forth by the Senate Park (McMillan) Commission of 1902 (see p. 116) in a new vision for Washington, served as poetic inspiration for transformations at Chicago, Cleveland, San Francisco, New Delhi, Canberra, and more. As we embark upon the twenty-first century, the powerful symbols of national unity and civic pride are visible in the compelling context of historic neighborhoods and civic urbanism of Washington, D.C.

INTRODUCTION—NOTES:

[1]As noted by Whayne S. Quin (author of Dermott Map text, p. 64), "To understand the relationship of these early maps [The Dermott Map, The L'Enfant Map, The Ellicott Map] to the founding of the original City of Washington, its land acquisition process, the government's role and agreements, and the relevance of the progressive changes in the maps' details it is appropriate to examine the text from the District of Columbia Superior Court case of Chesapeake and Potomac Telephone Company v. District of Columbia, (106 D. Wash. L. Rptr. 1065, 1070 (D.C. Supp. 1978), affirmed, 418 A.2d 114 (D.C. App. 1980)":

Detailed descriptions of the founding of the Nation's Capital and the laying out of the original squares, lots, streets and alleys are set forth in a number of cases....

Under the Charter of June 20, 1632 from Charles I, King of England, Cecilius Calvert, the first Lord Proprietary of the Province of Maryland, was granted all lands now embraced within the District of Columbia. After the American Revolution, Maryland became a state and succeeded to all rights of the Lord Proprietary, including the absolute right to the soil for use by the people of the new sovereignty.

The State of Maryland authorized cessation to the United States of the territory, which is now the District of Columbia.... [T]he cession was accepted and the President of the United States was authorized to appoint three commissioners whose duty was to "survey, and by proper metes and bounds define and limit" such district or territory. The Act empowered Commissioners "to purchase or accept such quantity of land on the eastern side of [the Potomac River] for the use of the United States, and according to such plans as the President shall approve." Additionally, the Commissioners were required to provide suitable buildings for the accommodation of public offices prior to the first Monday in December, 1800, at which time the seat of government would be transferred to the federal city.

On January 24, and March 30, 1791, the President by proclamation located and defined the limits of the District of Columbia and appointed the Commissioners who, with their successors, located and laid out the City of Washington.... The general boundaries of the proposed City, now called the "original" City, were the Eastern Branch, the Potomac River, Rock Creek to a point near P Street N.W. then following what is now Florida Avenue to 15th and H Streets, N.E., then south to C Street, N.E. then east to 20th Street and then south to the Eastern Branch. The land within was devoted mostly to farm purposes owned principally by 19 owners hereinafter sometimes referred to as "original proprietors".

While various maps and plans for the City were being prepared under the direction of the President and the Commissioners, negotiations were entered into between the Commissioners and the original proprietors which resulted in agreements executed by the parties providing for the disposition of the land within the original City pursuant to deeds of trust to be executed.

The Deeds of Trust ... were executed around June 30, 1791 by the original proprietors and provided for the disposition of land within the limits of the City of Washington in three different categories: (1) the fee title to streets was vested in the United States; (2) the land appropriations or reservations for the use of the United States were purchased by the Commissioners with fee title vesting in the United States at the rate of 25 pounds per acre; (3) the entire residue of the land, after being laid out in squares, parcels and lots was to be divided equally with one-half assigned to the Commissioners to be sold upon such terms and conditions as the President should deem proper with the proceeds from said sales to be first applied towards the payments due the original proprietors, with the remaining proceeds to the President of the United States to be applied for purposes under the Act of Congress (e.g., construction of new governmental buildings)-authorizing the acquisition or acceptance.

Quin concludes: "It can be seen that an official 'Plan' for the City of Washington was required to provide a mechanism for giving definition to the streets, the public reservations and the residue of the land to be divided between public and private ownership."

[2]"Bataille de Lawfeld." By Pierre L'Enfant; oil on canvas; oversize; Reunion des musées nationaux, Versailles; no. du cliché 79EN6745 (517); no. d'inventaire MV 212; 1747.

[3]"Bataille de Fontenoy." By Pierre L'Enfant; oil on canvas; oversize; Reunion des musées nationaux, Versailles; no. du cliché 79EN6744 (519); no. d'inventaire MV 188; 1745.

[4]"Siege de Fribourg et de ses Chateaux." By Pierre L'Enfant; ink de Chine and watercolor gouache over lead pencil sketch on paper; Musée de l'Armée, Paris; no. d'inventaire Eb1527; September 30 to November 25, 1744.

[5]"Vue de la Ville d'Amboise et du Chateau de Chanteloup." By Pierre L'Enfant; ink de Chine and watercolor gouache on lead pencil sketch on paper; 1,135x1,135 m, 1,025x1,043 m; Musée des Beaux-Arts, Tours; no. d'inventaire 794-1-42-LM 381-V425; provenance Chateau de Chanteloup; 1762 no. cliché LA-94.599.

[6]"Vue de Chateau de Chanteloup et de la Ville d'Amboise prise des portes de l'avenue d'Espagne." By Pierre L'Enfant; ink de Chine and watercolor gouache on lead pencil sketch on paper; 1,173x1,17 m, 1,067x1,079 m; Musée des Beaux-Arts, Tours; no. d'inventaire 794-1-41-LM 381-V424; provenance Chateau de Chanteloup; 1767 no. cliché LA-94.600.

[7]"Carte du Duché de Choiseul d'Amboise et de ses Environs, Levée Topographiquement a Six Lignes pour Cent Toises." By Charles Cozette (military scene), Jean-Baptiste Berthier, geographical engineer to king (map); Montage: ink de Chine with topographic indications, on paper (map), watercolor gouache on lead pencil sketch, on paper (military scene); 1,57x2,85 m; Musée des Beaux-Arts, Tours; no. d'inventaire D 11-3-1; provenance Chateau de Chanteloup; 1770.

[8]"Deux Lutteurs." By Pierre Charles L'Enfant; charcole on beige paper, figures enhanced with white chalk, annotation in pencil below: 2.M; 45.5x43.8 cm; Ecole Nationale Superieure des Beaux-Arts, Paris; no. d'inventaire EBA 3013; 1771.

[9]"Dépot de la guerre: Carte topographique des environs de Versailles, dites des chasses imperiales. Levée et dressée de 1764 & 1773 par les ingenieurs geographes des camps et armées commandes par feu M. Jean-Baptiste Berthier, Colonel, leur chef; terminée en 1807 par ordre de Napoleon, empereur des français, roi d'Italie et protecteur de la confederation du rhin, pendant le ministere de S.A.S.M. le Maréchal Alexandre Berthier, Prince de Neuchatel, Grand-Veneur, Grand Aigle de la Legion d'Honnor, etc. Sous la direction de General de Division Sanson, au depot general de la guerre." Versailles- Flle. 5, Paris - Flle. 6; drawn by Herault et Delahaye; engraved by Tardieu, Doudan et Bouclet l'aine; black and white engraving; scale = toises; Library of Congress, Geography and Map Division, G5834s .V4 29 .F7, 1807.

[10]"Plan geometral de la ville et Faubourg des Bordeaux avec tout les changements faits jusqua present." Attelier de M. Saige; published in London; black and white engraving; 23"x32.5", scale = toises; Library of Congress, Geography and Map Division, 1787.

[11]"Plan de Nancy, avec le tracé des anciennes fortifications et l'indication des monuments les plus remarquable." Cartographer unknown; colored lithograph; Musee Lorrain, Nancy; ca. 1850.

[12]"Federal Banquet Pavilion." Drawn from memory by David Grim; designed by Pierre Charles L'Enfant; ink and watercolor on paper, n.d.; 9.625x15.4375"; New-York Historical Society; Accession no. 1864.17; date depicted July 23, 1788.

FIRST AMERICAN MAP OF "GREATER" VIRGINIA AND CHESAPEAKE

The Captain John Smith Map

TITLE: Virginia/Discovered and Discribed by Captain John Smith, 1606
DATE DEPICTED: 1606
DATE ISSUED: 1624
ENGRAVER: William Hole
Engraved map, 6th State, 32 x 41 cm.
Library of Congress, Geography and Map Division, G3880 1624 .S541 Vault

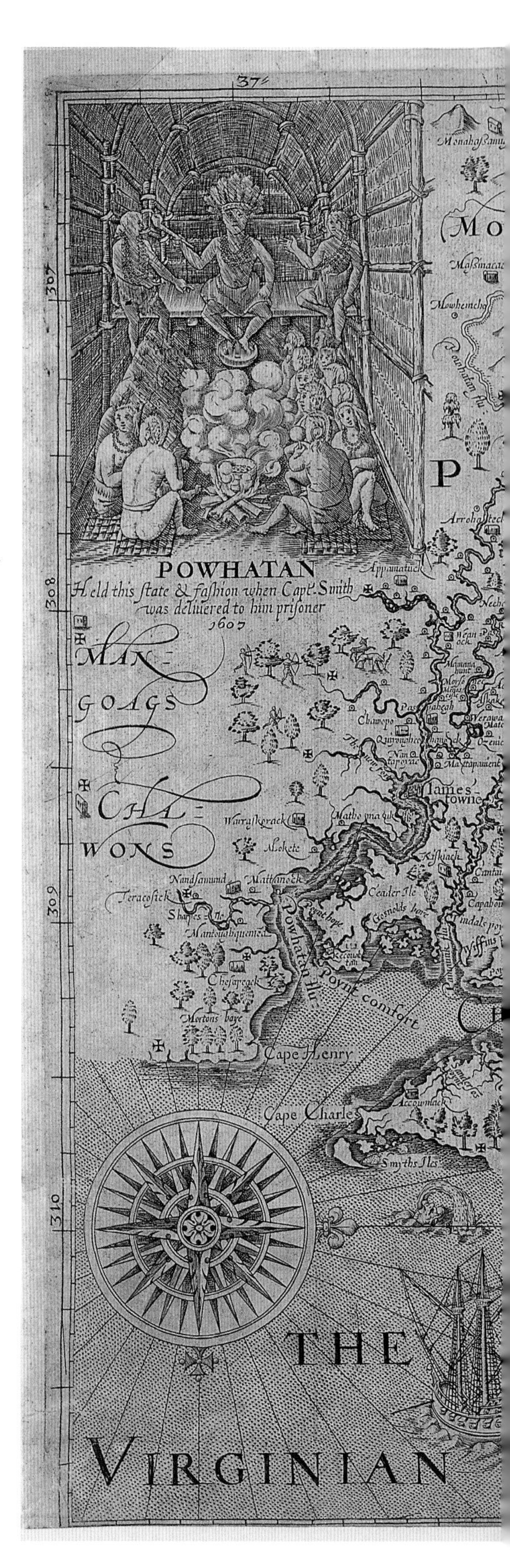

Spanning nearly 400 years, the cartographic history of the mid-Atlantic region dates from the maps of Captain John Smith. In 1606 he surveyed the Chesapeake Bay area and its principal estuaries with a group of 15 fellow Englishmen, publishing his first map in London in 1612. There were at least eleven revisions and publications of the Virginia Colony survey, which included Maryland and Washington, D.C. Typical of all colonial Virginia maps until 1721, the top of the map was oriented west, rather than north.

For 60 years Smith's work was the source for all subsequent printed maps. It continues to be a basic reference tool for locating sites of Native American settlements during the period of English colonization.

John Smith and his team began their surveyors' journey at Jamestown, Virginia. Reaching the mouth of the Potomac River in two weeks, they then continued upstream. Within three months their area survey was complete. Along the Potomac River in northern Virginia near where the City of Washington was to be located, Smith denoted five Indian villages between Aquia Creek and Great Falls. He spent two years exploring the Chesapeake Bay. Crosses on the map marked the upper limits of Smith's survey.

This intriguing map combines perspective images with other drawing techniques mixing plan and elevations. Overlaid on the base map are forests comprising several types of trees shown in elevation, while mounts and figures are drawn three dimensionally. A Susquehannock Indian, whom Smith described as "Gyant like," is illustrated in the upper right corner. In the upper left corner Chief Powhatan is seated in his lodge with tribesmen. A three-mast sailing vessel on the Virginian Sea is pictured at the lower left.

Many subsequent cartographers copied John Smith's map without crediting their source. A colorful, albeit inaccurate, map of the Chesapeake region was included in John Speed's 1670s atlas. Between 1629 and 1671 maps were published in Amsterdam and London, including those of Blaeu, the Hondius brothers, and derivatives of Arnoldus Montanus and John Ogilby. It was not until 1673 that Smith's map was replaced as a primary source of information on the Virginia and Maryland colonies by Augustine Herrman's map, "Virginia and Maryland As It is Planted and Inhabited This Present Year of 1670," published in London in four sheets. Herrman's map documented the extent of English settlement in the Chesapeake Bay region and Washington area. In the vicinity of the Potomac River, Creeks and secondary rivers, identified by Indian names, and counties and plantations, identified by English names, are accurately shown.

In 1719 John Senex's map, largely based on Herrman's work, broke with tradition by orienting north at the top. Another aspiring surveyor, the young George Washington at age seventeen drew a map of the small Virginia town, Alexandria, 1748-1749. Three years later Joshua Fry and Peter Jefferson published their significant map based upon a new survey that clearly delineated the Potomac River lands. From Smith's 1606 survey to Fry and Jefferson's 1755 map, con siderable development ensued in both cartographic skill and information gained about this region.

VIRGINIA
Massawomecks
Signification of these markes
To the crosses hath bin discouerd what beyond is by relation
Kings howses
Ordinary howses
MANNAHOACKS
The Sasquesahanougs are a Gyant like people & thus atyred
SASQUESAHANOUGH
TOCKWOGHS
KUSKARAWAOKS
Scale of Leagues
Discouered and Discribed by Captayn John Smith 1606
Grauen by William Hole
SEA

PERFECTING CARTOGRAPHIC PRECISION

The Fry and Jefferson Map

TITLE: A Map of the Most Inhabited Part of Virginia Containing the Whole Province of Maryland with Parts of Pennsylvania, New Jersey and North Carolina
DATE DEPICTED: 1775
DATE ISSUED: 1775 (1st edition 1755)
CARTOGRAPHER: Joshua Fry and Peter Jefferson
PUBLISHER: Thomas Jefferys
Colored engraving, 16 x 49 in
Courtesy, The Albert H. Small Washington Collection, 187.MP.ED.E.F.
(Also in Library of Congress, Geography and Map Division, G3880 .F72 1755 Vault)

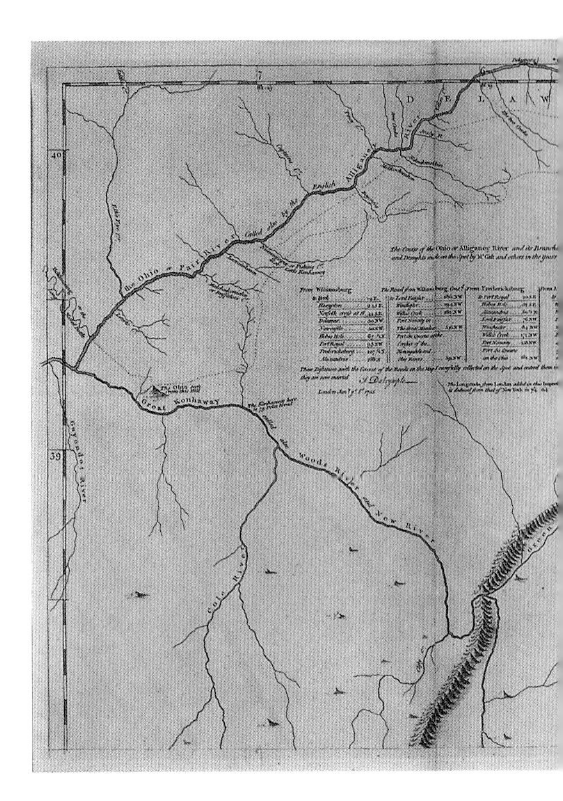

After more than 70 years had passed since Augustine Herrman's map, a new survey was undertaken in 1746 and 1749, and completed in 1751—following instructions by Virginia's governor to furnish a more accurate map for the colony. Joshua Fry, a mathematics professor at William and Mary College, and Peter Jefferson, an experienced surveyor and father of future president Thomas Jefferson, undertook this enormous effort.

The first versions of Fry and Jefferson's map were published on four sheets in London, between 1752 and 1754. The popular 1755 revised edition added a table of distances and roads linking the new port town of Alexandria to points south and to the fertile western Shenandoah Valley farms. This significant map was frequently reissued. It highlighted the critical location of Belhaven (also known as Alexandria) as a major market.

Subsequently published versions contained minor variations. The 1775 version, shown here, was published on two folio sheets; the 1794 map was issued on a single sheet. French editions were also published beginning in the 1750s. "Carte de la Virginia et du Maryland Dress sur la Grande Carte Anglaise de Mrs. Josue Fry et Pierre Jefferson," was issued by the prolific publisher, Georges Louis le Rouge in 1778. Another map edition, "Geographé ordinaire du Roi," by Gilles Robert de Vaugondy was engraved by E. Haussard (see article by Verner, Coolie, "The Fry and Jefferson Map," in *The Fry and Jefferson Map of Virginia and Maryland,* Princeton University Press, 1950).

Fry and Jefferson's map encompassed a vast amount of territory in the mid-Atlantic region, including the most inhabited portions. Part of northern North Carolina was delineated in the lower section, while the upper half showed Maryland, Delaware and parts of Pennsylvania and New Jersey. The barrier posed by the Allegheny Mountain Range to the west was formidable, underscoring the difficulties for transportation and communication to and from the fertile valleys between the mountain chains.

Features delineated in the vicinity that was to become Washington, D.C. still exist, although some names have changed. Among the plantations south of Belhaven and north of Occaquan River are "Washington" (plantation), now known as Mount Vernon, and Watson Landholding Plantation near the spectacular outcropping of rocky gorges above Georgetown at Great Falls.

Georgetown, which sits along the Potomac River at the southern boundary of Rock Creek, was only founded in 1751 and does not appear on early iterations of the map. Nor do Carrollsburgh and Hamburgh, the platted "paper" towns, which were not yet established. Hunting Creek and Four-Mile Run Creek, located in the flood plain, are identified south of Belhaven. Magees Ferry, Mason Island, Goose (Tiber) and Rock Creeks and Eastern Branch (Anacostia River) are named.

In the lower right a cartouche shows a slave rolling a hogshead to the wharf with planters seated nearby exemplifying the tobacco and slave economy of the Chesapeake region. The map was dedicated to Earl of Halifax and the Commissions for Trade and Plantations.

One might imagine the bond between the two generations of Jeffersons—Peter, father, and Thomas, son—and how the tradition of thinking about the land, spatial relationships, and map-making were passed from one to the other. These two men, who contributed significantly to the early formation of our nation and federal city, are seldom associated with each other in historical texts.

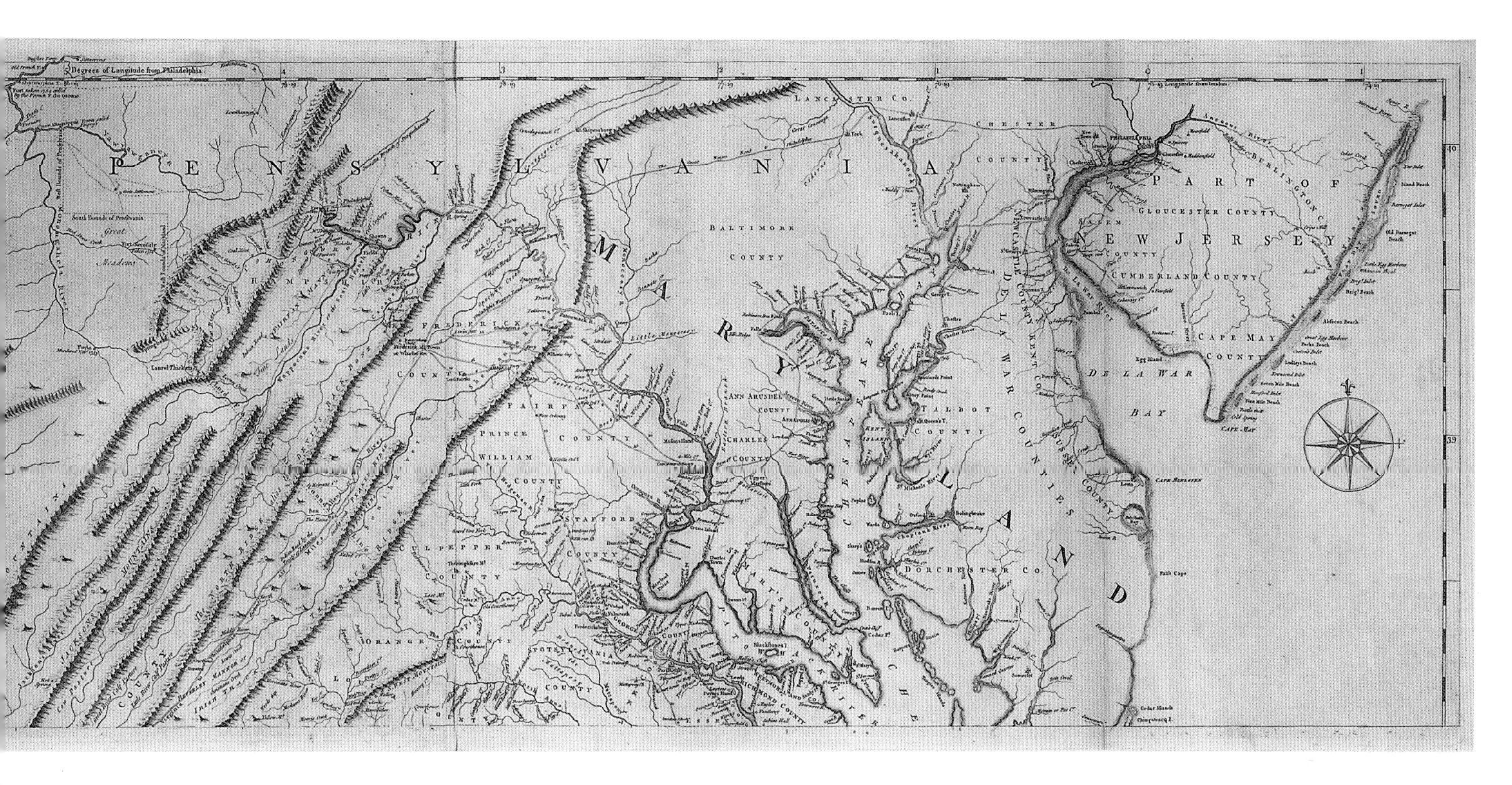
Degrees of Longitude from Philadelphia
P E N S Y L V A N I A
M A R Y L A N D
BALTIMORE COUNTY
LANCASTER CO.
CHESTER COUNTY
PHILADELPHIA
PART OF NEW JERSEY
GLOUCESTER COUNTY
CUMBERLAND COUNTY
BURLINGTON
SALEM COUNTY
CAPE MAY COUNTY
DE LA WAR BAY
CAPE MAY
DE LA WAR COUNTIES
TALBOT COUNTY
DORCHESTER CO.
ANN ARUNDEL COUNTY
CHARLES COUNTY
FREDERICK COUNTY
FAIRFAX COUNTY
PRINCE WILLIAM COUNTY
STAFFORD COUNTY
ORANGE COUNTY
CAPE HENLOPEN
Great Meadows
Laurel Thickets

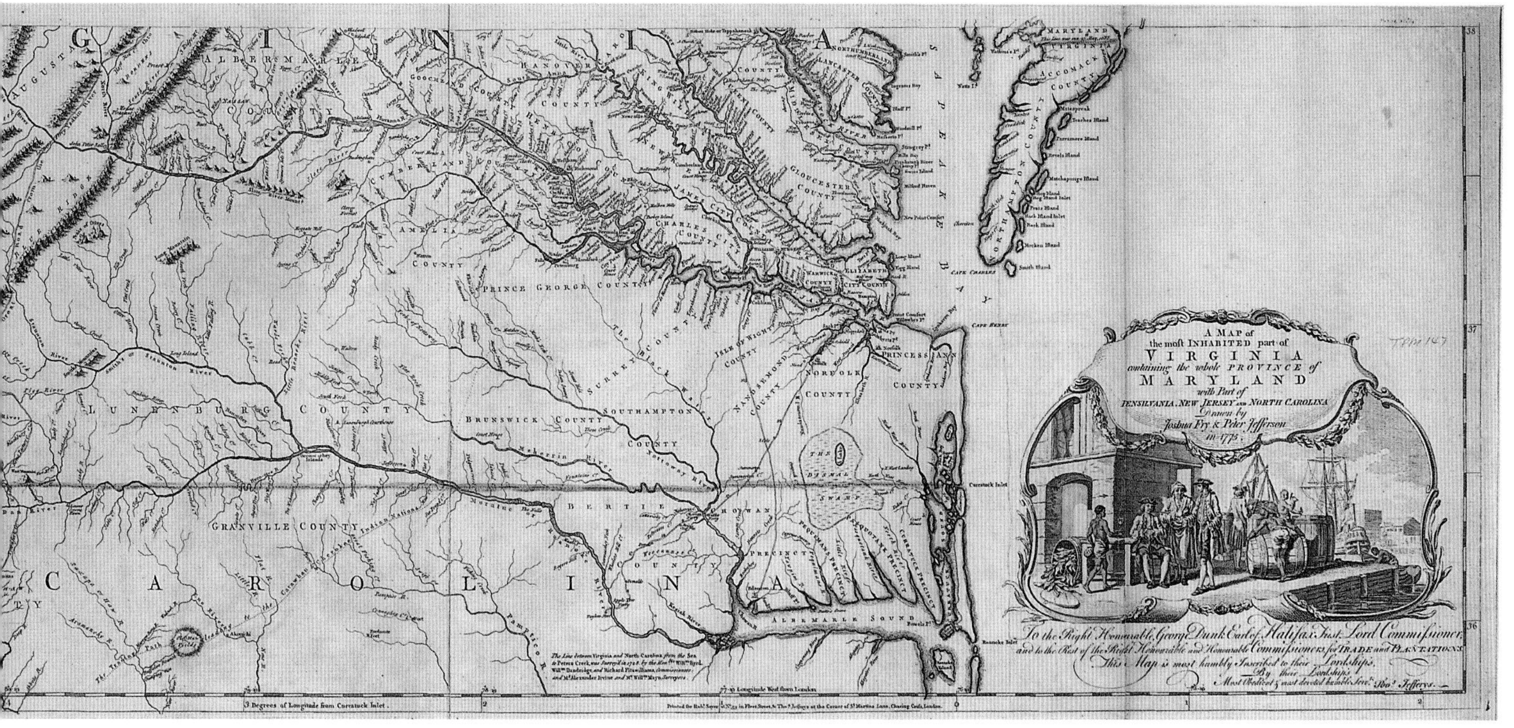
V I R G I N I A
ALBERMARLE
GOOCHLAND
CUMBERLAND
AMELIA COUNTY
PRINCE GEORGE COUNTY
GLOUCESTER COUNTY
CHARLES CITY COUNTY
ELIZABETH CITY COUNTY
NORFOLK COUNTY
PRINCESS ANN COUNTY
SURREY COUNTY
SOUTHAMPTON
BRUNSWICK COUNTY
LUNENBURG COUNTY
NANSEMOND COUNTY
ISLE OF WIGHT COUNTY
C A R O L I N A
GRANVILLE COUNTY
BERTIE
ALBEMARLE SOUND
CAPE CHARLES
CAPE HENRY
ACCOMACK
MARYLAND
3 Degrees of Longitude from Currituck Inlet
Longitude West from London
The Line between Virginia and North Carolina from the Sea to Peters Creek, was Surveyed in 1728 by the Hon. Wm. Byrd, Wm. Dandridge, and Richard Fitzwilliams, Commissioners; and Mr. Alexander Irvine and Mr. Wm. Mayo, Surveyors
A MAP of
the most INHABITED part of
VIRGINIA
containing the whole PROVINCE of
MARYLAND
with Part of
PENSILVANIA, NEW JERSEY and NORTH CAROLINA
Drawn by
Joshua Fry & Peter Jefferson
in 1775.
To the Right Honourable George Dunk Earl of Halifax First Lord Commissioner;
and to the Rest of the Right Honourable and Honourable Commissioners for TRADE and PLANTATIONS
This Map is most humbly Inscribed to their Lordships,
By their Lordships
Most Obedient & most devoted humble Servt. Thos. Jefferys.

REVOLUTIONARY WAR MILITARY CAMPS: FRENCH BATTALION

The Rochambeau Campaign Plans

TITLE: Amerique Campagne, 1782. Plans des Differents camps occupes par L'Armée aux Ordres de M. le Comte de Rochambeau
DATE DEPICTED: July - December 1782
DATE ISSUED: 1782
CARTOGRAPHER: Comte Jean Baptiste donatien de Vimeur Rochambeau (possibly drawn by a military engineer in his troop)
Bound manuscripts, pen and ink and watercolor, set of 45, 32 x 25 cm, 12.5 x 8 in per page
Library of Congress, Geography and Map Division, G1201.53R62 1782 Vault

A series of 45 watercolor drawings, numbered from 1 to 54, traces the Revolutionary War military route of Comte de Rochambeau's "Amerique Campagne 1782" from south to north as they encamped on their return march to Boston. These charming, quick illustrations of town plans and adjoining features expose the disposition of the territories under siege and opportunities for American triumph. The maps reveal street and road patterns, rivers, creeks, topography and location of structures.

The maps shown are the second part of a two-volume set. The first part, containing the drawings from the early campaign march from north to south, includes more drawings which are larger, more detailed and exuberant than the second set. Comparing the two sets, we might interpret the early drawings as reflecting the enthusiasm of a fresh regiment. The lack of similar detail in the drawings from south to north indicates familiarity with the sites, and perhaps some fatigue and desire for completion. The cover sheet, written by hand in fine calligraphy, entitled "Plans des differents camps occupés par L'Armée aux ordres de Mr. le Comte de Rochambeau" incorporates an eighteenth-century water color-coding convention for a key to elements on the plans. The key designates: red—"artillerie;" yellow—"Troupes Françaises;" blue—"Troupes Ameriquaines;" yellow-ochre—regiments of the Royal deux Ponts in Saintonge (which had camped in the same fields for the 1781 and 1782 campaigns); pale green-yellow—"woodlands;" blue-green—"water."

Plan 15, "Camp at Alexandria, July 17," 15 miles from Colchester, emphasizes the street grid of Alexandria—then a small village at the "Pootowmack" shore, bordered by two creeks at each end. Also highlighted are a number of structures, a military encampment, and several rural roads—one to "Clochester" and two leading to "George Town."

Plan 17, "Camp at Bladensburg, July 19, (20 and 21 sojourning)," eight miles from Georgetown, illustrates an L-shaped plan of the tiny community situated along the Eastern Branch (also known as Anacostia River). Also seen are the river tributaries, two bridges, the locations of three troop encampments, and three principal roads to "Anapolis," "George Town," and "Snowden Iron Works."

Drawing and graphics courses were an integral aspect of military school curricula in the seventeenth and eighteenth centuries. For military officers, engineers, cartographers, and battlefield artists engaing in battle, it was necessary to be accomplished in various drawing techniques. The drawings produced by these individuals were used like our instant polaroid photographs of today; they facilitated communication among the troops, recorded movement, provided a basis for making quick decisions, and chronicled the history of a battle. Documentation varied from notations and sketches to elaborately drawn maps, such as those of the Civil War (p. 88), or perspective paintings, such as those by Pierre L'Enfant (p. 16), the father of the designer of the City of Washington.

Rochambeau's 1782 maps permitted military field commanders, engineers, and communication specialists to grasp site conditions and battlefield positions rapidly in the midst of military maneuvers and stressful emergency conditions. In the theater of war, swift imaginative decision-making depended upon comprehension of the situation at hand. Eager to exploit an advantage, even under extreme weather conditions and hardships of terrain and water-crossings, field commanders required technical reports, such as these maps, to mount offensive or defensive positions, and to switch flanks abruptly.

From an eminent family, Count de Rochambeau was among a group of young, idealistic French officers who sided with the Americans in their struggle for independence from the British, joining in the service of the War of Revolution. He was later honored with a statue in his likeness, placed in Lafayette Square. Like Rochambeau, most French officers returned to France following the war. The Society of the Cincinnati was formed to honor and extend the bond between the French and American officers who had fought together—highlighting the critical role the French had played in the war. One French officer (lacking the status of French nobility) chose to remain in the new nation: Pierre Charles L'Enfant. Imbued with the ideals of freedom, justice and equality, he would become the designer of the plan for the federal City of Washington.

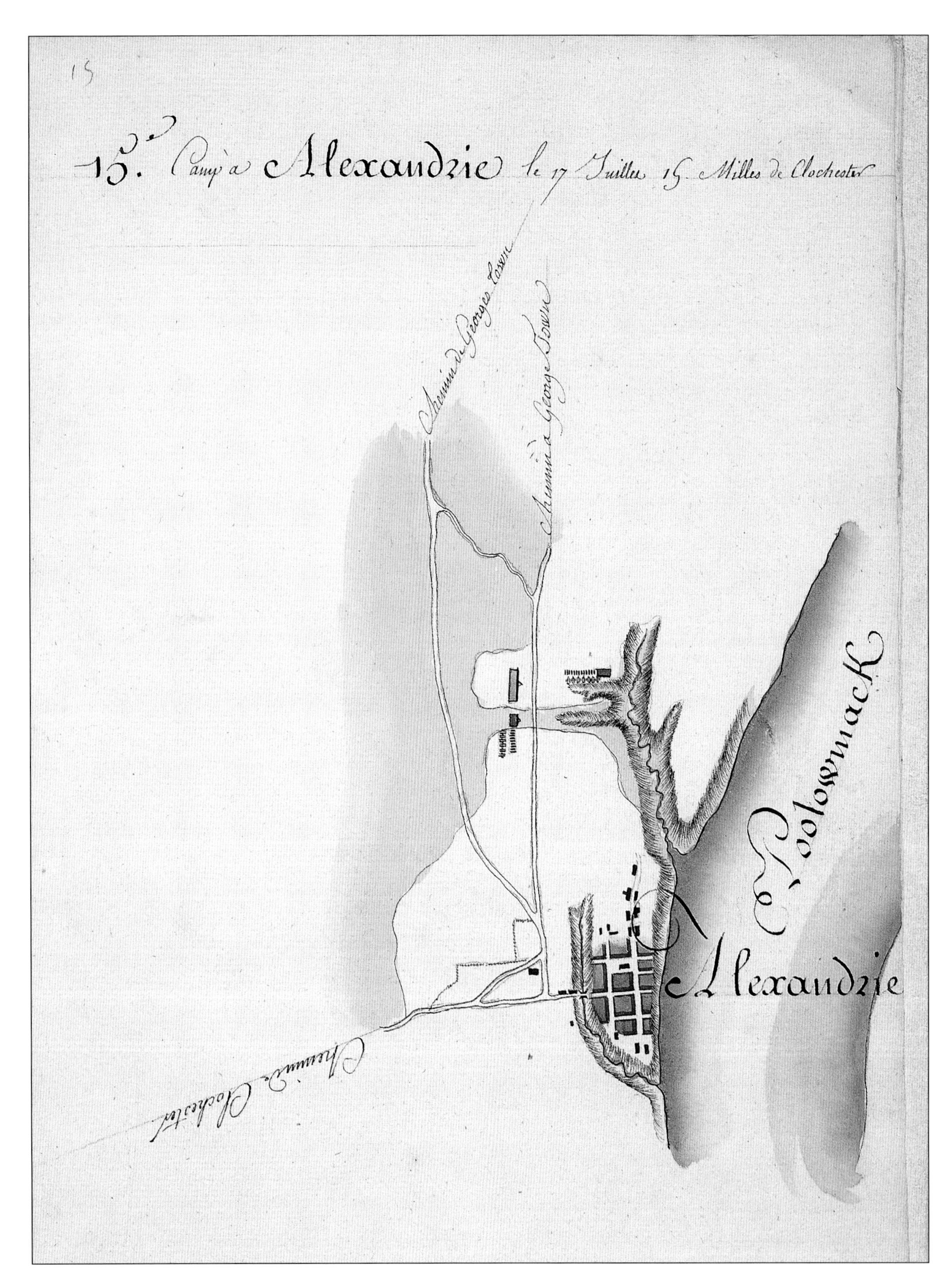
15
15.e Camp à Alexandrie le 17 Juillet 15 Milles de Colchester
Chemin de George Town
Chemin de George Town
Potowmack
Alexandrie
Chemin de Colchester

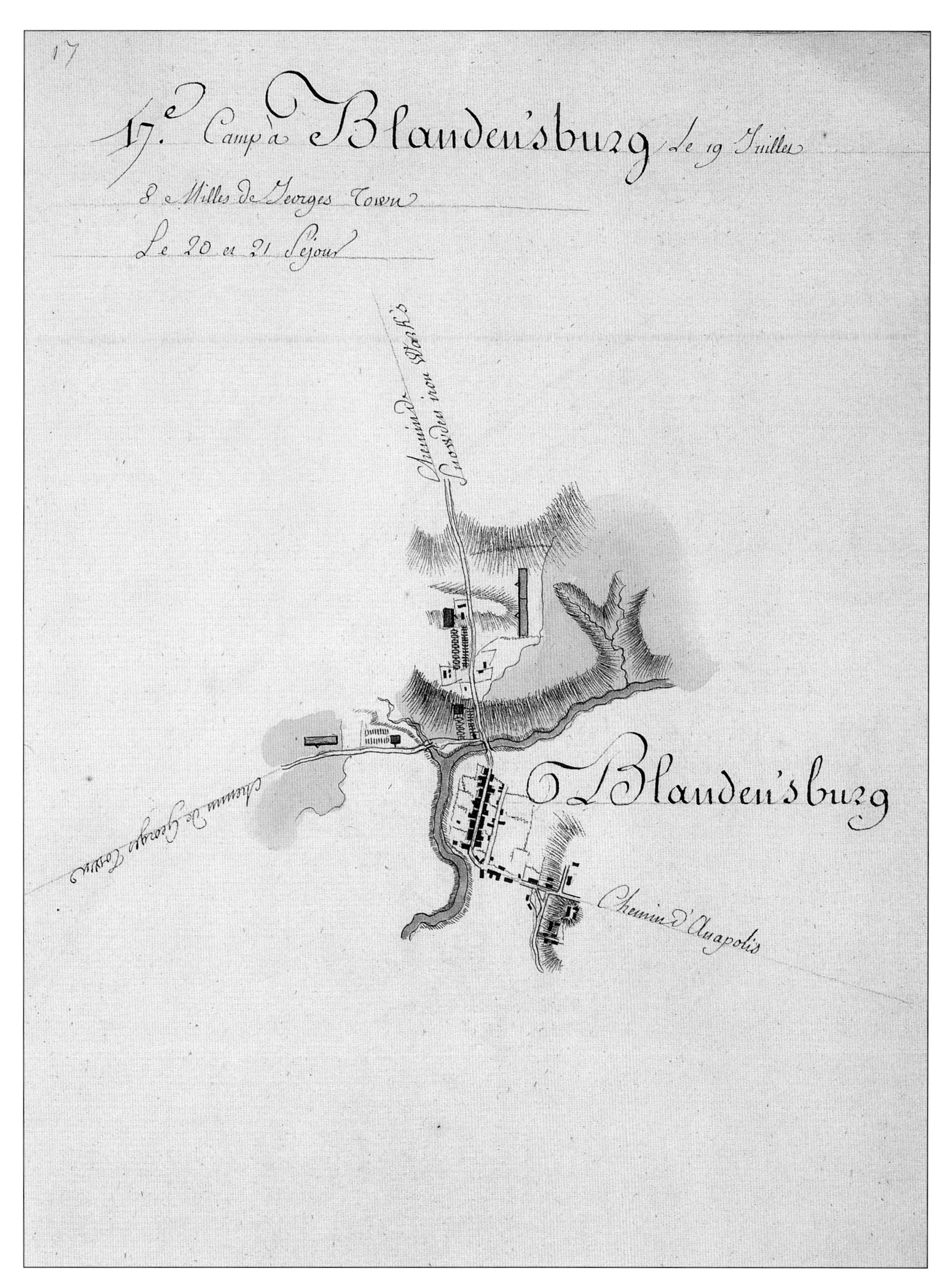
17
17e. Camp à Blandensburg le 19 Juillet
8 Milles de Georges Town
Le 20 et 21 Sejour
Chemin de Snowdens iron Works
Blandensburg
Chemin de Georges Town
Chemin d'Anapolis

FACILITATING TRANSPORTATION AND COMMUNICATION

The Prigs Map, a copy by Daniel Bell

TITLE: A map of the Eastern branch of Potomack river, St. James Creek, Goose Creek and the meanders of the Maryland side of Potomack river up to the mouth of Rock Creek
DATED DEPICTED: 1790
DATE ISSUED: 1790 original
CARTOGRAPHER: John Frederick Augustus Prigs
Manuscript, Ink on laid paper, watercolor and ink washes, 21.75 x 16.625 in.
Courtesy, The Albert H. Small Washington Collection, 455

ESSAY BY LUCINDA PROUT JANKE AND IRIS MILLER

An extremely important map surfaced recently during the decade of the bicentennial of the establishment of Washington, D.C. Drawn in 1790, one year before the visionary L'Enfant Plan, the Prigs Map exposes the foundation from which the city will arise. Viewed from today's vantage, one can envision this majestic city unfolding. As a rare document depicting the site before the arrival of the federal surveyors and planners, the significance of the Prigs Map cannot be overestimated.

As noted in its crown cartouche, this hand-drawn manuscript was the work of Daniel Bell—who copied a map drawn by "John Frederick Augustine Prigs," a Maryland surveyor. Prigs had been active in Prince Georges County and Georgetown as far back as 1758, the time during which he resurveyed the colonial town for its commissioners. In Library of Congress records, the surveyor is listed as "John F. A. Priggs." (Various spellings were typical of the period.) The laid paper by GB/Budgen contains a watermark.

The map depicts the land area between the Eastern Branch of the Potomac River and Rock Creek's mouth—the precise area that was soon to become Washington, D.C. Dated 1790, without specific reference to day or month, it is impossible to determine if the map was created before or after Congress passed legislation (on July 16 of the same year) placing the new federal city on the Potomac River. This legislation, or "Act of Residence," allowed President George Washington to select a specific site for the city along an eighty-mile stretch of the river. And, although Washington did not announce his decision until January 1791, it was well known that a site near Alexandria was his likely choice. Whether or not Prigs's map was created before the Act of Residence remains uncertain; but the fact of its creation certainly reflects the high interest in the site then current for navigation and overland opportunities.

This apparently prescient map seems less mysterious in light of a detailed article—published in Baltimore, early in 1789— that proposed this exact site for the federal city. Written under a pen name by George Walker, an enthusiastic promoter of the new city (as well as an investor in it), the article extolled the virtues of the Eastern Branch, calling it one of the safest and most commodious harbors in America—deep enough for the largest ships. (The Washington Navy Yard was located on the Eastern Branch in 1799, near St. Thomas Bay.) Walker would ultimately purchase a sizable tract in June 1791 in the eastern part of the city known as The Houpyard, just in time to turn it over to Trustees for the creation of the federal city. This map may even have been prepared for a client such as Walker. The Prigs-Bell invites study and detailed comparison with the later Toner (p. 56) and King plantation (p. 58) maps.

Included among the houses located near the water's edge of the Eastern Branch (and depicted on the map) were those of Walter Evans and William Young. Evans, who had patented his land in 1764, sold it to William's brother Abraham in 1791. The projection where the river turns northward is still known as Evans Point. The site of William's house, described as a 24 x 36-foot frame plus an 18-foot kitchen, is now incorporated in Congressional Cemetery. A sister, Elizabeth Young Wheeler, owned three small tracts on the river acquired from her brother William, in 1778. She married the operator of the Eastern Branch ferry, Aquila Wheeler and lived in the Ferry House depicted on the map— a frame house of modest dimension: it was only 14 x 21 feet. Nearby, a barn and cabin can be made out. Across the river, on the east side, two ferry houses were also indicated.

When the city plan was created, Virginia Avenue terminated near the ferry landing at the foot of 14th Street, S.E. At the time, most of the land in the city east of the future site of the Capitol was owned by the Young brothers and their sister; by Jonathan Slater (Slater's house also appears here on the Prigs Map); and by Daniel Carroll, of Duddington (p. 58). Slater, a small plantation owner who had acquired his tract in 1764, sold it to an Englishman named William Prout, in 1791. Prout ultimately rented-out the Slater property—near 8th and M Street, S.E.—for a tavern that might have been the first in the new city. Today, the Prout/Slater tract is the core of the historic Capitol Hill neighborhood.

Other houses of note on this map include: Notley Young's 1756 plantation house (p. 58), located near what became the intersection of 10th and G Street, S.W.; and, across the river, the substantial brick plantation house, Upper Giesborough—built by a member of the Addison family in 1735. Giesborough Point, where the rivers split, is the location today of Bolling Air Force Base.

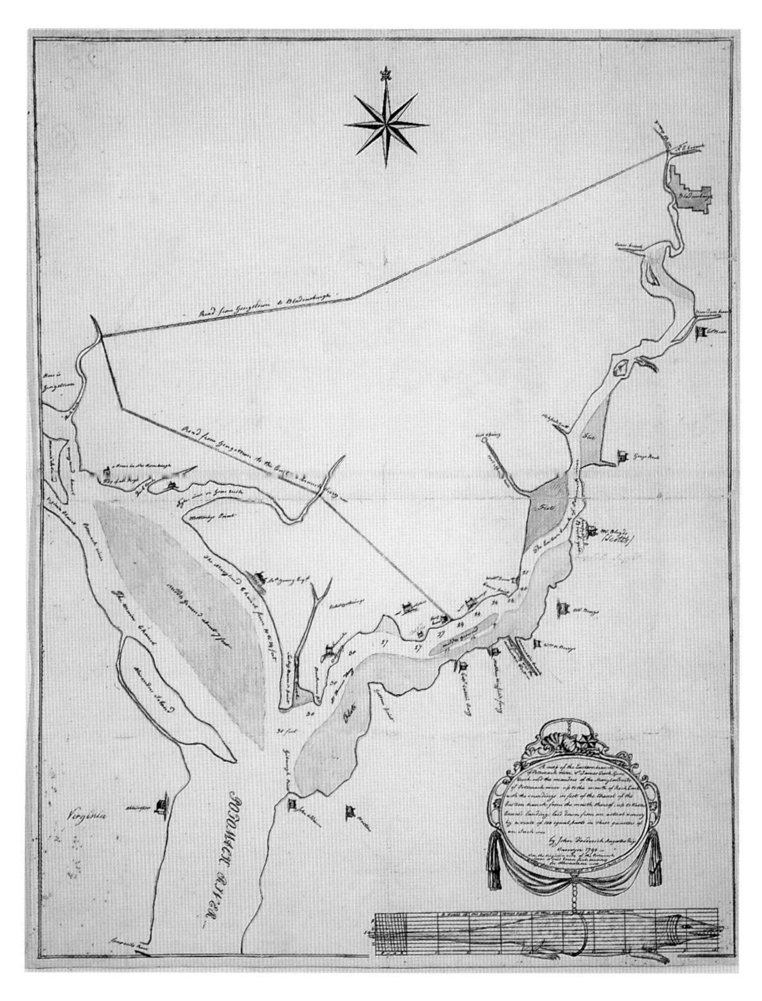

The full description or title of the map in the cartouche reads:

A map of the Eastern branch of Potomack river, St. James Creek, Goose Creek and the meanders of the Maryland side of Potomack river up to the mouth of Rock Creek with the soundings in feet of the Chanel of the Eastern branch, from the mouth thereof, up to Walter Evans's landing, laid down from an actual survey by a scale of 100 equal parts in three quarters of an inch.
by John Frederick Augustus Prigs
Surveyor 1790
Note, the Virginia side of the Potomack river is laid down from memory for Illustration.
A copy from the original—Daniel Bell.
A scale of one hundred equal parts in three quarters of an inch

The map also documented the use of now supplanted names, such as Turkey Buzzard Point (Hain's Point) and Alexander's Island (National Airport). In 1790, the Eastern Branch was not yet referred to as the Anacostia, the name in use today. Curiously, a branch flowing into the Eastern Branch was here designated as "Annacostin branch" with a reference to the "Annacostin Fort." Many creeks were named here, including Cool Spring Run—a creek the existence of which resulted in a jog in the city plan at today's C Street, N.E. at 15th. Its owner, Benjamin Stoddert, asked George Washington to exclude his property from the new city; Washington agreed! Hence, the strange configuration at the new city's eastern boundary.

The delineation of roads, the ferry, and the river surroundings suggest that whoever commissioned this map must have had a serious business venture in mind. Both the northern road from Bladensburg and the southern road from the ferry crossing led to Georgetown, one of the three existing colonial towns before the federal city.[1] Two earlier "paper" towns, platted but not developed, were included in the federal territory. This map showed "a House in New Hamburgh," just east of Rock Creek, but not Carrollsburg (p. 58), near Turkey Buzzard Point. Of particular interest, navigational markings are included: "The Maryland Chanel from 10 to 14 feet"; "Middle Ground about 7 feet"; "The Main Chanel, [on the Virginia side of the] Potomack river, 30 feet [at the confluence of the Potomac and] St. Thomas Bay [of the] Eastern branch of the Potomack river." These were noted along with shallow marsh areas and "Flats." Soundings indicate that the depth decreases toward the northeast to twenty-one feet and less. Because of silting, pollution and deteriorating navigational conditions, anticipated growth never occurred and Bladensburg has remained a village town.

No bridges existed across the rivers, thereby requiring crossing by ferry. Although the Maryland legislature, in which state this territory was located, had authorized bridges across the Eastern Branch in the late eighteenth century, none was built until the turn of the century.

Like L'Enfant a year later, Daniel Bell and John F. A. Prigs were, in 1790, clearly focused on the Eastern Branch waterfront and its possibilities. This map, recently acquired by Washington collector, Albert H. Small, may prove to be one of the seminal documents to convey the rationale for the site selection for the new federal city.

[1] Of the two primary roads spanning the territory from northeast and southeast toward the west side converging at Georgetown, the northern route from Bladensburg surely seems to have influenced how L'Enfant derived the commercial baseline—at K Street—to lay out the city (the other baseline being the ceremonial virtual axis of The Mall). Adjusted southward, K Street became an extra wide thoroughfare and the main connection between Bladensburg on the Anacosta River (anticipated as a major harbor) and Georgetown harbor and waterfront. On his plan, L'Enfant wrote: "New Road to Bladensburg." The southern road, as found on both the Prigs and L'Enfant maps, closely corresponds to Pennsylvania Avenue today.

AN ARTIST'S EYE: FRENCH AND AMERICAN INSPIRATION

THE "ORIGINAL" L'ENFANT PLAN

TITLE: Plan of the City intended for the Permanent Seat of the Government of the United States
DATE DEPICTED: Future projection
DATE ISSUED: 1791
CARTOGRAPHER: Peter (Pierre) Charles L'Enfant
Manuscript, colored map, ink, lead pencil and watercolor 70 x 81 cm, on sheet 73 x 104 cm
Library of Congress, Geography and Map Division, G3850 1791 .L4 Vault

L'ENFANT PLAN FACSIMILE: PRESERVING THE IMAGE

THE U.S. COAST AND GEODETIC SURVEY COPY

TITLE: Plan of the City intended for the Permanent Seat of the Government of the United States
DATE DEPICTED: 1791 projection
DATE ISSUED: 1887
CARTOGRAPHER: Peter (Pierre) Charles L'Enfant, FACSIMILE: B.A. Colonna, F.M. Thorn
PUBLISHER: U.S. Coast and Geodetic Survey Office; Julius Bien & Co., N.Y.
Photo lithograph, color, scale [ca. 1:16,000], 70 x 81 cm, on sheet 77 x 119 cm
Library of Congress, Geography and Map Division, G3850 1791 .L43 1887

The L'Enfant Plan—majestic in design, comprehensive in magnitude—is unique to its site. It follows the topography in a loose grid and gardenesque manner common to late-seventeenth- and eighteenth-century urban design modes. Walk the streets of Washington and feel the sensitivity its designer had for the terrain. Experience the two east-west baselines: the broad commercial K Street thoroughfare and the virtual arbitrary axis across The Mall. Throughout, an infusion of diagonals is apparent; they intersect to form *carrefours—a* sequence of open spaces in the pattern of stars—a pattern extolled by abbé Antoine Marc Laugier on the embellishment of towns (*Essai et Observations sur l'Architecture*, 1753, 1765). Inspiration and beauty, Laugier suggested in his writings, come not from perfect alignments of excessive regularity, but rather from variety and imagination, bizarre connections, streets entering squares at many angles, a bit of chaos and surprise within a sense of order.

In this spirit L'Enfant's plan does have derivatives, but none as direct imitation. Consider such urban designs throughout France in L'Enfant's time as LeNotre's Versailles and extension of the Tuilleries in Paris (p. 16). There's Heré's Nancy at Place Stanislas and Carriere (p. 20); Ledoux' Aix-en-Provence, in the medieval and baroque quarters; plans for alteration of Lyons, Nantes, Dijon, Reims, others; and such landscapes and hunting forests of chateaux as Marly-le-roi, Sceaux, Compiegne (p. 15); or Chanteloup, as painted by L'Enfant's father and L'Enfant's father's colleague, Cozette, for Duc de Choiseul (p. 16).

During the design process, Jefferson had provided L'Enfant with recently transformed European references: French plans of Lyons (reworked by Soufflot), Bordeaux (p. 20), Orleans, Montpelier, Marseilles; Italian plans of Milan and Turin; German plans of Karlsruhe and Frankfurt am Main; Netherlandish plans of Amsterdam. Similarly, L'Enfant's plan would influence subsequent city plans such as that of New Delhi, India (Lutyens); Canberra, Australia (Griffith); and Barcelona, Spain (Cerda).

Optimistic about L'Enfant's capabilities and vision, President George Washington chose him to design the new federal city. Washington was pleased to see the sweeping magnitude and symbolic grandeur of L'Enfant's plan. At first, work proceeded well. Thomas Jefferson coordinated communication among the President, Ellicott, and L'Enfant. Unfortunately, problems ultimately arose, and Washington was obliged to dismiss his talented designer. Surveyor Andrew Ellicott, who had worked with L'Enfant, redrafted the plan with numerous specific and objectionable morphological alterations (p. 44).

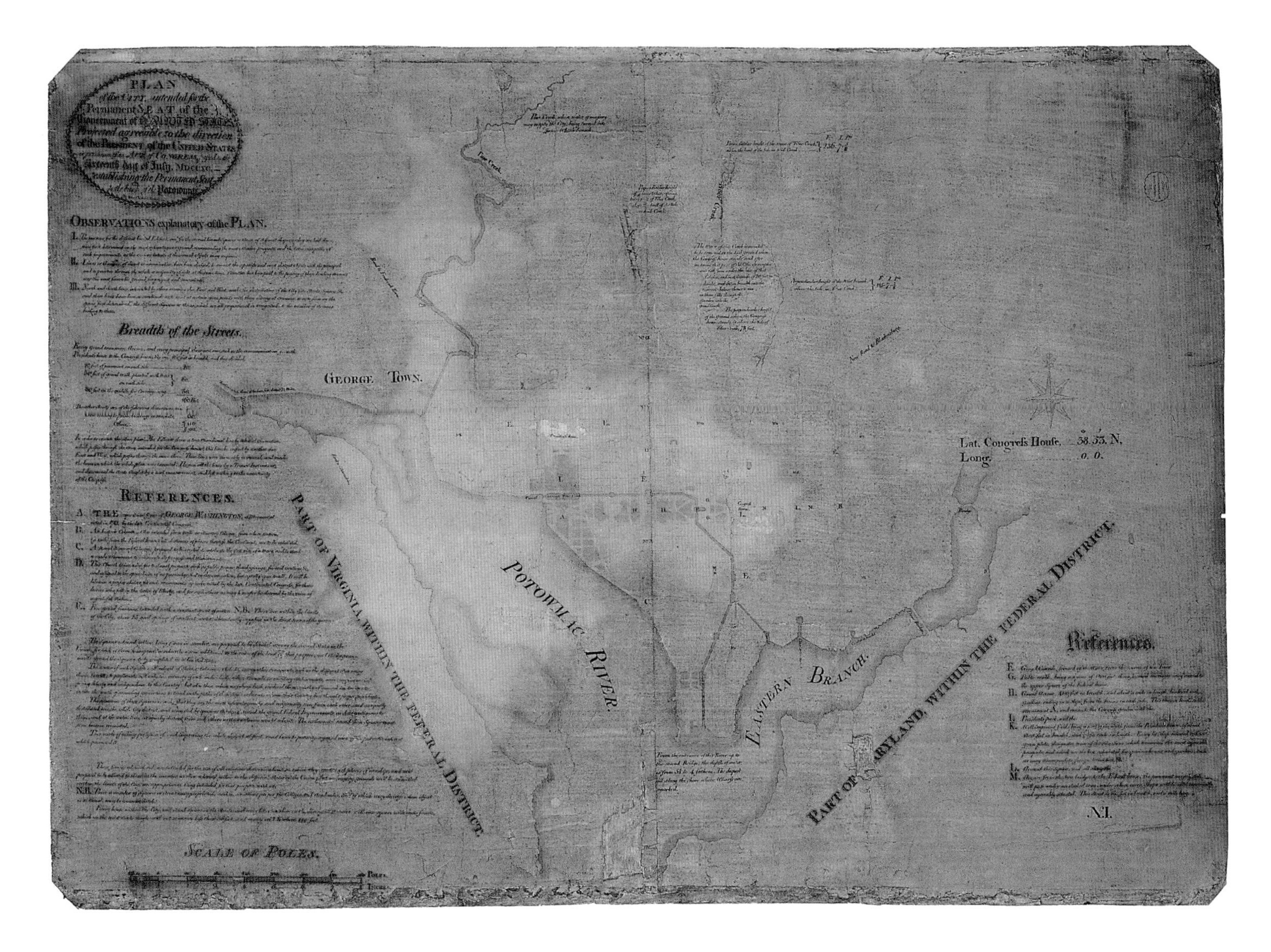

A brief account of the events leading to L'Enfant's replacement follows: On August 27, 1791 L'Enfant met with President Washington in Philadelphia. After this, he attempted to have his manuscript plan engraved in time for the federal city sale of lots on October 17. He chose N. Pigalle for the job. Pigalle was given a reduced plan that had been largely drawn by L'Enfant's assistant, Stephen Hallet. Inauspiciously, Pigalle was unable to obtain a copperplate in time to complete the work for the first sale of lots—resulting in a disappointed president and a discredited L'Enfant.

Distressed by the untimely work on his maps, and angered by a structure underway on a major avenue of his plan, L'Enfant assumed the authority of eminent domain, and effected the removal of the partially constructed home of Daniel Carroll of Duddington, which protruded seven feet into the roadway at New Jersey Avenue south of the Capitol. (This Daniel Carroll, son of Charles Carroll—proprietor of large landholdings in the original federal city—was born in "Duddington," his father's house near the Anacostia River. Members of this large, prominent family were thus identified by their land connections.) Soon thereafter, the president enjoined L'Enfant to continue his work—but under the "authority of the commissioners" who were familiar with the daily range of issues pertaining to the federal city—an unacceptable condition for L'Enfant, who had previously worked in a more direct way with the President.

In preparation for the second sale of lots, on November 28, 1791 Washington wrote to L'Enfant, requesting that he finish his map for publication. Toward this end, Benjamin Ellicott, was furnished with Hallet's small sketch plan. He was to draw a plan with accurate measurements appropriate for engraving. This, however, was not done quickly enough. To L'Enfant's consternation, he learned that in the meantime Thomas Jefferson had importuned Andrew Ellicott (Benjamin Ellicott's brother) to complete the work for engraving. The map, as completed by Andrew Ellicott, was received by President Washington on February 20, 1792 (p. 44)—ultimately without corrections by, nor attribution to, L'Enfant as its author, as had been the president's request. Thus, L'Enfant was the first, in a series of mapmakers in the service of the nation's capital, to be replaced.

By the late nineteenth century, the original L'Enfant Plan was in fragile, deteriorating condition. Impetus to make a reproducible copy arose among those who recognized its value and acknowledged the price of its constant use as court evidence. Between 1876-1884 it had been stored in the Capitol basement with records of the

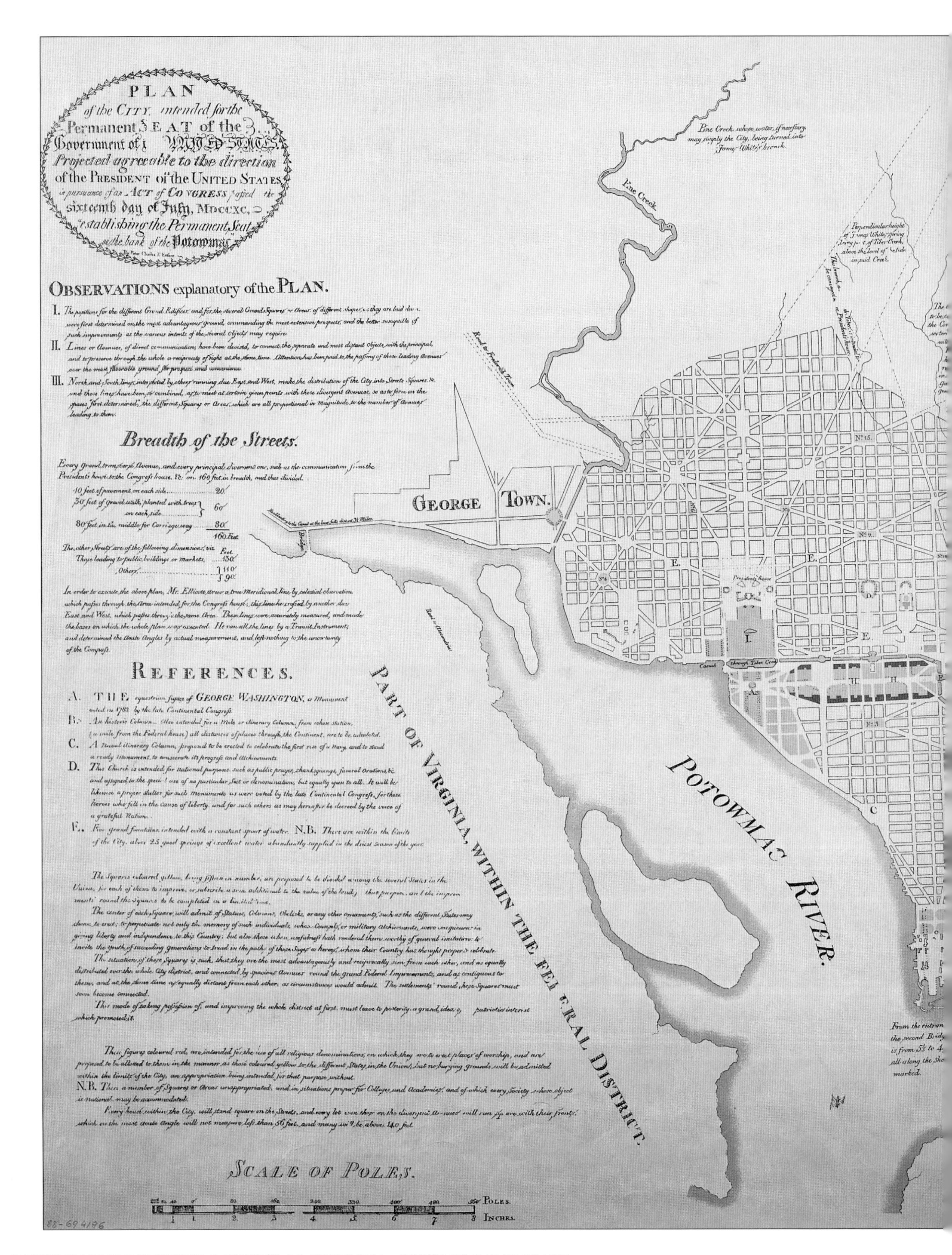
PLAN
of the CITY, intended for the
Permanent SEAT of the
Government of UNITED STATES:
Projected agreeable to the direction
of the PRESIDENT of the UNITED STATES,
in pursuance of an ACT of CONGRESS passed the
sixteenth day of July, MDCCXC,
"establishing the Permanent Seat
on the bank of the Potowmac"
OBSERVATIONS explanatory of the PLAN.
I. The positions for the different Grand Edifices, and for the several Grand Squares or Areas of different shapes, as they are laid down, were first determined on the most advantageous ground, commanding the most extensive prospects, and the better susceptible of such improvements as the various intents of the several objects may require.
II. Lines or Avenues, of direct communication have been devised, to connect the separate and most distant objects with the principal, and to preserve through the whole a reciprocity of sight at the same time. Attention has been paid to the passing of those leading Avenues over the most favorable ground for prospect and convenience.
III. North and South lines, intersected by others running due East and West, make the distribution of the City into Streets, Squares &c. and those lines have been so combined, as to meet at certain given points with those divergent Avenues, so as to form on the spaces "first determined," the different Squares or Areas, which are all proportional in magnitude to the number of Avenues leading to them.
Breadth of the Streets.
Every grand transverse Avenue, and every principal divergent one, such as the communication from the President's house to the Congress house, &c. are 160 feet in breadth, and thus divided.
10 feet of pavement on each side 20
30 feet of gravel walk, planted with trees on each side 60
80 feet in the middle for Carriage way 80
160 Feet.
The other Streets are of the following dimensions, viz. Feet
Those leading to public buildings or markets, 130
Others, 110 90
In order to execute the above plan, Mr. Ellicott drew a true Meridional line by celestial observation which passes through the Area intended for the Congress house; this line he crossed by another due East and West, which passes through the same Area. These lines were accurately measured, and made the bases on which the whole plan was executed. He ran all the lines by a Transit Instrument, and determined the Acute Angles by actual measurement, and left nothing to the uncertainty of the Compass.
REFERENCES.
A. THE equestrian figure of GEORGE WASHINGTON, a Monument voted in 1783 by the late Continental Congress.
B. An historic Column. Also intended for a Mile or itinerary Column, from whose station, (a mile from the Federal house) all distances of places through the Continent, are to be calculated.
C. A Naval itinerary Column, proposed to be erected to celebrate the first rise of a Navy, and to stand a ready Monument to consecrate its progress and Atchievements.
D. This Church is intended for national purposes, such as public prayer, thanksgivings, funeral Orations, &c. and assigned to the special use of no particular Sect or denomination, but equally open to all. It will be likewise a proper shelter for such monuments as were voted by the late Continental Congress, for those heroes who fell in the cause of liberty, and for such others as may hereafter be decreed by the voice of a grateful Nation.
E. Five grand fountains, intended with a constant spout of water. N.B. There are within the limits of the City, above 25 good springs of excellent water abundantly supplied in the driest season of the year.
The Squares coloured yellow, being fifteen in number, are proposed to be divided among the several States in the Union, for each of them to improve, or subscribe a sum additional to the value of the land; that purpose, and the improvements round the Squares to be completed in a limited time.
The center of each Square, will admit of Statues, Columns, Obelisks, or any other ornaments, such as the different States may choose to erect; to perpetuate not only the memory of such individuals, whose Counsels, or military Atchievements, were conspicuous in giving liberty and independence to this Country; but also those whose usefulness hath rendered them worthy of general imitation: to invite the youth of succeeding generations to tread in the paths of those Sages or heroes, whom their Country has thought proper to celebrate.
The situation of these Squares is such, that they are the most advantageously and reciprocally seen from each other, and as equally distributed over the whole City district, and connected by spacious Avenues round the grand Federal Improvements, and as contiguous to them, and at the same time as equally distant from each other, as circumstances would admit. The settlements round those Squares must soon become connected.
This mode of taking possession of, and improving the whole district at first, must leave to posterity a grand idea of the patriotic interest which promoted it.
These figures coloured red, are intended for the use of all religious denominations, on which they are to erect places of worship, and are proposed to be allotted to them in the manner as those coloured yellow to the different States in the Union; but no burying grounds will be admitted within the limits of the City, an appropriation being intended for that purpose without.
N.B. There a number of Squares or Areas unappropriated, and in situations proper for Colleges and Academies, and of which every Society whose object is national may be accommodated.
Every house within the City, will stand square on the Streets, and every lot even those on the divergent Avenues will run square with their fronts, which on the most acute Angle will not measure less than 56 feet, and many will be above 140 feet.
SCALE OF POLES.
POLES.
INCHES.
GEORGE TOWN.
Bridge
Road to Alexandria
Road to Frederick Town
Pine Creek
Pine Creek, whose water, if necessary, may supply the City, being turned into James White's branch.
PART OF VIRGINIA, WITHIN THE FEDERAL DISTRICT.
POTOWMAC RIVER.
President's house
Canal through Tiber Creek
From the entrance the second Bridge is from 5½ to 4 all along the Sho marked.
88-694196

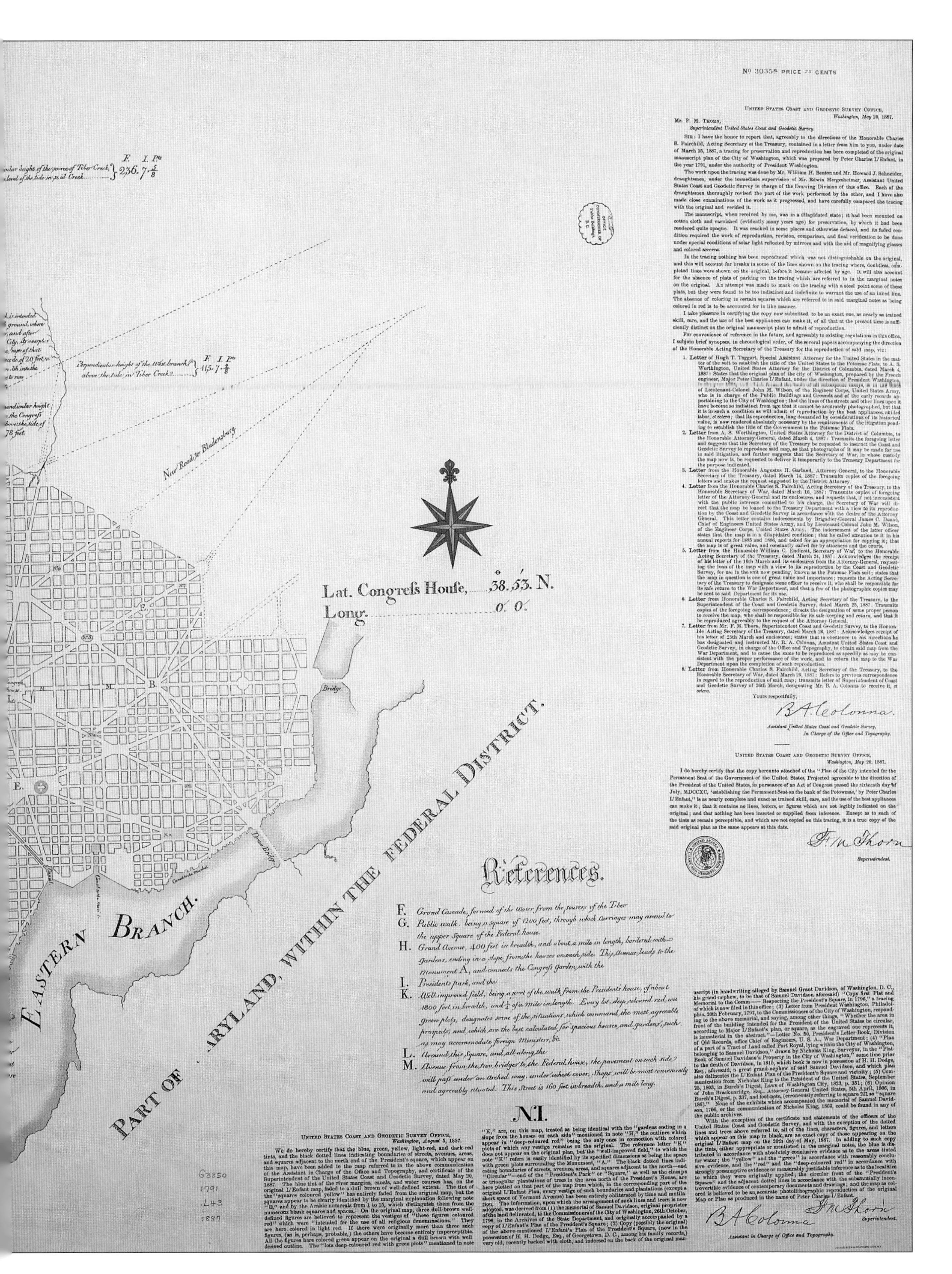

No 30358 PRICE 25 CENTS
New Road to Bladensburg
Lat. Congress House, 38.53. N.
Long. 0. 0.
Bridge
Eastern Branch
Part of Maryland, within the Federal District.
References.
F. Grand Cascade, formed of the Water from the sources of the Tiber
G. Public walk, being a square of 1200 feet, through which Carriages may ascend to the upper Square of the Federal house.
H. Grand Avenue, 400 feet in breadth, and about a mile in length, bordered with gardens, ending in a slope from the houses on each side. This Avenue leads to the Monument A, and connects the Congress Garden with the
I. Presidents park, and the
K. Well improved field, being a part of the walk from the Presidents house, of about 1800 feet in breadth, and ¼ of a mile in length. Every lot, deep coloured red, with green plots, designates some of the situations which command the most agreeable prospects, and which are the best calculated for spacious houses and gardens, such as may accommodate foreign Ministers, &c.
L. Around this Square, and all along the
M. Avenue from the two bridges to the Federal house, the pavement on each side will pass under an Arched way, under whose cover, Shops will be most conveniently and agreeably situated. This Street is 160 feet in breadth, and a mile long.
N.I.
United States Coast and Geodetic Survey Office, Washington, May 20, 1887.
Yours respectfully, B. A. Colonna.
F. M. Thorn, Superintendent.
United States Coast and Geodetic Survey Office, Washington, August 5, 1887.

Office of Public Buildings and Grounds, under chief Col. John Wilson. An ownership dispute over Potomac River Tidal Flats led to authorization of a facsimile and temporary transfer of the Plan to Coast and Geodetic Survey (National Ocean Survey), Office of Topography. The work was directed by B. A. Colonna, Drawings Division, and supervised by Edwin Hergesheimer. Two draughtsmen, William H. Benton and Howard J. Schneider, made the tracing.

On May 20, 1887 the first reproductions were printed in black and white. The tracing, mainly by Benton, was on two sheets of varnished French vegetable paper. Descriptions in the notes and bits of remaining color on the Plan show that color was subsequently added to the facsimile, which is still available for purchase. The original Plan was returned to storage on September 26, 1887.

On November 11, 1918 L'Enfant's plan was transferred to the Library of Congress—first to the Manuscript Division and later to the Geography and Map Division. In 1951 repair and restoration was attempted with the technology then available. A conservation evaluation was made again in 1981. To avoid losing details, only necessary minor repairs were undertaken and Conservation Officer Peter Waters recommended production of a new facsimile. In 1991 a like-size color facsimile and a computer-assisted reproduction were composed from the original manuscript utilizing computer-enhanced technology and photography. Thereafter, a hermetically sealed container was designed to protect the 200-year-old L'Enfant Plan.

Much has been written about the planning of Washington, District of Columbia, and the strengths and weaknesses of its urban landscape design. In truth, L'Enfant himself best describes his powerful ideas and profound principles. He wrote on the map (Note: clarifications by the author of this book are included below, enclosed in brackets):

OBSERVATIONS explanatory of the PLAN

I. The positions for the different Grand Edifices, and for the several Grand Squares or Areas as they are laid down were first determined on the most advantageous ground, commanding the most extensive prospects, and the better susceptible of such improvements as the various intents of the several objects may require.

II. Lines or Avenues of direct communication have been devised to connect the separate and most distant objects with the principal, and to preserve through the whole a reciprocity of sight at the same time. Attention has been paid to the passing of those leading Avenues over the most favorable ground for prospect and convenience.

III. North and South lines, intersected by others running due East and West, make the distribution of the City into Streets, Squares, etc., and those lines have been so combined as to meet at certain given points with those divergent Avenues, so as to form on the spaces "first determined," different Squares or Areas which are all proportional in magnitude to the number of Avenues leading to them.

Breadth of the Streets

Every grand transverse Avenue, and every principal divergent one, such as the communication from the President's house to the Congress house: 160 feet in breadth, and thus divided,

40 feet of pavement on each side 20
30 feet of gravel walk, planted
with tree on each side 60
80 feet in the middle for
Carriage way 80
160 Feet.

The other Streets are of the following dimensions,
Those leading to public
buildings or markets 130
Others 110
[or] 90

In order to execute the above plan, Mr. Ellicott drew a true meridional line by celestial observation which passes through the area intended for the Congress house (Capitol); this line he crossed by another one East and West, which passes through the same Area. These lines were accurately measured, and made the bases on which the whole plan was executed. He ran all the lines by a Transit Instrument, and determined the Acute Angles by actual measurement, and left nothing to the uncertainty of the Compass. [Note: nothing is mentioned here about the role of Benjamin Banneker, an astronomer and freed former slave, who is believed to have developed the position of the true north meridian.]

REFERENCES [Note: at this section, written in the margin in Jefferson's hand was a note, "All to be struck out" with a pencil line drawn through.]

A. THE equestrian figure of GEORGE WASHINGTON, a Monument voted in 1783 by the late Continental Congress.

B. An historic Column—also intended for a Mile or itinerary Column, from whose station (a mile from the Federal house) all distances of places throughout the Continent to be calculated.

C. A Naval itinerary Column, proposed to be erected to celebrate the first rise of a Navy, and to stand a ready Monument to consecrate its progress and achievements.

D. This Church is intended for national purposes, such as public prayer, thanksgiving, funeral orations etc. and assigned to the special use of no particular Sect or denomination, but equally open to all. It will be likewise a proper shelter for such monuments as were voted by the late Continental Congress for those heroes who fell in the cause of liberty, and for such other as may hereafter be decreed by the voice of a grateful Nation.

E. Five grand fountains intended with a constant spout of water. N.B. There are within the limits of the City, above 25 good springs of excellent water, abundantly supplied in the driest Season of the year.

F. Grand Cascade, formed of water from the sources of the Tiber.

G. Public walk, being a square of 1200 feet, through which carriages may ascend to the upper Square of the Federal house.

H. Grand Avenue, 400 feet in breadth, and about a mile in length, bordered with gardens, ending in a slope from the houses on each side. This Avenue leads to Monument A, and connects the Congress garden with the

I. President's park and the

K. Well-improved field, being a part of the walk from the President's house of about 1800 feet in breadth, and 3/4 of a mile in length. Every lot, deep-colored red with green plots, designates some of the situations which command the most agreeable prospects, and which are the best calculated for spacious houses and gardens, such as may accommodate foreign Ministers, etc.

L. Around this Square and all along the

M. Avenue from the two bridges to the Federal house, the pavement on each side will pass under an Arched way, under whose cover, Shops will be most conveniently and agreeably situated. This Street is 160 feet in breadth, and a mile in length.

The Squares colored yellow, being fifteen in number, are proposed to be divided among the several States of the Union, for each of them to improve, or subscribe a sum additional to the value of the land, that purpose, and the improvements round the Squares to be completed in a limited time.

The center of each Square will admit of Statues, Columns, Obelisks, or any other ornaments such as the different States may choose to erect; to perpetuate not only the memory of such individuals whose Counsels, or military achievements, were conspicuous in giving liberty and independence to this Country; but also those whose usefulness hath rendered them worthy of general imitations; to invite the youth of succeeding generations to tread in the paths of those Sages or heroes, whom their Country hath thought proper to celebrate.

The situation of these Squares is such, that they are the most advantageously and reciprocally seen from each other, and as equally distributed over the whole City district, and connected by spacious Avenues round the grand Federal Improvements, and as contiguous to them, and at the same time as equally distant from each other, as circumstances would admit. The settlements round those Squares must soon become connected.

This mode of taking possession, and improving the whole district at first, must leave to posterity a grand idea of patriotic interest that promoted it.

These figures colored red, are intended for the use of all religious denominations, on which they are to erect places of worship, and are proposed to be allotted to them in the manner as those colored yellow to the different States in the Union; but no burying grounds will be admitted within the limits of the City, an appropriation being intended for that purpose without.

[Note: in Paris at that period, an effort was made to develop cemeteries on the city's outskirts to improve inner-city health, stench, and tasteful aesthetics.]

N.B. There are a number of Squares or Areas unappropriated, and in situations proper for Colleges or Academies, and of which every society whose object is national may be accommodated.

Every house within the City, will stand square on the Streets, and every lot, even those on the divergent Avenues, will run square with their fronts, which on the most acute angle will not measure less than 56 feet, and many till be above 140 feet.

[In Jefferson's hand, recommended changes:]

Plan of the City of Washington in the territory of Columbia ceded by the states of Virginia, Mareland to the U.S. of America and by them as the seat of government after 1800.

[Note: L'Enfant placed pen-and-ink notes on the plan, some of which were struck out or edited by Jefferson, including notes at the northern creeks and southern wharf.]

The water of [Tiber] Creek is intended to be conveyed on the high ground, where the Congress house stands, and after watering that part of the City, its overplus will fall from under the base of that Edifice, and in a cascade of 20 feet in height, and 50 in breadth into the reservoir below; thence to run in three falls through the garden into the Grand Canal. The perpendicular height of the ground where the Congress house stands is above the tide of Tiber Creek, 78 feet.

From the entrance of this River up to the second Bridge, the depth of water is from 5 1/3 to 4 fathom. The deepest was all along the shore where Wharfs are marked.

[Upper left oval corner:]

PLAN of the City, intended for the Permanent SEAT of the Government of the UNITED STATES, Projected agreeable to the direction of the President of the United States, in pursuance of an ACT of CONGRESS, passed on the sixteenth day of July, MDCCXC, "establishing the Permanent Seat on the bank of the Potowmac."

ENVISIONING A FEDERAL CITY FOR A NEW NATION: LETTERS AND SKETCHES

The Jefferson Papers

TITLE: Jefferson Papers: Letters
DATE DEPICTED: Future projection
DATE ISSUED: 1790-1791: August 29, 1790; September 14, 1790; March 11, 1791; August 18, 1791
AUTHOR AND DESIGNER: Thomas Jefferson
Letters, ink on paper
Library of Congress, Manuscript Division, Neg - LC.MSS-27748-96

Thomas Jefferson played a major role in the design and development of the District of Columbia. This is evident in many of Jefferson's papers and much of his correspondence. Consider such writings as "Draft of Agenda for the Seat of Government," dated August 29, 1790; his "Report to Washington on Meeting Held at Georgetown," dated September 14, 1790; and a letter dated March 11, 1791 from Jefferson to President Washington. In these letters, Jefferson offered specifics for the layout of the District, the particular use of certain parcels, as well as the means to acquire land.

In the "Draft of Agenda," Jefferson determined a territory not to exceed "one hundred square miles in any form" to be located by "metes and bounds." According to Jefferson's Report to Washington, a position below Little Falls was designated to begin at ten miles. Another seven miles along the Eastern Branch was to be accepted from Maryland. An additional ten miles was to be accepted from Virginia beginning at the lower end of Alexandria. In locating the town, Jefferson proposed "give it double the extent on the Eastern branch of what it has on the river. The former will be for persons in commerce, the latter for those connected with the government." (See fig. b.)

In reporting to President Washington on conversations with local landowners, Jefferson determined that land for the proposed capital city could be obtained without the financial assistance of either the national government or local state assemblies. Jefferson proposed that affected property owners be paid twice the fair market value of their properties in exchange for a declaration of cessation to the new government. These purchases would be financed by the sale of lots; a $120,000 donation from Virginia; a possible like donation from Maryland. The market value was to be established without speculative consideration. Jefferson also proposed a maximization of the city's size by accepting equal portions of land from Maryland and Virginia. He explored several design and planning options, such as laying out the city in a rectangular pattern of lots with the exception of the lots obliquely angled along the river; and rotating the city boundaries from a squared orientation to a diamond to gain land advantage (and additional riverfront, if Bladensburg was to be included in the territory). It may be noted that George Walker had previously suggested this site.

To oversee this endeavor, Jefferson proposed the appointment of three commissioners. Their role was to purchase the land necessary for certain public buildings, right of way, and residential neighborhoods. The President's House, gardens, and offices required two consolidated squares. The Capitol and offices required one square, as did the land for a Market. In addition, Public Walks necessitated another nine consolidated squares. Jefferson intended each square to be "two hundred yards every way" or approximately eight acres. Squares were also reserved for a prison.

In his later letter of September 14, 1790, Jefferson determined that 1,500 acres were necessary for this design. Three hundred acres were to be allocated to the public buildings, walks and the like. The remaining 1,200 acres were to be subdivided into quarter-acre lots, from which street allotments were to be taken. This design would have resulted in approximately 2,000 lots. The lots and streets were to be situated at right angles, as in Philadelphia. Jefferson thought it best to "lay out the long streets parallel with the creek, and the others crossing them at right angles, so as to leave no oblique angled lots but the single row which shall be on the river." The streets were restricted to a width between 100 and 120 feet with 15-foot-wide "foot-ways".

According to Jefferson's letter of August 29, 1790, houses would have no setback restrictions. Those built at an even distance from the street would produce a "disgusting monotony," as in Philadelphia. "The contrary practice varies the appearance, and is much more convenient to the inhabitants." He specified, "the lots to be sold out in breadths of 50. Feet: their depths to extend to the diagonal of the square," which he illustrated with a diagram (see fig. c.). Today urban designers still debate the concept of maintaining a rigid street "wall," and the acceptable deviations permitting setbacks in the urban fabric.

Jefferson did admire the Parisian practice of height restrictions and encouraged its employment within the federal city to "keep houses low and convenient, streets light and airy, and fires more manageable." He admitted, of course, this was an "object of legislation." In his letter to President Washington dated March 11, 1791 Jefferson proposed that it be unlawful to build "any house with more than two floors between the common level of the earth and the eves." As the former ambassador to France, Jefferson was aware of French laws, in existence since the seven-

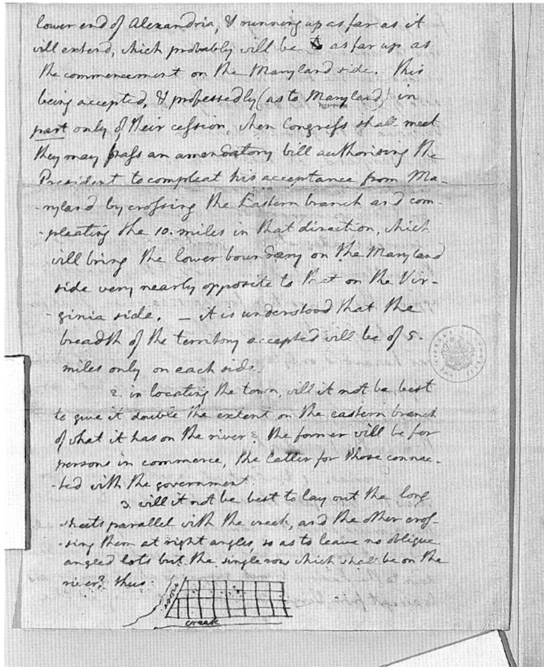

lower end of Alexandria, & running up as far as it will extend, which probably will be as far up as the commencement on the Maryland side. This being accepted, & professedly (as to Maryland) in part only of their cession, when Congress shall meet they may pass an amendatory bill authorising the President to compleat his acceptance from Maryland by crossing the Eastern branch and compleating the 10. miles in that direction, which will bring the lower boundary on the Maryland side very nearly opposite to that on the Virginia side. — it is understood that the breadth of the territory accepted will be of 5. miles only on each side.

2. in locating the town, will it not be best to give it double the extent on the eastern branch of what it has on the river? the former will be for persons in commerce, the latter for those connected with the government.

3. will it not be best to lay out the long streets parallel with the creek, and the other crossing them at right angles, so as to leave no oblique angled lots but the single row which shall be on the river? thus:

FIGURE B.

FIGURE A.

FIGURE C.

FIGURE D.

teenth century, limiting heights of buildings and designating the ratio of street width to building height. Such regulations aided in the avoidance of epidemics and promoted cleanliness, proper sewage systems, and trash collection.

In a letter to Pierre Charles L'Enfant of August 18, 1791, Jefferson urged L'Enfant to complete "laying out the lots…as a suggestion has been made here of arranging them in a particular manner which will probably make them more convenient to the purchasers, and more profitable to the sellers." He also informed L'Enfant of a request to engrave a map of the federal territory, which was the "best right" of "yourself or Mr. Ellicott." "If you do I would suggest to you the idea of doing it on a square sheet to hang corner upwards thus (see fig. d). The outlines being NW, NE, SE and SW. The meridians will be vertical as they ought to be; the streets of the city will be horizontal and vertical, and near the center; the Patowmac and Eastern branch will be nearly so also; there will be no waste in the square sheet of paper. This is suggested merely for your consideration." In 1793 at Jefferson's request, Andrew Ellicott prepared a manuscript of the Territory of Columbia with the north corner pointed upwards, and the following year he completed an engraving of the same image (p. 49). C. E. Arnold followed this layout theme in his 1862 Civil War map (p. 88) as a patriotic gesture.

AN ALTERNATE EYE: JEFFERSON'S CONCEPT OF EQUALITY

The Jefferson Plan

TITLE: Plan for the City of Washington
DATE DEPICTED: Future projection
DATE ISSUED: March 1791
DESIGNER: Thomas Jefferson
Manuscript, ink on paper, 24.75 x 39 cm, 9.75 x 15.5 in
Library of Congress, Manuscript Division, Neg - LC. MSS-27748-215;
Reel 14;# Box 1, oversized, #10805

From the initial planning process for a new capital as "seat of government," Thomas Jefferson's involvement was substantial. He consulted directly with President Washington regarding the precise site. He guided survey work, drew concept sketches, and coordinated the design process. Between 1790-1791 as the President's representative, he corresponded with designer Pierre Charles L'Enfant and surveyor Andrew Ellicott. Jefferson participated in formulating the map's title for the new federal capital. In addition to his numerous notations and corrections on L'Enfant's plan (p. 35), he offered his own alternate proposal, a celebratory design for the new capital that represented his pragmatic utilitarian value system.

Aspects of Jefferson's background may provide insight into the philosophical underpinnings of his city plan. Jefferson was a consummate renaissance man—third president of the United States, first Secretary of State, politician, diplomat, intellectual, author, designer, plantsman and landed gentleman. He was also the son of Peter Jefferson, a renowned cartographer, who, along with Joshua Fry, produced the most important mid-Atlantic regional map of the era (p. 26).

Having amassed a vast personal library, his books and maps included national and international works on history, politics, architecture, urbanisme, chinoiserie, and landscape. He designed his own home at Monticello, Virginia, laying out a *ferme ornee*, or ornamental farm, in the French-American tradition. Imbued with iconic meaning, his land furnished spiritual and optical pleasure. His daily journal described a broad array of plants, noting their habits, composition and species. Ultimately, he designed the preeminent university campus, the University of Virginia. Its construction, based on Jefferson's design, was authorized by the Virginia General Assembly on January 25, 1819.

Before departing for France as America's ambassador, to serve from 1785 to 1789, Jefferson introduced a plan to Congress for jurisdiction of the western territory ceded by the original states to the federal government. The following year, the Land Ordinance of 1785 was adopted, providing for surveys of public lands in divisions of six-mile-square plats for townships. A north-south meridian was established at the Ohio River, with parallel lines at six-mile intervals. Lines crossing at right angles formed 36-mile squares, which were subdivided into lots of one square mile (640 acres). Certain lands were reserved either for public school education or for the United States Government. Adjustments were devised for encroachments and special conditions. This Ordinance defining the standard for equitable land division was followed until the Civil War, and no doubt influenced the philosophical essence of Jefferson's Washington Plan—to distribute the squares in equal dimensions, thereby to represent a democratic ideal.

In early February 1791, Jefferson gave explicit instructions to Andrew Ellicott to survey and mark the Federal Territory's ten-square-mile boundary. In March he notified L'Enfant to begin his design plan, working first along the Eastern Branch, then continuing northward, marking an existing road between Georgetown and the Eastern Branch. L'Enfant was to note hills, valleys, all creeks, "Morasses," and accounting for and connecting fixed points.

As instructed, L'Enfant began working on the east, unaware that President Washington and Jefferson had been secretly collaborating with trusted landholders, Deakins and Stoddert, who were purchasing western land cheaply near Georgetown. Exchange of letters continued between Jefferson and L'Enfant during the design process.

Later in March, Jefferson prepared his own pen and ink plan based upon ideas he had shared with the President. More modest in scale than the inventive plan L'Enfant was to design, Jefferson's plan was superimposed on the paper town, Hamburg (laid out in 1768, of 234 lots). The total area, including the repetitive rectangular pattern of dots for anticipated future expansion, encompassed 2,000 acres. Jefferson's plan, similar in size to that of Philadelphia at that time, showed 600-foot-square blocks, to be divided into 60-foot-wide lots.

Reversing orientation from his prior sketch of 14 blocks along the Eastern Branch with four blocks at the Potomac (p. 41, fig. b), Jefferson's

Plan spread 11 blocks east-west and three blocks north-south. Conceived in a grid motif covering 273 acres, the centerpiece of his plan was a promenade, "public walks," one block wide, eight blocks (400 feet) long. The Capitol and President's House presided at opposite ends.

Other features were a dashed line representing Ferry Road, connecting Georgetown with the Eastern Branch (see also the Prigs-Bell Map p. 32). An intent to validate Washington as a major competitive shipping port to rival Baltimore was intrinsic to his plan. Hence, he noted water depths along the Potomac River from Georgetown to below Tiber Creek's mouth. We can only imagine how a port might have altered Washington's character.

The civic core of Jefferson's plan was laid out in the manner of Williamsburg, the capital of Virginia. In size it was less than the core of L'Enfant's plan; in sensibility, less glorious. Beyond that, it is astonishing that little consideration was given to distant views, elevations in terrain, designation of open space or parks outside the central area.

Looking back, Jefferson's gridiron Plan was noble in principle but retardataire in reality. Positioned between two worlds—sophisticated European Baroque-Classical philosophy, and a visceral American-utopian ideology of society, his American social vision prevailed. This was based upon a desire for human equality metaphorically rooted in a presence of nature extending unendingly westward.

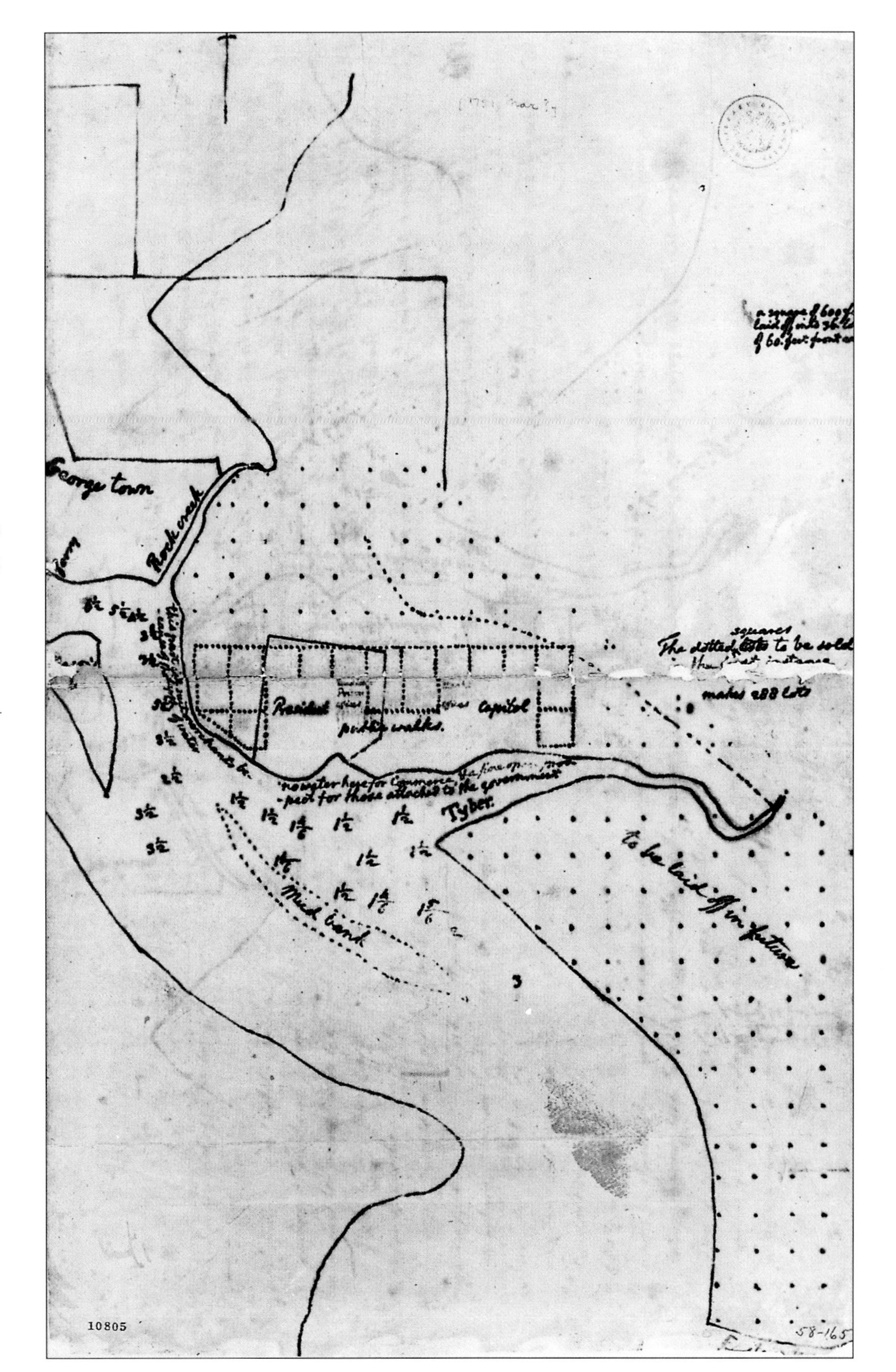

SURVEYOR'S REVISIONISM: FIRST PUBLISHED PLAN

The Ellicott Plan; Thackara & Vallance, Publisher

TITLE: Plan of the City of Washington
DATE DEPICTED: Future projection
DATE ISSUED: 1792
CARTOGRAPHER: Andrew Ellicott
PUBLISHER: Thackara & Vallance, Philadelphia, PA in "The Universal Asylum, and Columbian Magazine"
Engraving, 9.5 x 12.5 in. Courtesy, The Albert H. Small Washington Collection, 070.MP.ED.L.F.
(Also in Library of Congress, Geography and Map Division, G3850 1792 .L4 Vault)

ENGRAVED ELABORATION: FIRST "OFFICIAL" PLAN

The Ellicott Plan; Thackara & Vallance, Publisher

TITLE: Plan of the City of Washington, in the Territory of Columbia ceded by the States of Virginia and Maryland to the United States of America
DATE DEPICTED: Future projection
DATE ISSUED: 1792
CARTOGRAPHER: Andrew Ellicott
PUBLISHER: Thackara & Vallance, Philadelphia, PA
Engraved map, 20.25 x 27 in
Courtesy, The Albert H. Small Washington Collection, 064.MP.ED.E.F.
(Also in Library of Congress, Geography and Map Division, G3850 1792.L4 Vault)

Strange anomalies have frequently occurred throughout the history of the federal city. Among the unfortunate turn of events in 1792 was the replacement of the District of Columbia's urban designer, Pierre Charles L'Enfant, by a surveyor lacking in design qualifications. How was it possible that an extraordinary design of noble intent by a brilliant artist was reworked within a year by a surveyor who failed to understand precious nuances of his predecessor's versatile plan?

The second would-be designer, Andrew Ellicott, made subtle but quite significant changes to L'Enfant's exquisite urban plan. These alterations, seemingly rational and imperceptible, have affected the intrinsic character of the city, its buildings and streets. To L'Enfant, the re-draft "unmercifully spoiled" the experiential premise of the original. In truth, it defrauded the nation and urban architects of a superb model from which to learn and emulate.

Examine elements from L'Enfant's plan that Ellicott declined to grasp. One finds that Ellicott's engravings reflect the hand and mentality of an American-trained surveyor—not a European urban artist. In L'Enfant's late-seventeenth and early-eighteenth-century France, inventive composition of loosely gridded streets and radiating diagonal avenues was perfected by Andre Le Notre based upon philosophies of Descartes, Pascal and others. These responded to variations in topography that influenced spatial perception. Ellicott's regularized application of this plan neglected to account for design illusions that enhance urbanism.

Selected avenues, especially Massachusetts and Pennsylvania, were considerably shifted and straightened by Ellicott, although L'Enfant had positioned them for reciprocity of views between contiguous public reservations. The repetitive system of open spaces set out by L'Enfant was determined by the visual effect of grade changes spanning the gently rolling hills, a morphology developed in baroque gardens and urban landscape.

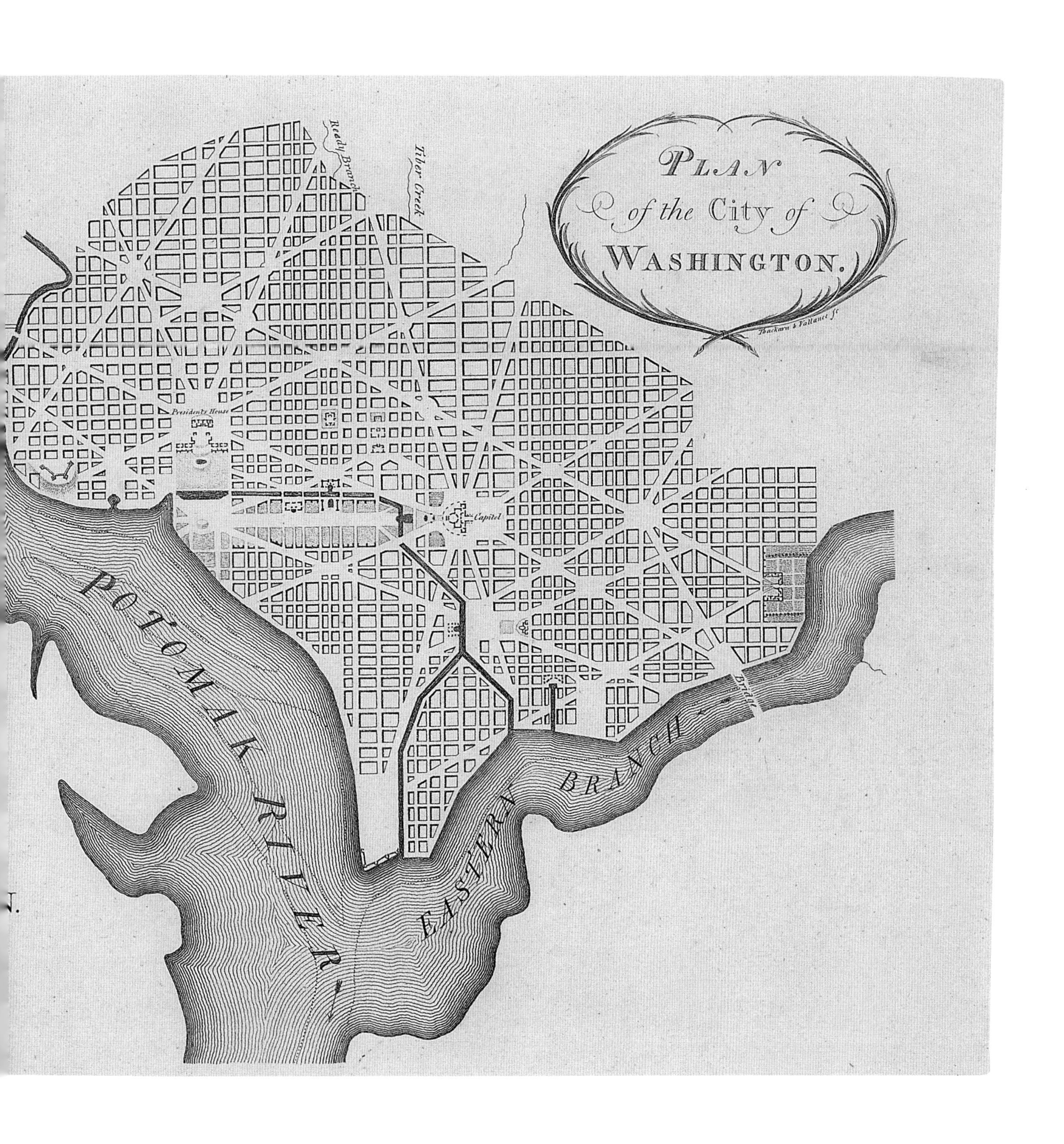
PLAN
of the City of
WASHINGTON.
Thackara & Vallance sc
Ready Branch
Tiber Creek
Presidents House
Capitol
POTOMAK RIVER
EASTERN BRANCH

Connecting avenues, aligned and spaced for comfortable site and walking distances, were meant to go to (not through) the reservations, adjusting for localized situations at each segment. Reservations were to be places for people, the center of identifiable neighborhoods. The egregious alterations of L'Enfant's design vocabulary preclude a true reading of Baroque space at the onset of the age of enlightenment.

Worse, the Ellicott engraving, first submitted as a manuscript on February 20, 1792, was printed without L'Enfant's name despite the President's authorization. Further, twelve civic reservations and five radial avenues were eliminated; public building sites were reshaped; and five stately fountains, placed at strategic "gateways," were deleted—never to be implemented in the ensuing 200 years.

Comparison of L'Enfant's pencil-and-ink erasure-filled manuscript (p.- 35) with Ellicott's engraved plan at the 8th Street corridor reveals consequential differences in the division of blocks between Mount Vernon Square (#2 on L'Enfant's plan) and Pennsylvania Avenue: Ellicott drew three block pairs north and south of the proposed national church while L'Enfant showed only two each. Reconfiguring sites of the "Congress House" (Capitol), President's House, and Pennsylvania Avenue connection trivialized L'Enfant's drawing, which was still in the process of refinement.

Time provided advantage for Ellicott to realize his plan following L'Enfant's dismissal. Benjamin Ellicott, Andrew's brother and a competent draftsman who had worked with L'Enfant, had in his possession survey material and drawings, including a L'Enfant manuscript drawn partly by Stephen Hallet. From these he could adapt components not shown on the L'Enfant Plan, such as numbering the squares throughout the city, listing soundings for river channel depths, and naming the avenues dedicated to the 15 states of the union. On a missing sketch, L'Enfant presumably had designated names of the grand network of avenues to reflect the relative position and direction of states according to their regional geographic locations, i.e., northern, middle and southern states.

Ellicott's engravings imposed adjustments to the seven foremost civic structure sites; fortifications flanking the Potomac River; the Washington City Canal congruent with Tiber Creek, before turning southward to meet St. James Creek; and along The Mall's central "Grand Avenue" and greensward.

Ellicott's manuscript was presented to President Washington and was subsequently given to Thackara and Vallance of Philadelphia to engrave for use at the second sale of lots scheduled for October 8, 1792. The engravers, requiring eight weeks to complete a large map, immediately prepared a reduced version for wide circulation as the frontispiece for the March 1792 publication of *The Universal Asylum,* and *Columbia Magazine.* This map is referred to as the "first" published map of Washington (p. 41).

Because of continuing intrigue related to the timing and production of an official map, the president secretly asked Secretary of State Thomas Jefferson to have another engraving prepared in a different city. Samuel Hill of Boston printed 4000 copies by July 20, 1792, albeit lacking certain useful data, in time for the second sale of lots. Not until November 13, 1792 was the Thackara and Vallance engraving delivered. Elegantly produced on a large copper sheet (52.75 x 71.5 cm versus Hill's plate at 42 x 51.5 cm), it included depths by soundings and 1,146 designated squares (ten more than Hill's).

Map details were finalized at a meeting between the Commissioners, Thomas Jefferson, and James Madison on September 8, 1791. The federal city was named "the City of Washington" and the federal district was called "The Territory of Columbia." The map was titled "A Map of the City of Washington in the Territory of Columbia." It was agreed that the city be divided into four quadrants projecting from the Capitol—north, south, east, and west—naming the streets alphabetically in one direction and numerically in the other.

Ellicott's large map by Thackara and Vallance was a work of handsome cartography, incorporating a simplified version of text from the L'Enfant Plan. For many years, it was regarded as the official map of Washington, until the U.S. Supreme Court ruling in Morris v. United States, 174 U.S. 196 (1899), determined that the Dermott Map (p. 65), the first map of the federal city signed in 1797 by Presidents George Washington and John Adams, was the "official" map of Washington, D.C.

Alas, Andrew Ellicott was soon discharged from service to the city, meeting a similar daunting fate as Pierre Charles L'Enfant who had preceded him, and James Dermott who was to follow.

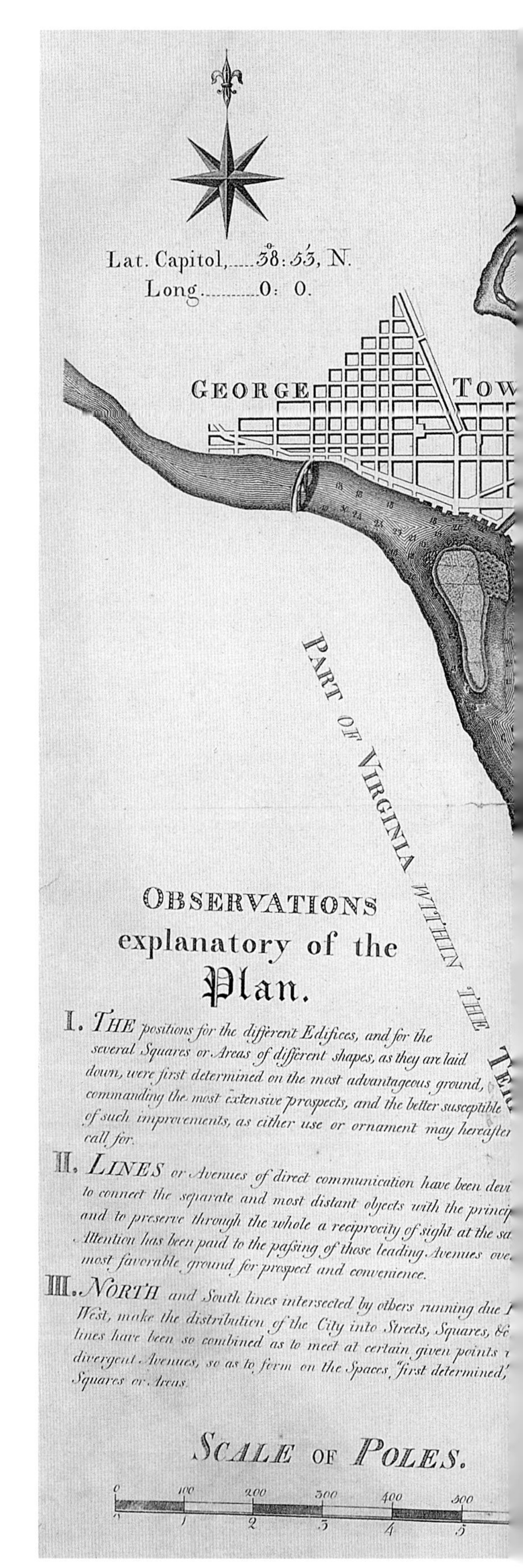

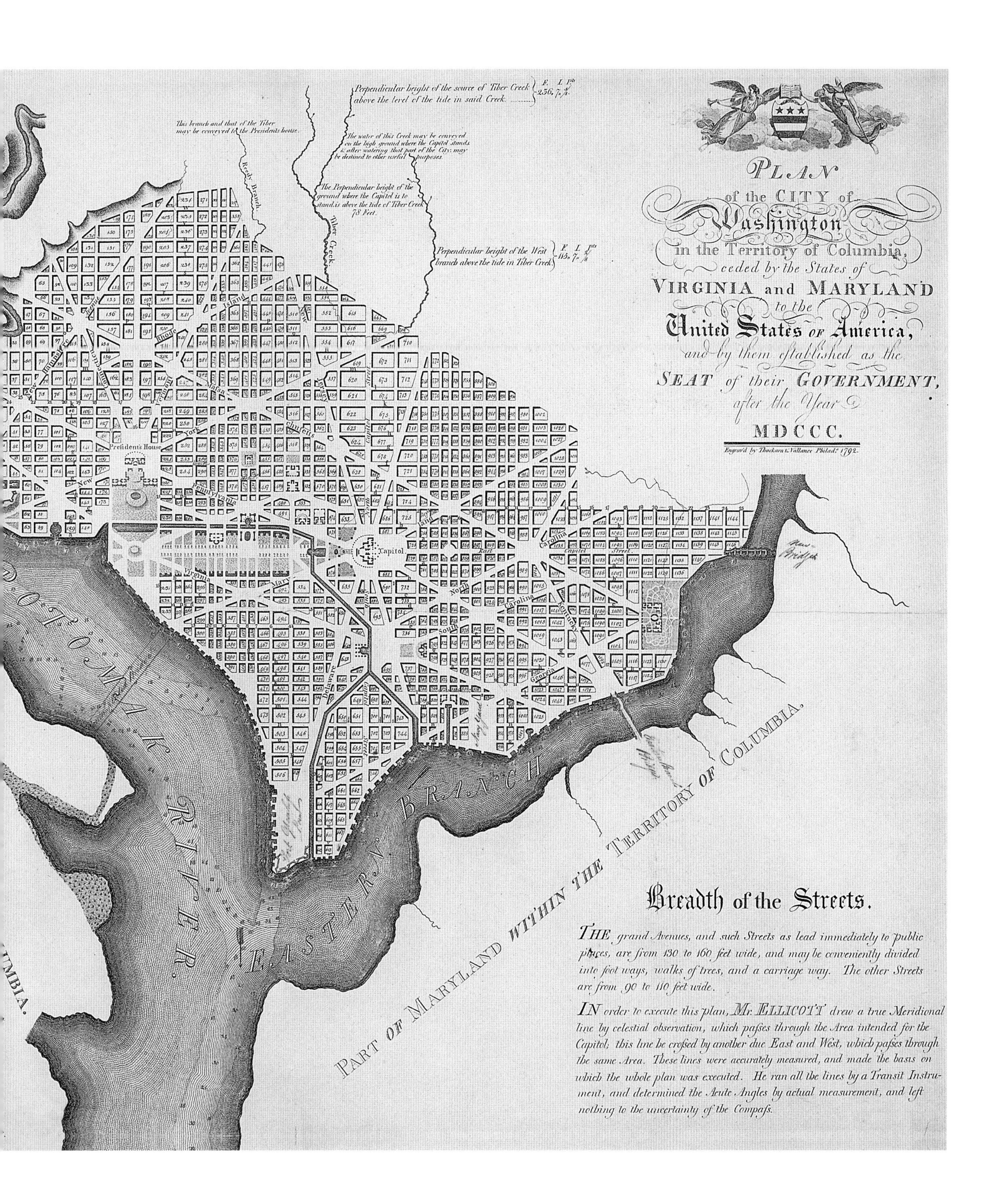
PLAN
of the CITY of
Washington
in the Territory of Columbia,
ceded by the States of
VIRGINIA and MARYLAND
to the
United States of America,
and by them established as the
SEAT of their GOVERNMENT,
after the Year
MDCCC.
Engrav'd by Thackara & Vallance Philad.a 1792.
Perpendicular height of the source of Tiber Creek above the level of the tide in said Creek
This branch and that of the Tiber may be conveyed to the Presidents house.
The water of this Creek may be conveyed on the high ground where the Capitol stands & after watering that part of the City may be destined to other useful purposes.
The Perpendicular height of the ground where the Capitol is to stand, is above the tide of Tiber Creek 78 Feet.
Perpendicular height of the West branch above the tide in Tiber Creek
Reedy Branch
Tiber Creek
President's House
Capitol
POTOMAK RIVER
EASTERN BRANCH
PART OF MARYLAND WITHIN THE TERRITORY OF COLUMBIA.
Breadth of the Streets.
THE grand Avenues, and such Streets as lead immediately to public places, are from 130 to 160 feet wide, and may be conveniently divided into foot ways, walks of trees, and a carriage way. The other Streets are from 90 to 110 feet wide.
IN order to execute this plan, Mr. ELLICOTT drew a true Meridional line by celestial observation, which passes through the Area intended for the Capitol; this line he crossed by another due East and West, which passes through the same Area. These lines were accurately measured, and made the basis on which the whole plan was executed. He ran all the lines by a Transit Instrument, and determined the Acute Angles by actual measurement, and left nothing to the uncertainty of the Compass.

INSERTING PLAN AND TERRITORY INTO BOUNDARIES: CITY OF WASHINGTON AND WASHINGTON COUNTY

The Ellicott "Ten-Mile Square" Map

TITLE: Territory of Columbia
DATE DEPICTED: 1794
DATE ISSUED: 1794 (based on 1793 manuscript)
CARTOGRAPHER: Andrew Ellicott
PUBLISHER: Thackara & Vallance, Philadelphia, PA
Engraved, hand colored map, scale [1:32,500] 55 x 55 cm
Library of Congress, Geography and Map Division, G3850 1794 .E4 Vault

The "10 Mile Square," or 100 square miles, boundary of the Territory of Columbia was surveyed by Andrew Ellicott, who subsequently drew two plans at the request of Thomas Jefferson: a manuscript pen and ink drawing in 1793, and an engraving in 1794 (seen here). The territory included 64 square miles ceded from Maryland in 1788 and 36 square miles ceded from Virginia in 1789.

The first boundary stone was consecrated in a Masonic ceremony on April 15, 1791. The initial boundary point from which the site was laid out was in Virginia at Jones Point on Hunting Creek at the Potomac River south of Alexandria. A line was drawn due northwest 10 miles to Falls Church, at which point a line was drawn due northeast 10 miles crossing through Maryland to the northern border at Woodside (Sixteenth Street); a line was then drawn 10 miles due southeast to "Ches. Beach Junction," followed by a line 10 miles due southwest back to Jones Point. Congress approved the survey with the stipulation that all public buildings must be constructed on the Maryland side of the Potomac.

As late as March 3, 1791 the boundaries were still being adjusted southward from an earlier point at the Eastern Branch and the Potomac to include Alexandria and portions of Virginia. On March 30 after Ellicott completed his survey, President Washington with the three district commissioners, Major Pierre Charles L'Enfant, and Andrew Ellicott rode along the territory boundary. That evening Washington met with most of the original 19 proprietors who signed the "Articles of Agreement." The Commissioners were directed to have boundary stones placed one mile apart to permanently mark the boundary lines. These mile markers are labeled along the edges of this map.

Spanning the extant District of Columbia portion of the 10 Mile Square, the boundary stones' locations and names, from the Potomac River northwest to the Potomac River southeast are: Chain Bridge, Near Reservoir, American University Park, Chevy Chase Circle, Pinehurst, Rock Creek Park, Woodside; Blair Lee, Takoma Park, Stotta Station, Sargent Road, Queen's Chapel Road, Brentwood Road, Reform School, Kenilworth, Burrville, Ches. Beach Junction; Central Avenue, Ridge Road, Suitland Road, Walker Road, Oxon Run, Wheeler Road, Livingston Road, Blue Plains, Fox's Ferry. Since 1916 the Daughters of the American Revolution have preserved the boundary stones. Some of the small original stones are preserved today within black iron fences. Taller stones, to replace the missing originals, were set out in the mid-twentieth century.

The map's image within the boundary highlights the City of Washington at the center of gently rolling lowland adjacent to the Potomac River, Eastern Branch Annakostia, Rock Creek and Boundary Street (Florida Avenue) just below the hills to the north. Surrounding the central City of Washington, the steeply graded land was known as Washington County. Major roads and streams are seen traversing the valleys. Clearly visible, the north arrow is placed near the tight street grid of Alexandria. (Alexandria was not originally included in the proposed Territory of Columbia for the federal Ten-Mile Square; Congress approved Washington's request that it be added in 1791.)

Ellicott's map, with some inaccuracies, was the first of several topographical drawings of the "Territory of Columbia" following President Washington's proclamation of the selected site on January 24, 1791. It is thought that Benjamin Banneker, an astronomical researcher who helped establish true north, may have assisted with this early survey. Periodically, topographical studies to correct inaccuracies have been undertaken by others. Among those are Arnold's Civil War map (p. 89), the handsome pastel-colored 1880 statistical topographical map revisions by Lieutenant Greene (p. 94-99), and Don Hawkins' recent studies.

TERRITORY of COLUMBIA
CITY of WASHINGTON
George Town
Capitol
POTOMAK RIVER
EASTERN BRANCH
Alexanders Island
Alexandria
Four mile Run
West Corner
East Corner

POPULAR PLAN ABROAD: LONDON, FRANCE

The Ellicott Plan; based on Samuel Hill, Publisher

TITLE: Plan of the City of Washington
DATE DEPICTED: Future projection
DATE ISSUED: February 1, 1793
CARTOGRAPHER: based on Andrew Ellicott
PUBLISHER: in *Literary Magazine*, J. Good, Bond Street, London
Hand colored map, scale [ca. 1:40,000], 21 x25 cm, on sheet 27 x 31 cm
Library of Congress, Geography and Map Division, G3850 1793 .G6 Vault

TITLE: Plan de la Ville de Washington
DATE DEPICTED: 1794/1800
CARTOGRAPHER: based on Andrew Ellicott
PUBLISHER: Bibliotheque Nationale en Brumaire An XI, Paris, France

MARKETING FASHION: "BANDANNA" MAP

The Ellicott Plan; based on Samuel Hill, Publisher

TITLE: Plan of the City of Washington in the Territory of Columbia, ceded by the States of Virginia and Maryland to the United States of America
DATE DEPICTED: Future projection
DATE ISSUED: ca. 1792-1794
CARTOGRAPHER: Andrew Ellicott
PUBLISHER: based on Samuel Hill, Boston, MA
Red ink on cloth, 23.5 x 27 in
Courtesy, The Albert H. Small Washington Collection, 274.MP.ED.E.F.

How extraordinary it is to realize the fascination Americans and Europeans felt for the new American democracy and its capital. Would this experiment in republican government work? Soon after the Ellicott Plan was issued, a plethora of maps of Washington began to appear in this country and abroad. European journals and individual publishers distributed maps in a number languages and world capitals. In addition to those shown here—from London and Paris—others were produced in Amsterdam, Russia, Germany, and Spain.

These plans were based upon those published in 1792 by James Thackara and John Vallance in Philadelphia, and on those by Samuel Hill in Boston. Although they were issued between 1792 and 1797, many appear to be from 1800. Confusion arises from the title: "Plan of the City of Washington … Seat of their Government after the Year 1800." The title anticipates the transfer of the capital from its temporary location in New York City to the City of Washington in the Territory of Columbia. Close scrutiny of a map may reveal the true issue date. However, the casual reader, indeed some museum curators, may be fooled by the 1800 date.

The original Thackara and Vallance and the Hill engravings had been beautifully crafted with great clarity, thus enabling the European copies to retain a high quality. Often text was eliminated. Most were black and white engravings although some had added colored tints, as does this one from London. Whereas L'Enfant had been annoyed that his name was deleted from Ellicott's map, these European maps typically have omitted the names of both L'Enfant and Ellicott!

With meticulous study one can decipher the differences in workmanship between the Ellicott Plan's first two engravers. One can also differentiate these from the L'Enfant Plan. First, look at rivers. Hill's map shows channels but no soundings. Next, study details on The Mall and nearby civic buildings. Finally,

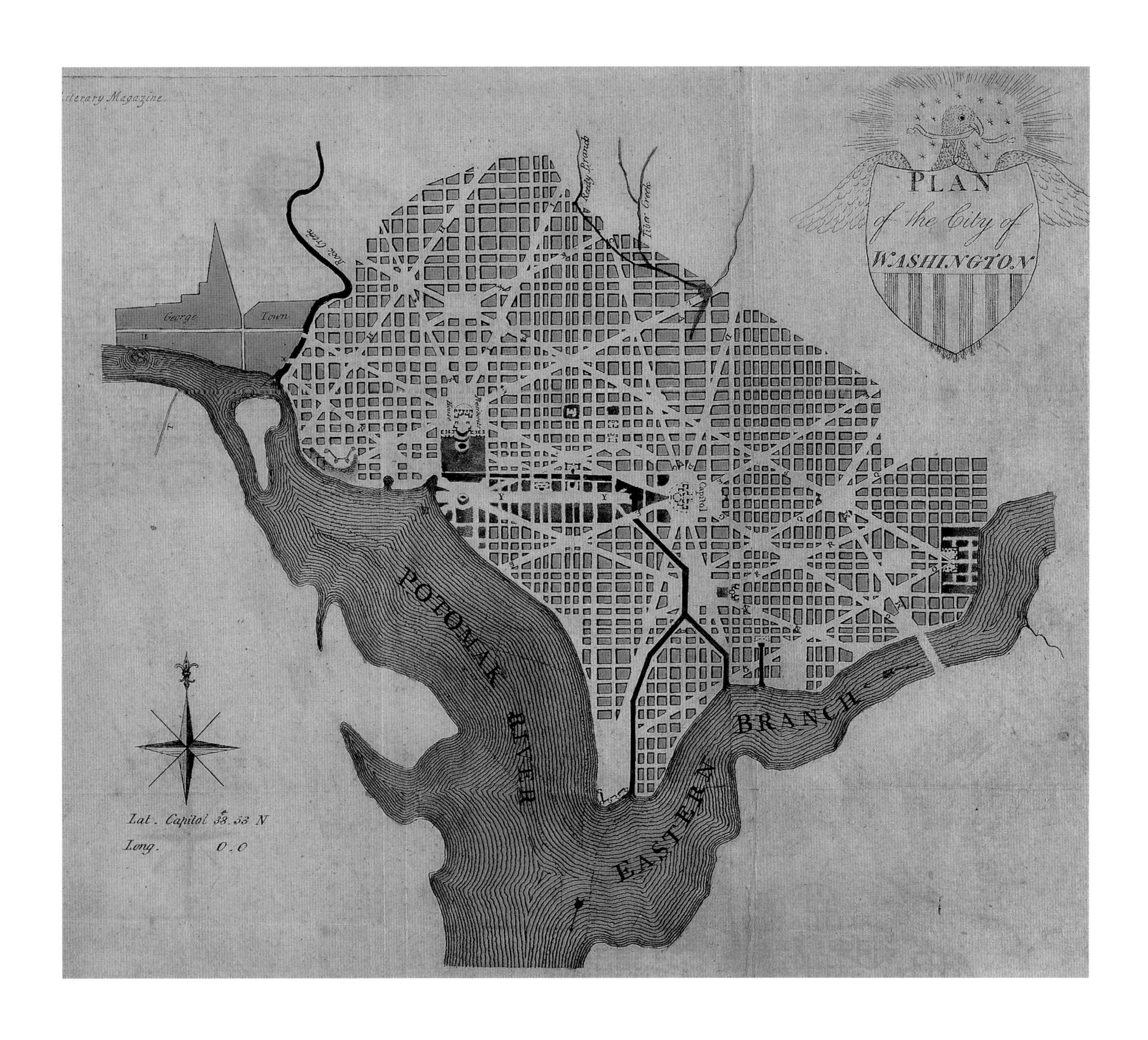
Literary Magazine.
PLAN
of the City of
WASHINGTON
George
Town
Rock Creek
Reedy Branch
Tiber Creek
Capitol
POTOMAK
RIVER
EASTERN
BRANCH
Lat. Capitol 38.53 N
Long. 0.0

compare the morphology of open space, particularly at intersections along Massachusetts Avenue, which Ellicott altered considerably. L'Enfant's streets did not converge as straight lines as they entered and exited squares and circles. They followed the terrain, being concerned with vistas between localized segments. Some historians have said the end result is virtually the same. To those who discern the nuances of the original urban landscape concept, a great deal has been lost. To L'Enfant, the difference was humiliating.

The "Bandanna" map was produced on cloth, presumably used as a table covering or woman's shawl. It has since become an artistic phenomenon to be found framed on gallery and office walls. Its soft deep rose-bronze tone enhances the exquisite delicacy of the corners and border design.

The location of many L'Enfant drawings, sketches, and papers remains a mystery. It is known that he made several versions of his plan, with tracings, sketches, and so forth—as designers always do during the design process. Some of his materials were stolen. He gave one drawing to Benjamin Ellicott, Andrew's brother, to prepare for the first official engraving. It is thought that one of his drawings may have made its way to Europe, possibly with Tobias Lear, while other papers and drawings remain in private collections. These recurring questions will be left to others to solve.

Breadth of the Streets.
THE Grand avenues and such Streets as lead immediately to pubic places are from 130 to 160 feet wide and may be conveniently divided into foot ways walks of trees and a carriage way. The other Streets are from 90 to 110 feet Wide
The Water of Tiber Creek may be conveyd on the high ground where the Capitol stands and after watering that part of the City, may be destined to other usefull Purposes.
The perpendicular height of the ground where the Capitol is to stand, is above the tide of Tiber Creek 78 Feet
PLAN
of the City of Washington
in the Territory of Columbia.
ceded by the States of
VIRGINIA and MARYLAND
to the United States of America
and by them established as the
SEAT of their GOVERNMENT
after the Year
MDCCC.
Perpendicular height of the west branch above the tide of Tiber Creek
F. I. P.
115.7.2/8
Latitude Capitol 38.53.N.
Longitude 0. 0.
Rock Creek
GEORGE TOWN
Tiber Creek
Capitol
Observations explanatory of the Plan.
I. THE positions for the different Edifices, and for the severall squares or Areas of different shapes as they are laid down, were first determined on the most advantageous ground, commanding the most extensive prospects, and the better susceptable of such improvements as either use or Ornament may hereafter call for.
III. NORTH and South lines, intersected by others running due East and West, make the distribution of the City into Streets Squares &c
different Squares or Areas.
PART OF VIRGINIA WITHIN THE TERRITORY OF COLUMBIA
POTOMAK RIVER
EASTERN BRANCH
PART OF MARYLAND WITHIN THE TERRITORY OF COLUMBIA
In order to execute this plan, Mr. ELLICOTT drew a true Meridional line by Celestial Observation, which passes through the Area intended for the Capitol; this line he crossed by another due East and West, which passes through the same Area. These lines were accurately Measured, and made the bases on which the whole plan was executed. He ran all the lines by a Transit Instrument, and determined the acute Angles by Actuall Measurement, and left nothing to the uncertainty of the Compass
SCALE OF POLES.
100
200
300
400
500 Poles
1
2
3
4
5 Inches

A TOWN PLAN CULLED THROUGH LETTERS AND DOCUMENTS

The Loftin Drawing

TITLE: City of Washington 1800 ©
DATE DEPICTED: 1800
DATE ISSUED: 1982
CARTOGRAPHER: T. L. Loftin
PUBLISHER: T. L. Loftin ©
Map, 24.5 x 30.5 in
Courtesy, The Albert H. Small Washington Collection, 152.MP.TD.E.F.

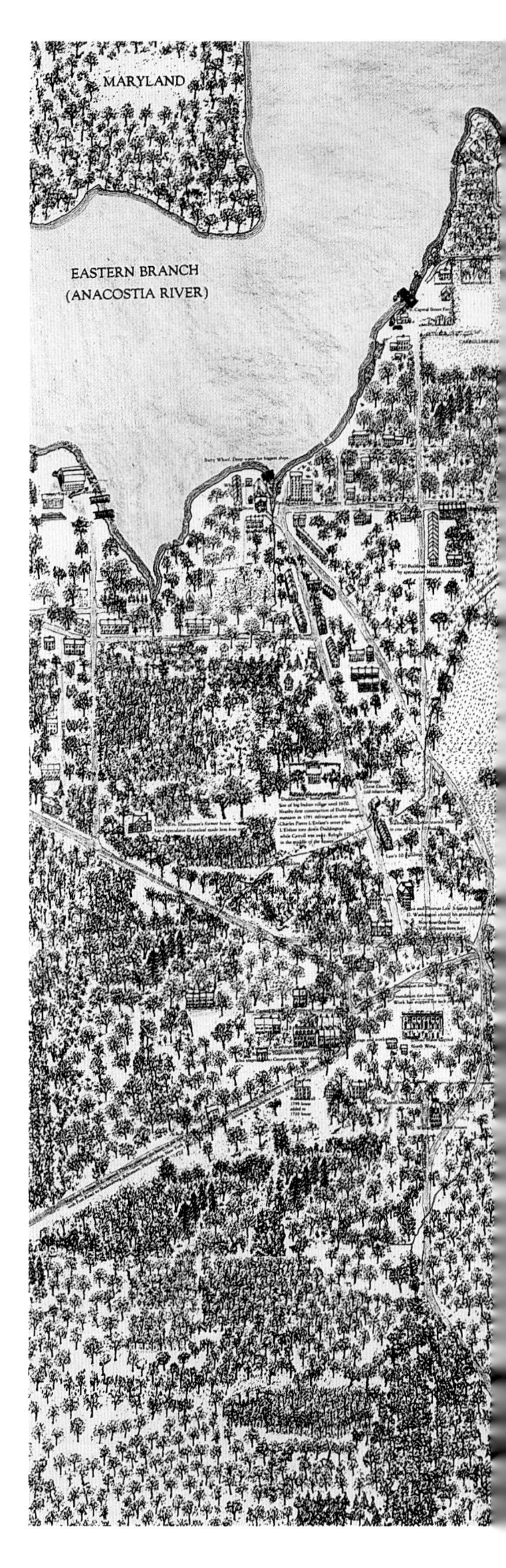

Picturing Washington in its infancy renders surprising views of a young city in the process of being carved out of a wilderness and farmland. This map-view, a reconstruction of Washington as it would have looked in 1800, was prepared in the late twentieth century. Clustered buildings scattered among a forest of trees present a disarming image of a small rural village. Structures are largely dispersed between the President's Palace and the Capitol north of Pennsylvania Avenue, and south of the Capitol toward the river. Streams, roads and patches of farmland are portrayed with often amusing explanatory notes throughout.

How was the information culled for such a map, one is inclined to ask. What of its accuracy? T.L. Loftin, who worked at National Geographic, painstakingly researched the data. One primary source was an 1801 building census requested by President Thomas Jefferson and completed under the directive of the city Commissioners. This invaluable register listed the structures on each block, including materials (brick, stone, or wood) of each.

Two private sources also offered significant information. Written accounts and recollections of Christian Hines, resident for 71 years—his residency there beginning in 1795, when age 14—made mention of almost every street and building. Historian Wilhelmus Bryan's History of the National Capital, Volume One, provided additional details about building type (hotel, house, log cabin, brewery, tobacco barn, tavern, church), apple orchards, flowing springs, and placement of buildings on city blocks. Other material was sifted from books and from articles in Records of the Columbia Historical Society (now the Historical Society of Greater Washington), including George Washington's rental houses and Potomac ferryboats.

Hines noted that stone posts, carved with the address and square number of each building lot and each city block or square, were secured deep in the ground at corners. Tiber or "Goose" Creek, proposed on L'Enfant's plan as a canal, was said to be a temporary home for millions of migrating noisy geese in spring and fall, turning the creek snowy white.

Greenleaf, or Turkey Buzzard Point, is now Fort McNair, the Army War College. Hains Point and the Tidal Basin would be developed almost 100 years later from river channel dredgings and earth dug from the marshlands.

This charming view of the City of Washington resulted from a series of tracings, transfers, and creative imagination. Surveyor James Dermott's relatively accurate "Tin Case Map" of 1797 (p. 65), the first signed map of Washington, signed by Presidents Washington and Adams, was traced by the author on heavy transparent paper, using a light table at the Library of Congress, Geography and Map Division. Each building was located at the correct city block and address on a second transparent trace overlay copy. Trees (sycamore, apple, chestnut, oak, maple) mentioned by Bryan and Hines were drawn by artist/ethnobotanist, Barrie Kavasch, who also drew period houses employing engravings, photographs, and written descriptions. Through a process of photo-transfers trees and houses were imprinted on the map by the author.

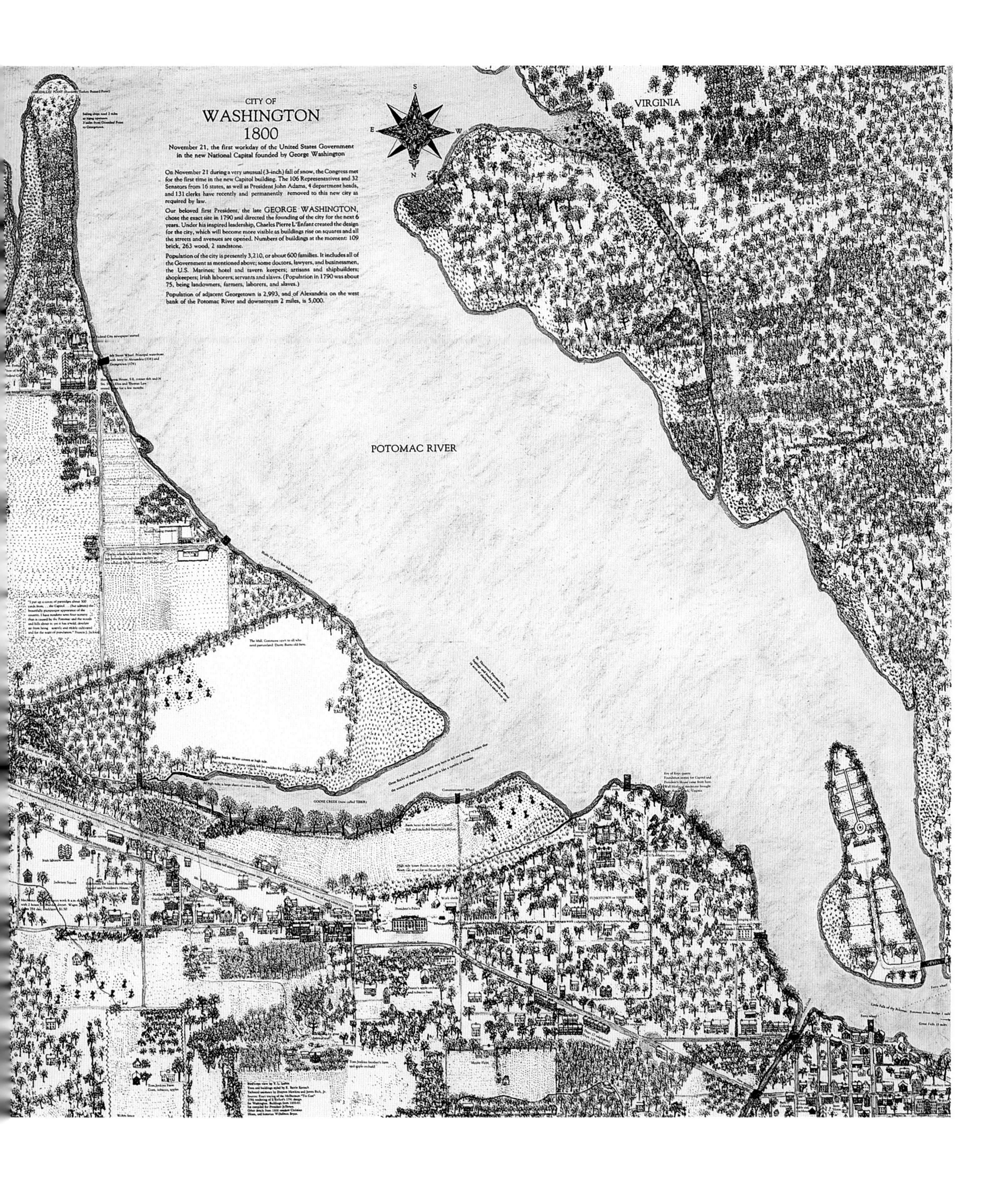
CITY OF
WASHINGTON
1800
November 21, the first workday of the United States Government
in the new National Capital founded by George Washington
On November 21 during a very unusual (3-inch) fall of snow, the Congress met for the first time in the new Capitol building. The 106 Representatives and 32 Senators from 16 states, as well as President John Adams, 4 department heads, and 131 clerks have recently and permanently removed to this new city as required by law.
Our beloved first President, the late GEORGE WASHINGTON, chose the exact site in 1790 and directed the founding of the city for the next 6 years. Under his inspired leadership, Charles Pierre L'Enfant created the design for the city, which will become more visible as buildings rise on squares and all the streets and avenues are opened. Numbers of buildings at the moment: 109 brick, 263 wood, 2 sandstone.
Population of the city is presently 3,210, or about 600 families. It includes all of the Government as mentioned above; some doctors, lawyers, and businessmen, the U.S. Marines; hotel and tavern keepers; artisans and shipbuilders; shopkeepers; Irish laborers; servants and slaves. (Population in 1790 was about 75, being landowners, farmers, laborers, and slaves.)
Population of adjacent Georgetown is 2,993, and of Alexandria on the west bank of the Potomac River and downstream 2 miles, is 5,000.
S
E
W
N
VIRGINIA
POTOMAC RIVER

EARLY LANDHOLDERS, "IN EMBRYO"

The Toner Sketch

TITLE: Sketch of Washington in Embryo, viz.: Previous to its Survey by Major L'ENFANT...
DATE DEPICTED: 1792
DATE ISSUED: 1893 (facsimile of 1874 map)
CARTOGRAPHER: Ernest F. M. Faehtz and F.W. Pratt, compilers; S.R. Seibert C.E., engineer
PUBLISHER: By authority of the Capitol Centennial Committee, Norris Peters Co., Washington, D.C.
Hand colored map, 20 x 26 cm
Library of Congress, Geography and Map Division, G3851.G46 1792 .F3 1893

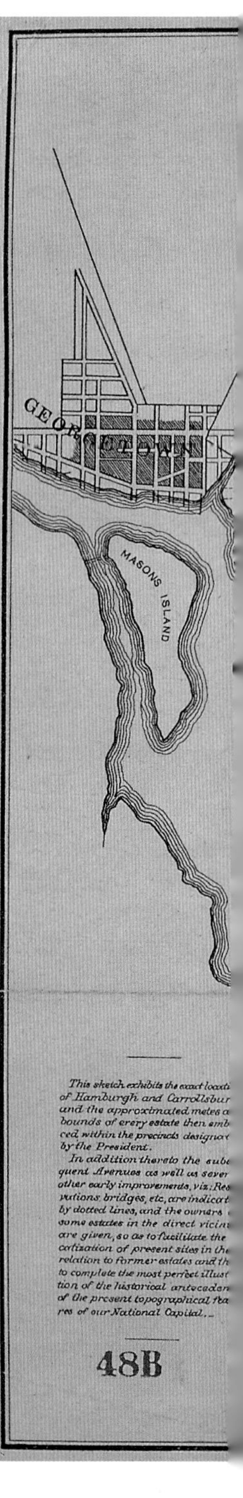

Three-quarters of a century after the founding of the City of Washington, Dr. Joseph Toner, a nineteenth-century collector of Washingtoniana, undertook a remarkable historical research project. He identified names and landholdings of original proprietors at the moment that the city was established. For many years his map offered the best source of landholders and existing farms within the District of Columbia prior to the L'Enfant Plan.

Dr. Toner provided his research information to associates who assisted in creating a retrospective map and booklet, "Sketch of WASHINGTON IN EMBRYO viz: Previous to its Survey by Major L'ENFANT. Compiled from the rare historical researches of Dr. JOSEPH M. TONER, who by special favor has permitted the use of his labor and materials for the publication of a grand historical map of this District now in progress by his efforts combined with the skill of S.R. Seibert, C.E., Compilers, E.F.M. Faehtz, F.W. Pratt." (This complete document also exists in the Washington Kiplinger Collection.)

A note at lower left states, "This sketch exhibits the exact location of Hamburgh and Carrollsburg, and the approximated metes and bounds of every estate then embraced within the precincts designated by the President."

"In addition thereto the subsequent Avenues as well as several other early improvements, viz: Reservations, bridges, etc, are indicated by dotted lines, and the owners of some estates in the direct vicinity are given, so as to facilitate the localization of present sites in their relation to former estates and thus to complete the most perfect illustration of the historical antecedents of the present topographical features of our National Capital."

Besides the grid towns of "Georgetown, Hamburgh, and Carrollsburg," the map includes "Masons Island, Alexanders Island (Boundary Channel), Fort (west of Hamburgh), Patawmack River, Anacostia River or Eastern Branch, St. James Creek, Goose Creek or Tyber Cr.," and extensions to Goose and Rock Creek. For each estate a sketch elevation shows approximate locations of the main house plus principal and secondary structures. Bridges, avenues, and public reservations are overlaid in dotted lines, without names, calling attention to the plan's *parti* (concept diagram) penetrating across the landscape.

The original proprietors listed from east to west are: "Robt. Peter (3 parcels), J. M. Lingan, Sam Blodget, Sam. Davidson, John Davidson, Lynch Sands, David Burns (2 parcels), Wm. Deakins, Ben. Oden, Daniel Carroll, N. Young (2 parcels), Isherwood, Geo Walker, Wm. Prout, A. Young, Wm. Young, Abraham Young & George Walker." George Walker, an ambitious proprietor, published a broadside, "A description of the Situation and Plan," in London dated March 12, 1793. Walker intended to promote the city and enhance the value of his own property. Many of the landowners hoped to profit from the placement of public buildings on their property.

Among possible sources for this research might have been the following: the special-purpose "colonial plantation" maps, drawn for some original proprietors, and prepared by Nicholas King; James Dermott's plats for Carrollsburgh and Hamburgh (manuscript copies were made by William Forsyth, 1858 and John Stewart, 1894, respectively); Dermott's block plans; and the Commissioner's records. Toner's research is essentially accurate, except for minor discrepancies related to property in transition and time sequences of land ownership.

Several other researchers have since compiled similar studies of the same land area or portions thereof. James M. Stewart's 1884 map, based upon Toner's, "Showing the Lines of the Various Properties at the Division with the Original Proprietors in 1792" adds References A-Z reflecting some property transfers. Arthur B. Cutter, Army Corps of Engineers, used Dr. Toner's work for his 1952 map. Recent painstaking research by Lucinda Prout Janke and Ruth Ann Overbeck was published as an article by the Historical Society of Greater Washington. Patricia McNeil's research resulted in a 1991 map (not shown) delineated by Don Hawkins, with some differences from Toner's. T. Loftin's map-view (p. 54) creates an engaging pictorial image of the period a few years after Toner's map.

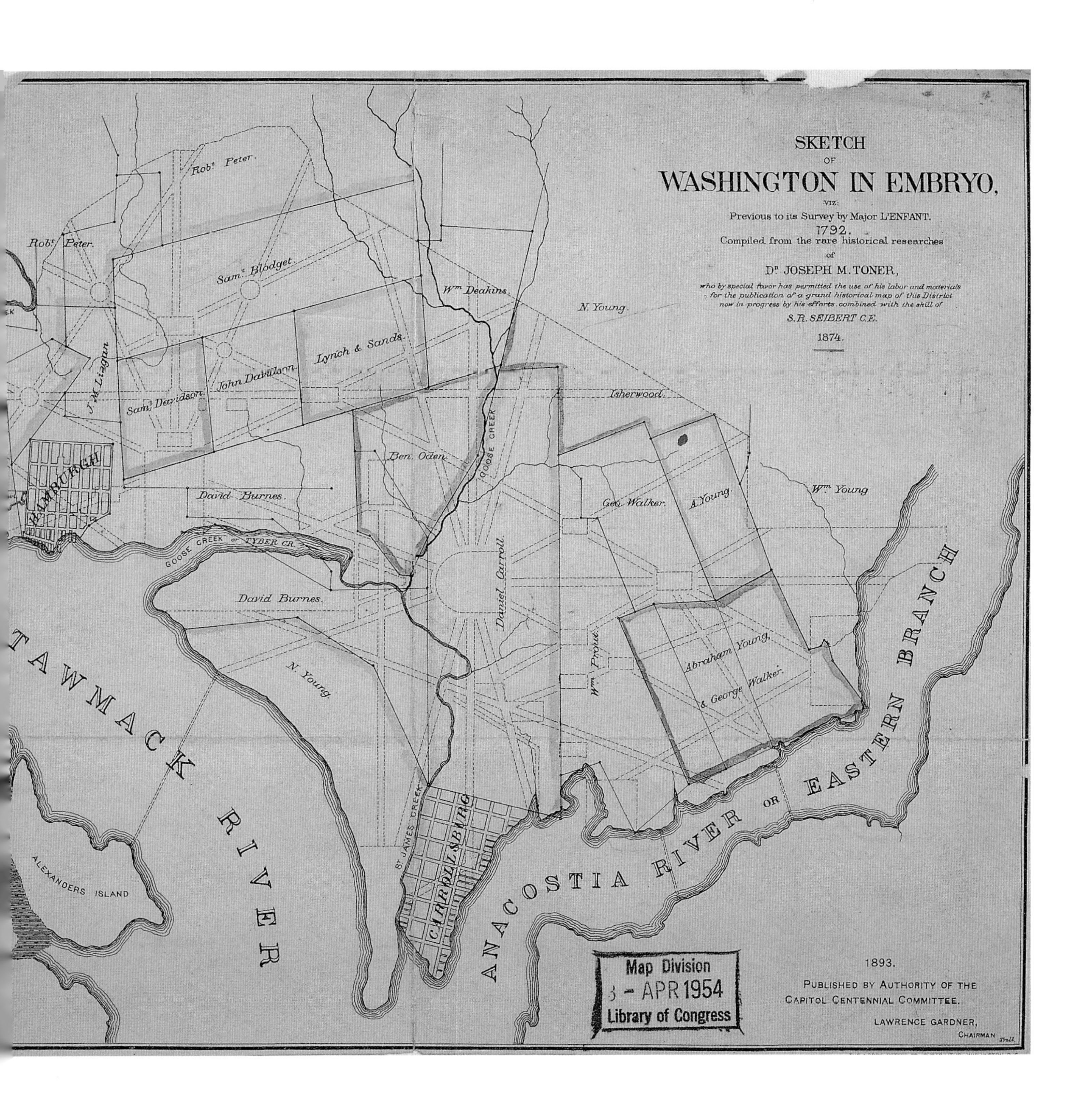
SKETCH
OF
WASHINGTON IN EMBRYO,
VIZ:
Previous to its Survey by Major L'ENFANT.
1792.
Compiled from the rare historical researches
of
Dr JOSEPH M. TONER,
who by special favor has permitted the use of his labor and materials
for the publication of a grand historical map of this District
now in progress by his efforts combined with the skill of
S.R. SEIBERT C.E.
1874.
Robt Peter.
Robt Peter.
Saml Blodget.
Wm Deakins.
N. Young.
Lynch & Sands.
John Davidson.
Saml Davidson.
Isherwood.
HAMBURGH
Ben. Oden.
GOOSE CREEK
David Burnes.
Geo. Walker.
A. Young.
Wm Young
GOOSE CREEK or TYBER CR.
Daniel Carroll.
David Burnes.
Wm Prout.
Abraham Young,
& George Walker.
N. Young.
RIVER
ALEXANDERS ISLAND
ST JAMES CREEK
CARROLLSBURG
ANACOSTIA RIVER OR EASTERN BRANCH
1893.
Published by Authority of the
Capitol Centennial Committee.
Lawrence Gardner,
Chairman

TYPICAL "SURVEYOR'S GRID": CARROLLSBURG PLAT

The Francis Deakins Survey

TITLE: Plat of Carrollsburg
DATE DEPICTED: Surveyed October 1770
DATE ISSUED: 1770
CARTOGRAPHER: Francis Deakins
PUBLISHER:
Cadastral survey map, brown and red inks, scale [1:2400], 38 x 71 cm
Library of Congress, Geography and Map Division, G3852.C35G46.D4 1770 Vault

SITE PLANS FOR TWO PROPRIETORS

The Plantations of Notley Young, Daniel Carroll; N. King Surveys

TITLE: Map of Part of the City of Washington shewing the situation of the Mansion House, Graveyard & Buildings belonging to Mr. Notley Young
DATED DEPICTED: 1791
DATE ISSUED: 1796
CARTOGRAPHER: Nicholas King
PUBLISHER:
Manuscript, black and brown inks, lead pencil, watercolors, scale [1:1200], 33 x 84 cm
Library of Congress, Georgraphy and Map Division, G3851.G46 1796 .K58 Vault

TITLE: Plan of Part of the City of Washington Shewing the situation of the Buildings belonging to Mr. Dan'l Carroll of Duddington
DATED DEPICTED: 1791
DATE ISSUED: 1796 (watermarked: 1794)
CARTOGRAPHER: Nicholas King
Manuscript, pen and ink, watercolor, scale [ca. 1:1200], 37 x 46 cm
Library of Congress, Georgraphy and Map Division, G3852.G32 1796 .K5 Vault

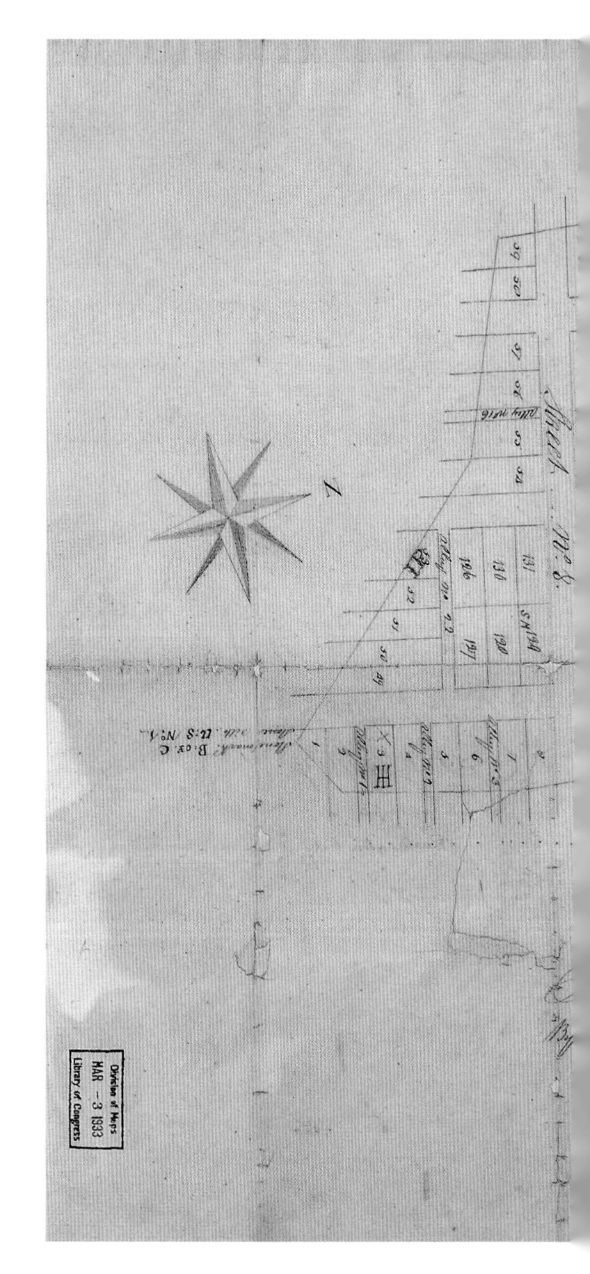

Two "paper" towns, Carrollsburg and Hamburgh (not shown), designed twenty years before L'Enfant's plan, were planned in the customary American grid pattern, introduced in cities such as Philadelphia and Savannah. These two towns were planned along the Potomac to gain advantage from the deeper river channels below Georgetown harbor, a major port for tobacco inspection (see Tones, p. 56).

Carrollsburg, 160 acres at the confluence of the Eastern Branch and Potomac Rivers, was named for its prominent Maryland developer, Charles Carroll. This cadastral map, signed by the surveyor Francis Deakins in 1770, shows lot numbers, some with initials of landholders. Street names differ from those of today. Hamburgh, 130 acres at the mouth of Tiber Creek, sometimes called Funkstown, was named by its founder, Jacob Funk, for his German ancestral hometown.

The streets and squares of these two towns were platted into lots in a manner remarkably different from that proposed by L'Enfant. To reconcile the discrepancies and compensate the

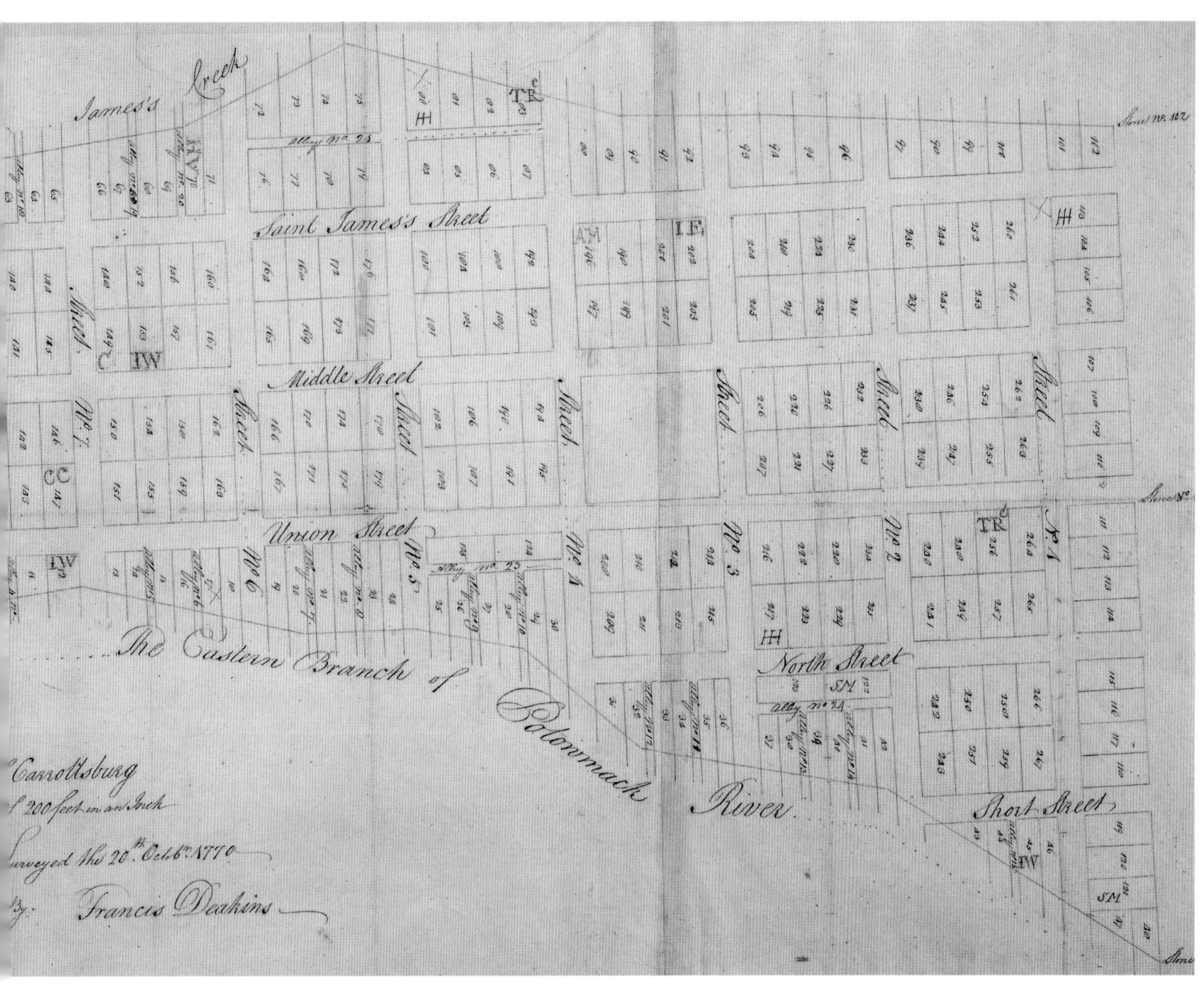
James's Creek
Saint James's Street
Middle Street
Union Street
North Street
Short Street
The Eastern Branch of Potowmack River
Carrollsburg
Francis Deakins

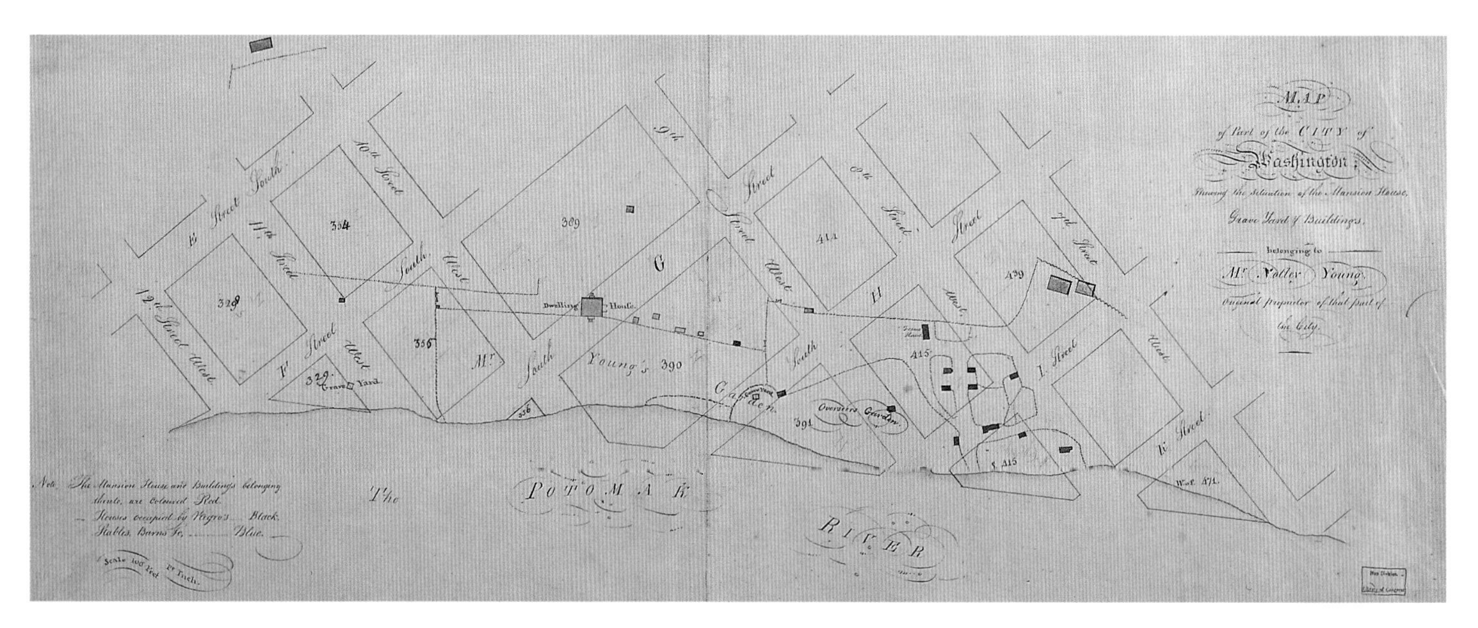
MAP
of Part of the CITY of
Washington;
Shewing the Situation of the Mansion House, Grave Yard & Buildings,
belonging to
Mr. Notley Young,
Original proprietor of that part of the City.
POTOMAK
RIVER
The
Dwelling House
Mr. Young's
Garden
Overseers Garden
Grave Yard

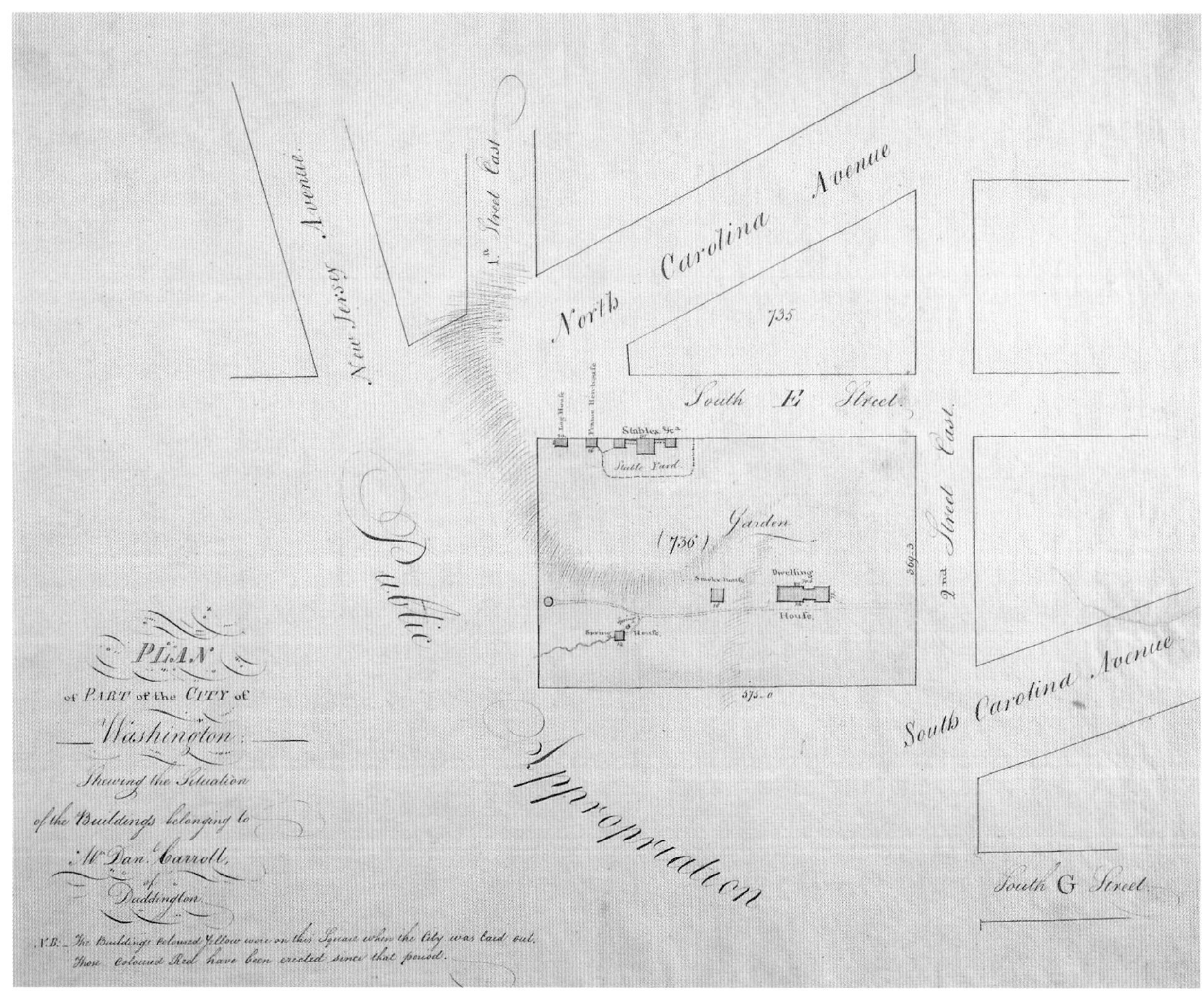
PLAN
of PART of the CITY of
Washington;
Shewing the Situation
of the Buildings belonging to
Mr. Danl. Carroll,
of
Duddington
N.B. — The Buildings coloured Yellow were on this Square when the City was laid out.
Those Coloured Red have been erected since that period.
New Jersey Avenue
1st Street East
North Carolina Avenue
735
South E Street
Stables &c.
Stable Yard
(736) Garden
Smoke House
Dwelling House
2nd Street East
Public
Appropriation
South Carolina Avenue
South G Street

lot's owners, James Dermott later surveyed and produced new plats beginning in April 1793. These manuscript maps portray both the original proposed subdivisions and Dermott's adjustments of the squares. Although the original plats were lost, manuscript copies exist. These were made by William Forsyth in 1858 for Carrollsburg (now in Library of Congress, Geography and Maps) and by John Stewart in 1894 for Hamburgh (now in National Archives).

Specific "special-purpose" maps, which advanced the state of mapping in Washington at that time, were also produced in these early years. Eminent surveyor Robert King, Sr. and his sons, Nicholas and Robert, Jr., were responsible for most of the maps in this category. Nicholas King was the skilled surveyor for the City of Washington from 1796-1797, and again from 1803-1812. His elegant special-purpose maps were among the more notable early descriptive works of the young federal capital.

In the fall of 1796, King prepared a historically significant series of seventeen large pen-and-ink and watercolor wash drawings of colonial plantations located within the city, including plans for Daniel Carroll and Notley Young (shown here). To the landowners these renderings were of great consequence. The city Commissioners, "agreeably to the Deeds of Trust," had granted permission to retain existing property features as the new city was constructed—except for those structures sited on or interfering with proposed streets.

These site plan delineations typically situate the original grounds and buildings: "Mansion House" (main dwelling), slave quarters, overseer's house, gardens, water features, fence lines, and graveyards. Dimensions can often be found for large structures. Thus, King's maps served as a fascinating graphic inventory and guide as to which items could be retained. As the largest collection of colonial plantation maps for a locality, they provide indispensable detailed historical data.

Among other important King family maps were Robert, Sr.'s "surface-profile" street topo maps (not shown), consisting of six manuscripts (two sheets executed by Thomas Freeman). These included vertical cross-sections and linear measurements, drawn 40 feet to 1 inch at the centerline of each street, to permit proper grading, re-leveling, and alignments of streets.

Maximum slope of streets was maintained at eight degrees to enable loaded wagons and carriages ease of use. The degree of slope was also important to facilitate water runoff and alignment of streets with existing adjacent structures (National Archives and Library of Congress, Geography and Maps).

Nicholas King's "wharfing" plans manuscript of 1797 (not shown), produced on twelve sheets at 100 feet to 1 inch, extend from the mouth of Rock Creek along the waterfront to the Eastern Branch at G Street. Existing conditions (original shoreline, wharves, Water Street) adhere to the 1793-1795 Andrew Ellicott-James Dermott resurvey overlaid with a proposed adjusted location for Water Street, wharves, new "water lots," and a market adjoining the canal between 20th and 21st Streets near the Potomac.

The Commissioners had directed preparation of these wharfing plats to resolve discrepancies with earlier maps. King both enlarged Water Street to 80 feet wide and created more space between wharves to promote a healthier, cleaner environment—a growing concern in eighteenth-century America and Western Europe. The wharfing plans, while never published or built because of expense, nevertheless had great influence upon later parks, waterfront and federal land development (National Archives).

This extraordinary family of mapmakers, father and sons, compiled another set of large-scale plans between 1802-1804 (not shown). In sixteen sheets at 200 feet to 1 inch and bound in an atlas, they were known as the "King Plats." Begun by Robert King, Sr., his extensive previous experience in England enabled him to set high standards for accuracy, graphic quality of linework and watercolor wash—skills that he passed on to his sons. These comprehensive King Plats would become base data for the popular 1818 map (p. 71) by son Robert King Jr., City Surveyor with Benjamin Latrobe from 1812-1817.

APPORTIONING BLOCKS, LOTS, ALLEYS; SQUARES 37 AND 70

THE DERMOTT SURVEYS; PETER MANUSCRIPT ATLAS

TITLE: Transactions of Commissioners with Robert Peter in re: Mount Pleasant and Mexico Tracts, (D.C.) 1791
DATE DEPICTED: 1794-1796
DATE ISSUED: Square 37—1794; Square 70—1796
CARTOGRAPHER: James Reed Dermott
PUBLISHER:
Bound manuscript, hand colored, ink and watercolor,145 leaves, atlas 18.5 x 11.5 in, 47 x 28 cm
Library of Congress, Georgraphy and Map Division, G1275.K52 1809 Vault

During the eighteenth century most of the land in the District of Columbia was farmland, owned by 19 proprietors. Most families were acquainted and many were related by marriage. New land generally had been acquired through inheritance or by sale within this small group.

George Washington had designated some land for public buildings and open space, while other property was to be sold to cover improvements and other expenses of the new federal city. Among these expenses was the purchase of land both for Federal property and for resale as lots to would-be buyers. Within a year of Washington's declaration, some land was bought by land speculators, but anticipated sales did not take place as hoped for, and city growth remained slow until after the Civil War.

Among the original lots sold at auction in the District of Columbia in October 1791 was Lot #30 in Square 127, located on 17th Street, Northwest, between H and I Streets. The purchaser was Peter (Pierre) Charles L'Enfant—the very same who had designed the plan of Washington. A certificate of purchase indicates the percentage paid and that which is owed. Signed by the city Commissioners, this example of certificates, conserved in the Library of Congress, Manuscript Division, states:

At a public Sale of Lots in the City of Washington, Peter Charles L'Enfant of Georgetown Maryland became purchaser of Lot number thirty—In Square number one hundred twenty seven for the consideration of ninety nine of Maryland on the terms and conditions published at the same sale: And he hath accordingly paid one-fourth part of the said consideration money, and given Bond, with security, for the payment of the residue; on the payment whereof, with interest, according to the said Bond the said Peter Charles L'Enfant—or his assigns will be entitled to a conveyance in fee.

18th October 1791

Square No 127. Lot No 30

A little known chapter in the history of the United States and Washington City, except among purveyors of local lore, is the intrigue, competition, disputes with the Commissioners, and charges of negligence between its early designers and surveyors. Eventually, this led to the dismissal of many, including Pierre Charles L'Enfant, Andrew Ellicott and his brother Benjamin, a colleague Isaac Briggs, and James Dermott.

Six "section maps" of the southern portion of Washington were submitted in October 1793 by the Ellicott brothers and Briggs shortly before their dismissal. These surveys, scaled at 200 feet to the inch, varied from 100 to 300 squares (city blocks) each. Streets, squares with dimensions, and water elements were marked, enabling Dermott to use them to verify dimensional discrepancies with other maps.

Dermott also was assigned the task to register and divide the squares into lots, resulting in a set of 1,100 cadastral plats scaled 40 feet to the inch. For each square, or city block, manuscript maps were drawn in triplicate and distributed to: Clerk of the District, the Commissioners Office, and the original proprietors. The maps were signed and dated by the surveyors, including transcription of "Returns."

Lot and alley configurations varied according to differing shapes of the squares—rectangular or transfigured by a diagonal avenue. Alleys at the center of squares were typically 30 feet wide while the entry lanes from the street were generally 15 feet wide. Most squares had at least two entry and exit lanes.

All lots were numbered with dimensions given for all boundary lines by Dermott, 1794—both the square and lots, in addition to alley widths. Lots "to remain to the proprietors" were tinted with yellow watercolor wash, setting them apart from those "subject to be sold agreeably to the Deeds of Trust concerning Lands in the City," or the city government. The agreement between the Commissioners and proprietors noted the Square and Lot numbers assigned to each party. It concluded with "Witness our hands this (date)" and signatures by witnesses, commissioners, and proprietors, or their attorneys.

Examples shown are from the Peter Manuscript Atlas of "Transactions of Commissioners with Robert Peter" regarding the Mount Pleasant and Mexico Tracts, 1788-1809, for Squares No. 1-416 in the City of Washington. The U.S. National Archives holds the original cadastral plats for Squares 1-1149 in 28 volumes, "Division Sheets," compiled for the Commissioners, and squares 1-1305 in 4 volumes, "Records of Squares," compiled for the Clerk of the District. Robert Peter, Georgetown's pioneer businessman, was owner of Rock Creek Store, which sold tea. He was one of the largest landowners in northwest Washington, and in Maryland with holdings as far as Frederick.

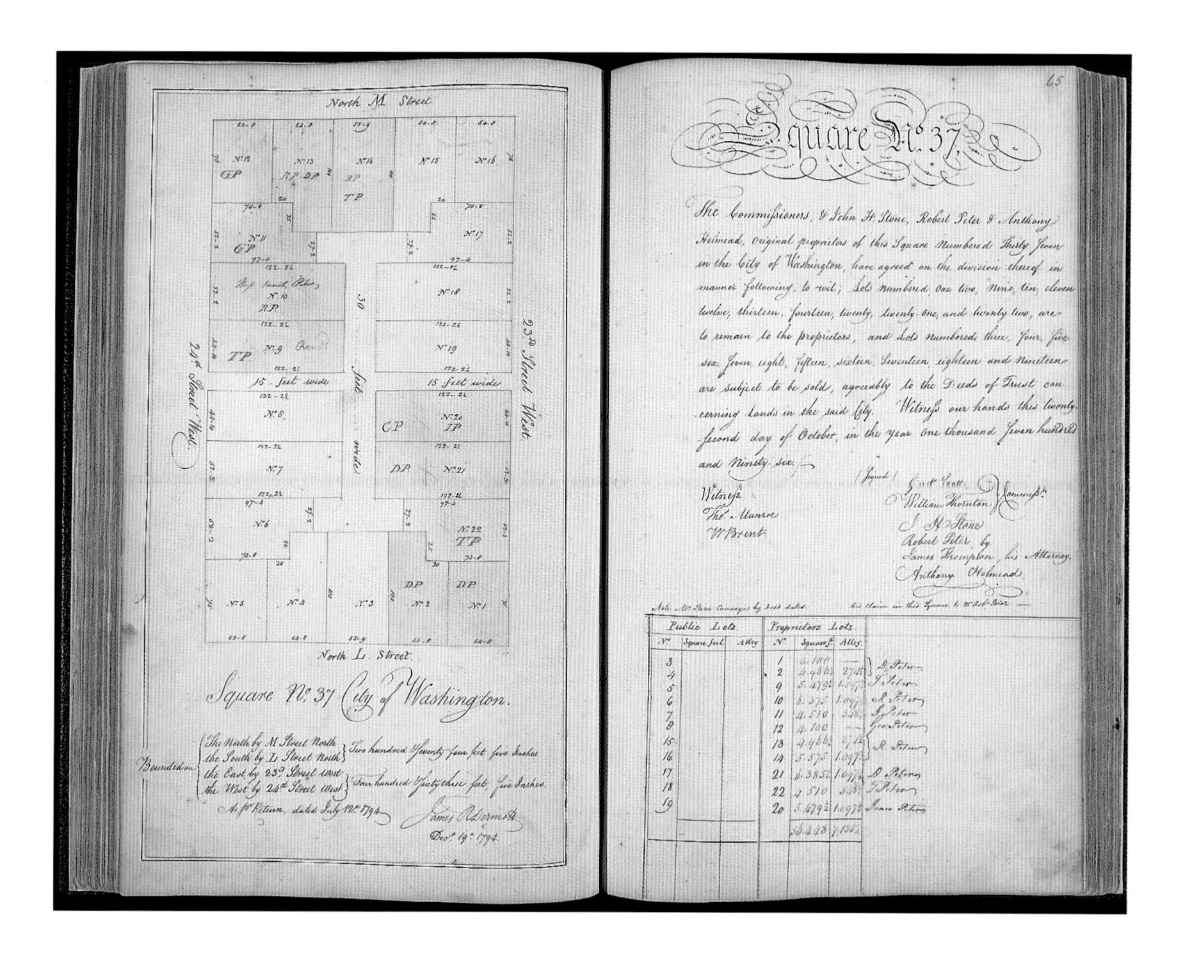

Square No. 37.

The Commissioners, & John H. Stone, Robert Peter & Anthony Holmead, original proprietors of this Square Numbered Thirty seven in the City of Washington, have agreed on the division thereof in manner following, to wit; Lots numbered one two, nine, ten, eleven twelve, thirteen, fourteen, twenty, twenty one, and twenty two, are to remain to the proprietors, and Lots numbered three, four, five six, seven eight, fifteen, sixteen, seventeen eighteen and nineteen are subject to be sold, agreeably to the Deeds of Trust concerning Lands in the said City. Witness our hands this twenty second day of October, in the year One thousand seven hundred and Ninety six.

Witness Thos. Munroe, W. Brent

(Signed) Gust. Scott, William Thornton, Commissrs.; J. H. Stone; Robert Peter by James Thompson his Attorney; Anthony Holmead

Note Mr. Stone conveyed by deed dated [illegible] his claim in this Square to Mr. Rob. Peter

Public Lots No.	Square feet	Alley	Proprietors Lots No.	Square ft.	Alley	
3			1	4.100	—	D. Peter
4			2	4.466½	274½	D. Peter
5			9	5.479½	1.097½	J. Peter
6			10	5.375	1.097½	R. Peter
7			11	4.510	548¼	J. Peter
8			12	4.100	—	Geo. Peter
15			13	4.466½	274½	R. Peter
16			14	5.575	1.097½	R. Peter
17			21	6.385½	1.097½	D. Peter
18			22	4.510	548¼	T. Peter
19			20	5.479½	1.097½	James Peter
				56.448	7.135½	

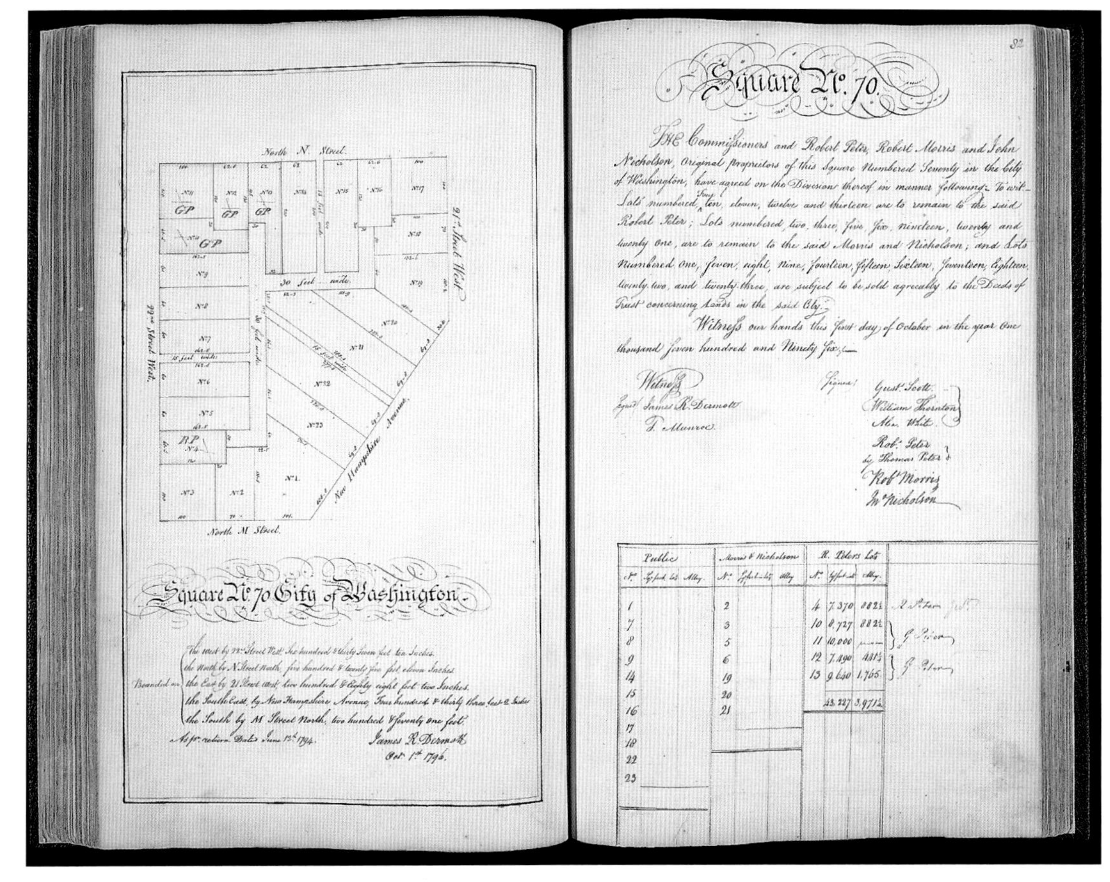

Square No. 70.

THE Commissioners and Robert Peter, Robert Morris and John Nicholson, Original proprietors of this Square Numbered Seventy in the City of Washington, have agreed on the Division thereof in manner following. To wit. Lots numbered four, ten, eleven, twelve and thirteen are to remain to the said Robert Peter; Lots numbered two, three, five six, nineteen, twenty and twenty One, are to remain to the said Morris and Nicholson; and Lots Numbered One, seven, eight, Nine, fourteen, fifteen, Sixteen, seventeen, Eighteen, twenty two, and twenty three, are subject to be sold agreeably to the Deeds of Trust concerning Lands in the said City.

Witness our hands this first day of October in the year One thousand seven hundred and Ninety six.

Witness (Signed) James R. Dermott, T. Munroe

(Signed) Gust. Scott, William Thornton, Alex. White; Robt. Peter by Thomas Peter; Robt. Morris; Jn. Nicholson

Public No.	Sq. feet lots	Alley	Morris & Nicholson No.	Sq. feet lots	Alley	R. Peters Lots No.	Sq. feet lots	Alley	
1			2			4	7.370	882½	R. Peter
7			3			10	8.727	882½	G. Peter
8			5			11	10,000	—	G. Peter
9			6			12	7.490	441¼	G. Peter
14			19			13	9.640	1.765	G. Peter
15			20				43.227	3.971½	
16			21						
17									
18									
22									
23									

EVOLVING ACCURACY

The Dermott "Tin Case" Map

TITLE: The Dermott or Tin Case Map of the City of Washington, 1797-1798
DATE DEPICTED: 1797
DATE ISSUED: May 4, 1888 (facsimile)
CARTOGRAPHER: James Reed Dermott
PUBLISHER: U.S. Coast and Geodetic Survey Office, Washington, D.C.
Colored map on 4 sheets; scale [ca. 1:7000], 125 x 135 cm, sheets 89 x 73 cm
Library of Congress, Geography and Map Division, G3850 1798 .D4 1888

ESSAY BY WAYNE S. QUIN

Following the dismissals of first Pierre Charles L'Enfant and then Andrew Ellicott—whose maps of Washington were considered not yet complete—President Washington, in June of 1795, called on James R. Dermott. Washington instructed Dermott to draw a plan of the City; Dermott would eventually complete the map in 1797, and he would incorporate the field resurveys completed by him and other surveyors between 1793-1795.

The Dermott Map was the third of the early historic maps depicting the plan of the City of Washington—following the L'Enfant and Ellicott Plans. It is sometimes referred to as the "Tin Case" Map because it was for many years stored in a tin case among the records of the Surveyor of the City of Washington. It was likewise named for the large tin case in which it was transported to Philadelphia in 1797. The map was to be used for, among other things, the establishment of boundaries and the clarification of the ownership of the streets, public reservations (term for federal public open space), and usable land vested in the public domain and the original private sector proprietors. Accuracy of identification of the extent of the squares (term for city blocks) to be divided between the original proprietors and the city was another goal. (The specific divisions of lots and alleys within the squares were denominated in separate large volumes.)

President Washington sent a letter (March 2, 1797) to the city Trustees, who were holding land in trust while the federal city was being planned, instructing them to convey all city streets as they were laid out and shown on Dermott's plan. This included as well the squares, parcels, and lots appropriated to the use of the United States for seventeen separate sites. However, President John Adams stated in a letter of July 23, 1798 that "in the press of business" the Dermott Map had been omitted as an attachment to President Washington's letter, and that it was then being resubmitted to the Trustees with the same instructions of the late President (Morris at 254). Thereby, the Dermott Map became the first map of the national capital signed by United States presidents.

In Morris v. United States (1899) (174 U.S. 196, 208-209, at 256), the U.S. Supreme Court determined that the Dermott Map, of the first three historic Washington plans, was the most complete for purposes of implementing the Deeds of Trust (which provided for the disposition of land within the limits of the City of Washington). However, as the Supreme Court stated:

> *...while we regard the Dermott Map as officially authenticated, we do not accept the contention that it is to be considered as the completed and final map of the city...*
>
> *On the contrary, we think it plain, upon the facts shown by this record, that the President, the Commissioners, and the surveyors proceeded, step-by-step, in evolving a plan of the City. Under each of the plans mentioned [the L'Enfant, Ellicott, and Dermott Plans], lots were sold and private rights acquired. Changes were, from time to time, made to suit the demands of interested parties, and additions were made as the surveys were perfected. Even the last map as approved by President Washington [the Dermott Plan], as was said by President Jefferson in 1804, left many things unfinished, some of which still remain to be declared.*
>
> *In short, we think that these several maps are to be taken together as representing the intentions of the founders of the City, and, so far as possible, are to be reconciled as parts of one scheme or plan.*

The Court went on to illustrate the importance and shortcomings of each of the plans. It said the L'Enfant Plan, "Contains all the essential features of the City of Washington as they exist today" (Id. at 258). However, there was significant open space on the plan, "undoubtedly intended as a thoroughfare and for public purposes," which was not named as a street, and none of the other streets or avenues were named.

The Ellicott Plan, frequently referenced as the "engraved" plan, "shows the squares numbered, the avenues named, and the numbered streets all designated." But as the court pointed out, it neither showed all the squares nor correctly placed the public reservations and, indeed it was made before the completion of the surveys of blocks and boundaries (Id. At 257).

The Dermott Map showed the progress that had been made since 1792. It indicated the location and extent of the public reservations or appropriations and also certain new squares not previously shown on the Ellicott Plan which were laid off at the intersection of the streets that did appear on the engraved plan. But Dermott's plan, according to President Jefferson's letter of July†14, 1804, "left many things unfinished, some of which still remain to be declared" (Id. At 258).

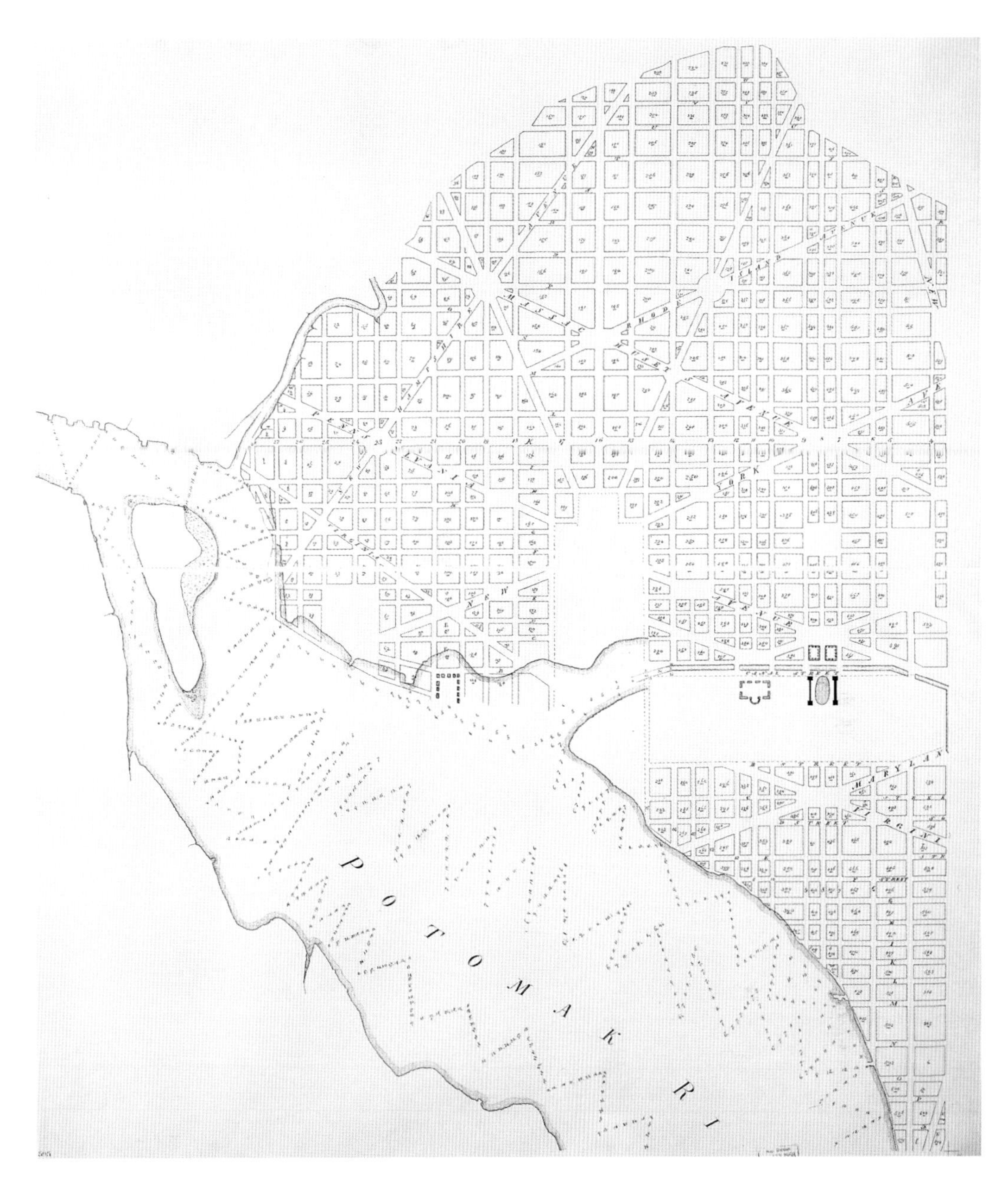

In summary, comparison of the three plans shows a significant progression in accuracy and redrafting of the location of squares and streets as well as the reservations. Of the three plans, the Dermott Map was the most complete and was the one that President Washington intended to be annexed to his instructions (as stated in his letter of March 2, 1797) as the basis for the Trustees to approve the location and boundaries of the streets, reservations and squares.

[Note: At the time of its publication as a photolithograph by the United States Coast and Geodetic Survey Office in 1888—nearly ninety years after its completion—several items were added to the bottom of the map. These items include: letters originally attached to the map (to Thomas Beall of George and John M. Gantt), signed by Presidents George Washington and John Adams; an 1854 letter by Joseph Elgar verifying that the original signed presidential letters had been attached to the map; and an accounting of Dermott's connection with the city as a surveyor, in a letter by B.A. Colonna, Assistant in Charge, United States Coast and Geodetic Survey.

A copy of a third note, written in 1854 by the then Commissioner of Public Buildings, Joseph Elgar, was also attached attesting to the letters by both Washington and Adams. The letter reads:

This Plan of the City of Washington I recognize as the plan of the City to which were attached two papers; one signed by George Washington, March 2nd 1797—the other by John Adams, July 23rd 1798, which papers I have this day identified as having been attached to this plan; this being the plan referred to in the first paragraph of the paper first above mentioned. The said papers having become very much disfigured and torn by the unrolling and rolling up of the plan, I detached them, in order to prevent their entire destruction.

Joseph Elgar (signed)
Late Commissioner of Public Buildings
Witness
B. B. French, Commr. Of P.B.

VIEW FROM GEORGETOWN: THE PROMISE OF POLIS

The Cartwright Image, after the Beck Drawing

TITLE: George Town and Federal City, or City of Washington
DATE DEPICTED: 1801
DATE ISSUED: 1801
CARTOGRAPHER: T. Cartwright (London), after George Beck (Philadelphia)
PUBLISHER: Atkins & Nightingale, London & Philadelphia
Aquatint, 57.5 x 45 cm
Library of Congress, Prints and Photographs Division,
Neg LC-USZC4-530, #PGA-C-Cartwright George Town

Forest and farms marked the scene English landscape painter George Beck encountered when he arrived in America in 1795, anticipating the transfer of the seat of government from Philadelphia to the City of Washington. By 1800, urban growth, while still disappointingly slow, had begun near the Potomac, down river from bustling Georgetown. Yet the impression of a "patchwork" town of clustered scattered buildings persisted.

This view appears to be taken from nearly the same station point above Georgetown as one taken five years earlier by George Isham Parkyns (published by James Harrison, New York, in Library of Congress, Prints and Photographs Division, not shown). However, Beck's is a more accurate portrayal of the scene. The unpaved road in the foreground portends a problem that still exists on many Washington streets two hundred years later—ruts, potholes, and unrepaired roads.

From this vantage point Mason's Island in the middle ground has one road traversing the farmland leading to two small structures. An arched bridge at Rock Creek basin connects Georgetown to Washington. Although many vessels are shown in Beck's painting, silting and sandbars would block the channel, soon giving greater advantage to the port at Baltimore.

Streets are laid out according to the undulating land, crossing at high points and natural distances. Two principal civic buildings, the President's House and Capitol, share reciprocal views, connected by a stately avenue. At Greenleaf Point near the Navy Yard, the greatest number of buildings bunched in one location form an important neighborhood at the mouth of the Eastern Branch.

The charming image presents a sense of uniformity of materials and scale, which was the intention of the city's founding fathers. Regulations established a building height at 30 feet along streets (avenues could be 40 feet). It was required that structures be built of brick or stone, aligning parallel to the street. Existing wood buildings were considered "temporary"—soon to be torn down!

Publish'd June 1st 1801, by Atkins & Nightingale, No. 143, Leadenhall Street, London, & No. 35, North Front Street, Philadelphia.
Engraved by J. Cartwright, London
FEDERAL CITY, or CITY of Washington.

URBAN AMENITY: CONVERTING A CREEK TO WASHINGTON CITY CANAL

The Latrobe Plan

TITLE: Plan of the Washington Canal: No. I and No. II
DATE DEPICTED: Future projection
DATE ISSUED: February 5, 1804
CARTOGRAPHER: Benjamin Henry Latrobe, engineer
Manuscript, ink and watercolor, 2 maps; scale [ca. 1:2370], 53 x 82 cm and 54 x 84 cm
Library of Congress, Geography and Map Division, G3852.W28 S2 L3 Vault

Two existing creeks flowing through Washington's monumental core were to be transformed into a canal creating a new city amenity conforming to the L'Enfant Plan. Tiber (Goose) Creek at the northwestern edge of The Mall (formerly also the location of B Street, now called Constitution Avenue) and St. James Creek southwest of the Capitol were to be connected, thus linking the Potomac and Eastern Branch. When the Washington Canal ultimately opened on November 21, 1815, except for a few alterations, it was substantially the same as it had been positioned by L'Enfant.

It was foreseen that this new waterway would share similarities with the delightful urban canals of St. Martin and St. Denis of L'Enfant's Paris, which even to this day offer agreeable prospects and occasions for strolling.

The canal's path eastward turned 90 degrees at the base of the Capitol, crossing The Mall until it was to divide into two branches connecting the Anacostia and St. James Creek. The Dermott Appropriation Map contained two changes from L'Enfant's plan. First proposed in the 1793-1795 resurvey, these included moving the branch point of convergence northward and realigning the western leg at St. James Creek. This latter portion appears still unbuilt on Albert Boschke's maps of 1857 and 1861 (p. -84-87). To the west, the canal terminated at 17th Street. In 1831 Robert Mills was selected to redesign this junction of the canal where the Potomac River and Tiber Creek estuary frequently overflowed onto The Mall.

Numerous construction delays occurred following establishment of a privately owned Washington Canal Company in 1802. Incorporation of another canal company in 1809 and appointment of Benjamin Henry Latrobe in 1810 as chief engineer were followed by more delays due to lack of funds, the 1812 War and British invasion of the federal city. A public lottery finally enabled completion of the canal.

Benjamin Latrobe prepared this first plan for the Washington Canal between 1802-1804 for the first canal company. It locates and shows details for locks, bridges, a tollhouse and channel depths. A longitudinal section notes the "Ultimate Level of a thorough Baycraft Navigation," while the plan states "This Bridge must be a Drawbridge if Baycraft be admitted." About water depth and seasonal tides, he records "the tide is so much affected by the Wind as to render the Navigation always uncertain and difficult above the President's Square." This splendid watercolor drawing also indicates a lumberyard, Market Square and House, Public Wharf, streets and private properties.

On a later 1815 drawing Latrobe, by now the city surveyor, included additional information, obtained from Nicholas King's surface-profile maps, regarding street levels at canal crossings. This plan showed a proposal (never built) for a circular basin linking the head-waters of Tiber Creek below the Capitol to a basin "mud lock" where The Mall and canal intersect.

By 1871 the canal was no longer viewed as an urban asset. Limited initial funding had caused the structure to be built primarily of wood rather than stone. Soon the wood rotted. Worse, it became a collector of sewage and a dumping area for debris and refuse from the abutting Central Market, attracting mosquitoes and becoming a health hazard. The odor was foul. What was envisioned as a delightful urban amenity was filled in, moved underground into a culvert—by directive of Alexander Shepherd, Board of Public Works—to be replaced by a roadway, Constitution Avenue.

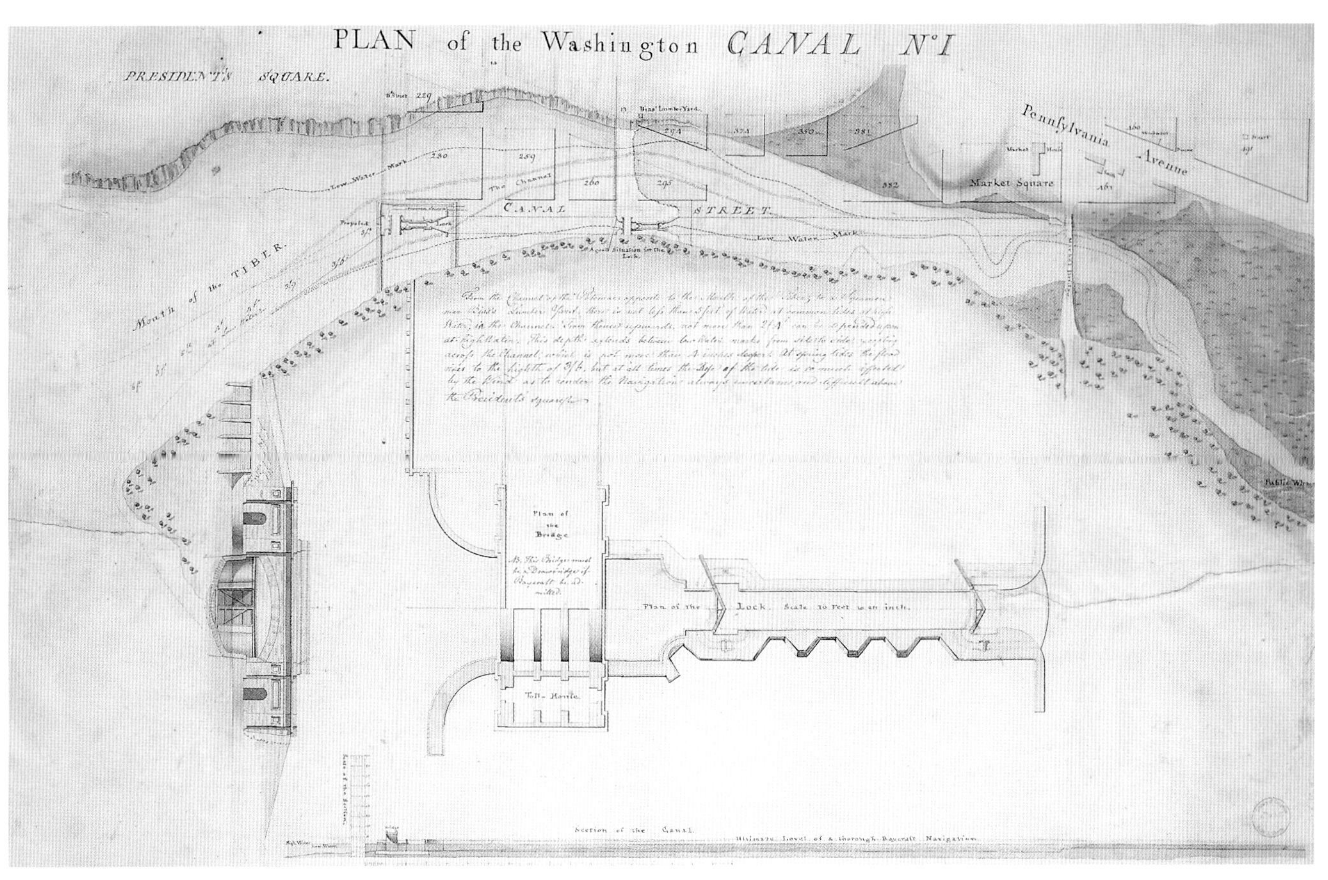
PLAN of the Washington CANAL N°I
PRESIDENT'S SQUARE.
CANAL STREET.
Mouth of the TIBER.
Pennsylvania Avenue
Market Square
Plan of the Bridge
Plan of the Lock. Scale 10 Feet to an inch.
Section of the CANAL.

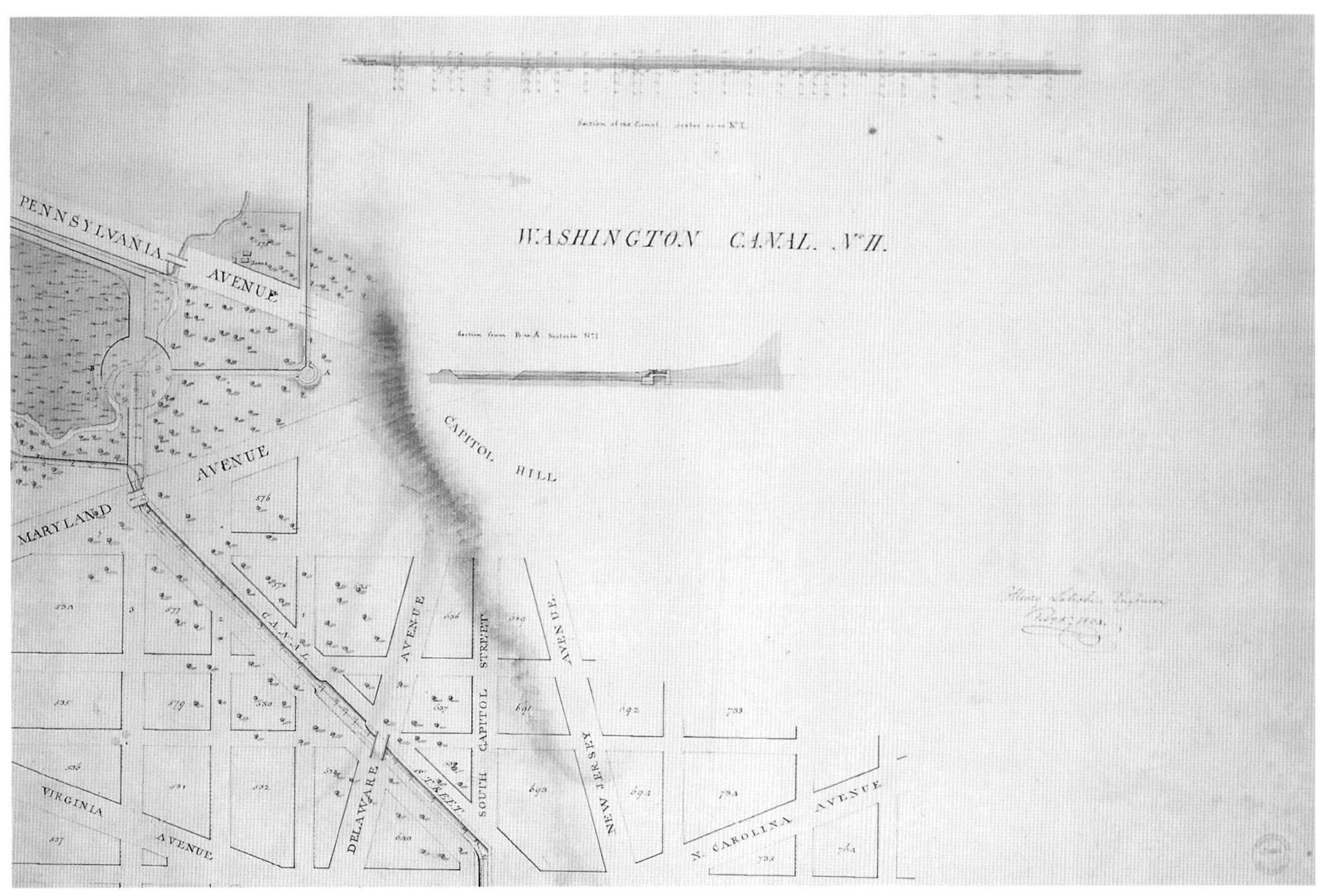
WASHINGTON CANAL. N°II.
PENNSYLVANIA AVENUE
CAPITOL HILL
MARYLAND AVENUE
CANAL
DELAWARE AVENUE
SOUTH CAPITOL STREET
NEW JERSEY AVENUE
VIRGINIA AVENUE
N. CAROLINA AVENUE

FACILITATING NAVIGATION TO MARKETS: C&O CANAL

The Latrobe Plan; Nicholas King Engraving

TITLE: Plans and Sections of the Proposed Continuation of the Washington Canal from Rock Creek to Little Falls of the Potomak to the Navy Yard in the City of Washington
DATE DEPICTED: Future projection
DATE ISSUED: 1802-1803
CARTOGRAPHER: Nicholas King, Benjamin Henry Latrobe, engineer
PUBLISHER: Washington, D.C.
Manuscript, pen and ink and watercolor, 4 maps; scale [ca. 1:792] 56 x 83 cm or smaller
Library of Congress, Geography and Map Division, G3852 W28G45 507 .L3 Vault

Picture a crystal clear sun-drenched autumn day, blue sky, gentle breezes, golden leaves shimmering, trees forming a tectonic-like wall on either side framing a view of a slender band of water that is the C&O Canal National Historic Park. The towpath along the canal becomes a recreational parade route in all seasons for people and pets—strolling, jogging, bicycling. This treasured linear park, a cultural landscape stretching alongside the Potomac River, was almost lost to the automobile as a paved scenic highway in the early 1950s as part of a proposed interstate highway system connecting to Ohio and Pennsylvania through Cumberland, Maryland.

Protests by environmentalists, spearheaded by Supreme Court Justice William O. Douglas in 1954, and joined by Interior Secretary Stewart Udall and others in a publicized canal trek in 1962, succeeded in stopping the highway momentum. In 1972 Congress was persuaded to pass a bill creating a national historic park, maintained by the National Park Service under the Secretary of Interior with advice from an appointed C&O Canal National Historic Park Advisory Commission.

In the early years of this country there were a great many canal "ventures." Limited-length canals in tandem with rivers were often used to get around obstacles. Continuous lateral canals parallel to the river were typical of a later stage of canal development.

George Washington envisioned the Potomac River as a great avenue to the west to facilitate transportation and trade, with Alexandria and Georgetown as major ports. Hence, he wanted to locate the federal capital here where it might become both the administrative and commercial heart of the country. The Potomac Canal, completed by 1802 on the Virginia side of the Potomac, skirted Great Falls with the intention of extending well into Virginia by providing navigation on the river above the Falls.

Simultaneously, a lateral canal on the Maryland side, Chesapeake and Ohio (C&O) Canal, was conceived to move goods to and from Virginia and Maryland's "Great Valley" and lands west of the District of Columbia. The first design of 1802, a proposed extension to Washington Canal, aimed to connect from Little Falls to Georgetown, crossing Rock Creek to the Washington Canal along the northern edge of The Mall (terminating at the Navy Yard) as shown on L'Enfant's plan.

Rendered on four sheets of plans and cross-sections, the project was designed and signed by Benjamin Latrobe. The exquisite watercolor and ink drawings, however, are now largely attributed to the skilled hand of Nicholas King who had prepared two maps of Georgetown in 1800—a plan and a surface-profile of streets.

The first two sheets of plans and sections depict the "Probable Line of the proposed Canal" west of Georgetown from Little Falls, locating the road, views of gorges, inlets, forest, tunnels, bridges, toll house, farm, mill and furnace. The third and fourth sheets portray lower Georgetown terminating at the mouth of Rock Creek to the east. Taverns, banks, structures (some landowners are named), a churchyard, street names, a lock and the aqueduct over Rock Creek are indicated.

A note on sheet No. III states: "On this and the following streets, are laid down the Natural Levels as well as the proposed Graduation of the Streets of George Town, which were accurately ascertained by Mr. Nicholas King in the Year 1800; the buildings then Standing are also noticed. But, as a considerable number of Houses have been since erected, the situation of which could not be laid down on these plans, for want of sufficient time, it is probable that the line marked for the course of the Canal may require alteration whenever it shall be executed. It does not however appear from an attentive view that such alteration will affect the expense in any material degree."

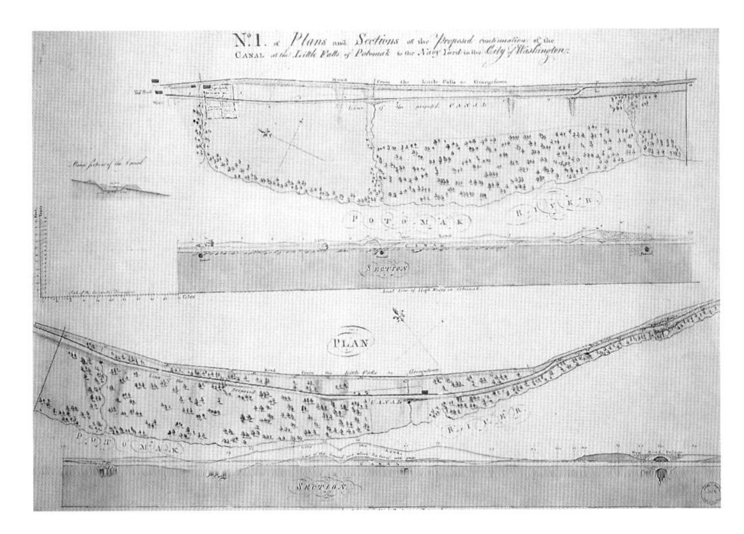

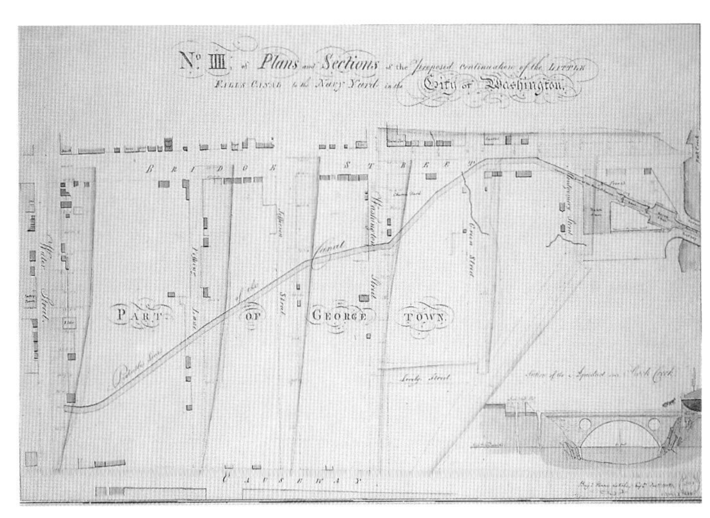

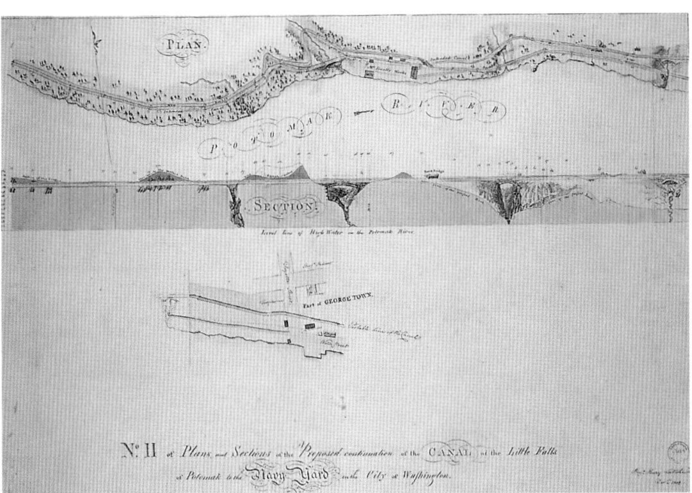

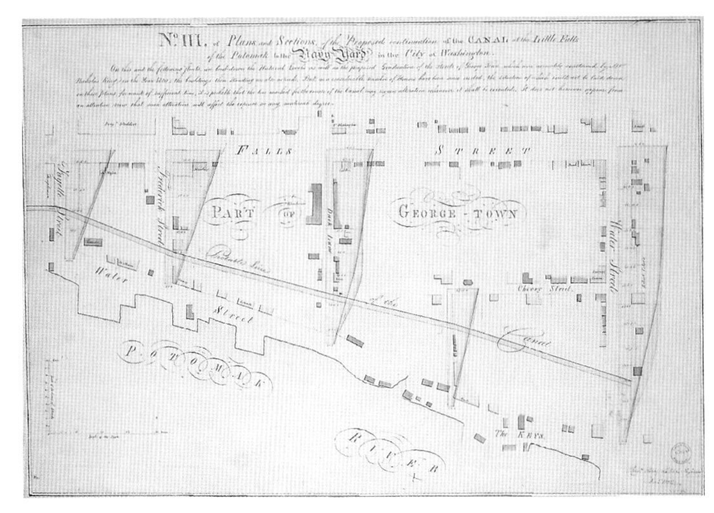

Chartered by Congress in 1825, the C&O Canal was inaugurated on Independence Day (July 4th) during the same year by President John Quincy Adams in a well-attended ground breaking ceremony, which has inspired routine "reenactments," still occurring today. Commercial mule-barge traffic began in 1830. Aqueduct Bridge, constructed in 1843, crossed the Potomac River to connect to the Alexandria Canal, begun five years earlier. The bridge, called "water-trough" bridge, enabled barge cargo (mostly coal and bulky freight) to reach the deep-water port of Alexandria, often bypassing Georgetown harbor to the dismay of Georgetown entrepreneurs. Almost twenty years later the bridge was modified to carry traffic for the war effort. Albert Boschke's 1857 and 1861 maps (p. 84-87) show the locks across Rock Creek channel connecting to the Washington Canal.

As a link westward, the C&O Canal promised to be as successful as New York State's Erie Canal. However, lack of investor financial support precluded rapid completion. By the time the canal reached Cumberland in 1850, the Baltimore & Ohio (B&0) Railroad had already been in place for eight years, rendering the canal obsolete. B&O purchased the C&O Canal Company, continuing its operation for limited use. But a series of floods in 1889 and 1924 wiped it out. In 1938 the federal government acquired the property. Adhering to a 1901 Senate Park (McMillan) Commission recommendation the canal was made into a park. The public now has access to one of the most extraordinary cultural landscapes in America.

SYMBOLISM AND CIVIC PRIDE

The R. King Map

TITLE: A Map of the City of Washington in the District of Columbia
DATE DEPICTED: 1818
DATE ISSUED: 1818
CARTOGRAPHER: Robert King, surveyor of the City of Washington established as the permanent Seat of the Government of the United States of America
ENGRAVER: C. Schwarz; Washington, D.C.
PUBLISHER: W. Cooper, Washington, D.C.
Engraving, scale [ca. 1;12,599], 61 x 79 cm
Library of Congress, Geography and Map Division, G3850 1818 .K5 Vault

Among the first, beautifully embellished, nineteenth-century maps with facades of Washington's two principal civic buildings is that by Richard King, Jr. Elevations of the "South Front of the President's House as designed and executed by James Hoban" and "East Front of the Capitol of the United States as originally designed by William Thorton and adopted by General Washington—President of the United States" are meticulously rendered.

Companion buildings to the President's House (White House) are shown in plan only. First, positioned 450 feet east of the President's House were the Departments of State and the Treasury. To the west at about the same distance were the War and Navy Departments. These buildings, predominately red brick, were similar in height and design.

Following his resignation as city surveyor, King published this popular map on one single engraved sheet on fine paper. It became available in sheet form at $3, or book or roller form at $5. Easy for the user to carry, city topography could be readily understood from a distance. Named streets and avenues and numbered squares offered additional advantage. This was probably the first Washington map published in Washington. Until then, American map trade was centered in Philadelphia, with other leading publishers in Baltimore and Boston.

King's map is significant for its accuracy. Finally, as a result of thorough ground surveys, the exact position of streets and depiction of contours in the city landscape appear correctly. A glance at the map allows the reader to know its water courses and perceive the high points of the terrain. It was more precise than any previous map, including those by L'Enfant, Ellicott, and Dermott. Its accuracy, in large part, is dependent upon three prior surveys: the King Plats (Robert King, Sr. with aid from sons Nicholas and Robert, Jr.), elements of Dermott's Appropriation Map, and canal surveys by Benjamin Latrobe and Nicholas King.

The map indicates the shoreline of rivers, headwaters of streams, and the Washington Canal as completed in 1815 (Tiber Creek's former configuration is also drawn). Other features include the landscape and site plan of Mason's Island (also called Analostan, now Teddy Roosevelt Island), and designation of public appropriations: "Marine Hospital," "Navy Yard," "Greenleaf Point," "Judiciary" (square).

This handsome map became a collector's item. One such collector and printer, Peter Force, copied and re-published King's map. The result was a lawsuit by Robert King, Jr. in 1820 for copyright violation. His copyright information was on the back of the map. Although King lost the suit because he neglected to mark the copyright date on the map's face, the landmark suit set a precedent for future mapmakers. King's map was the last of an era of intensive surveying and mapping of Washington until the Civil War.

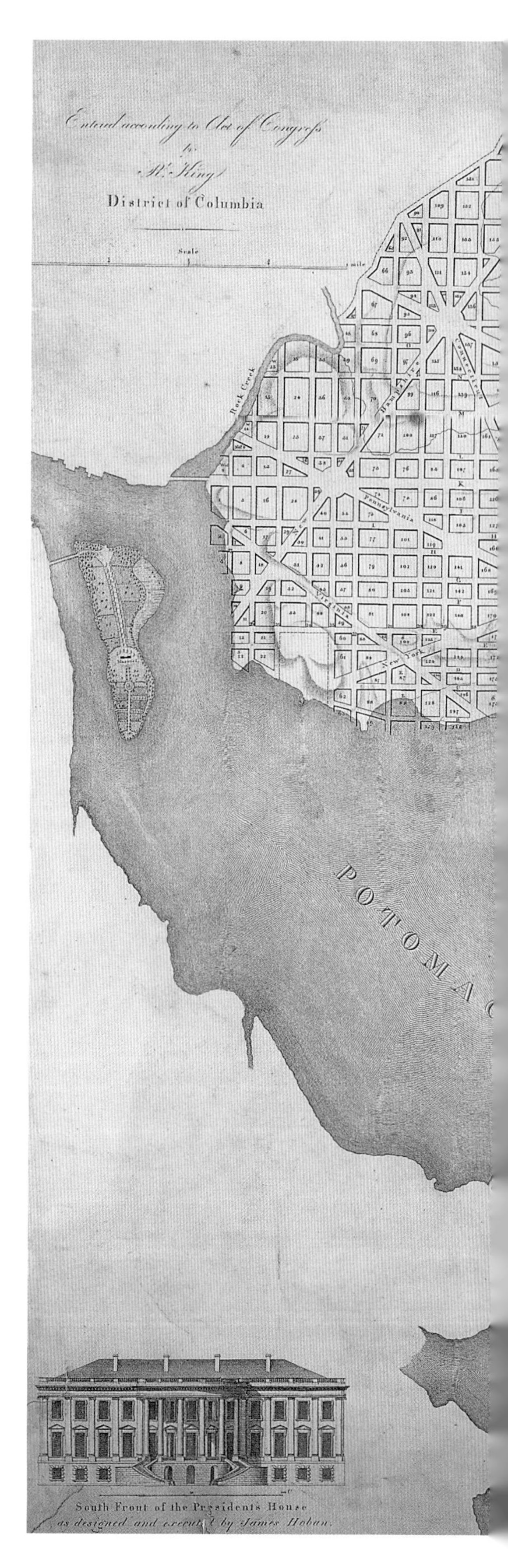

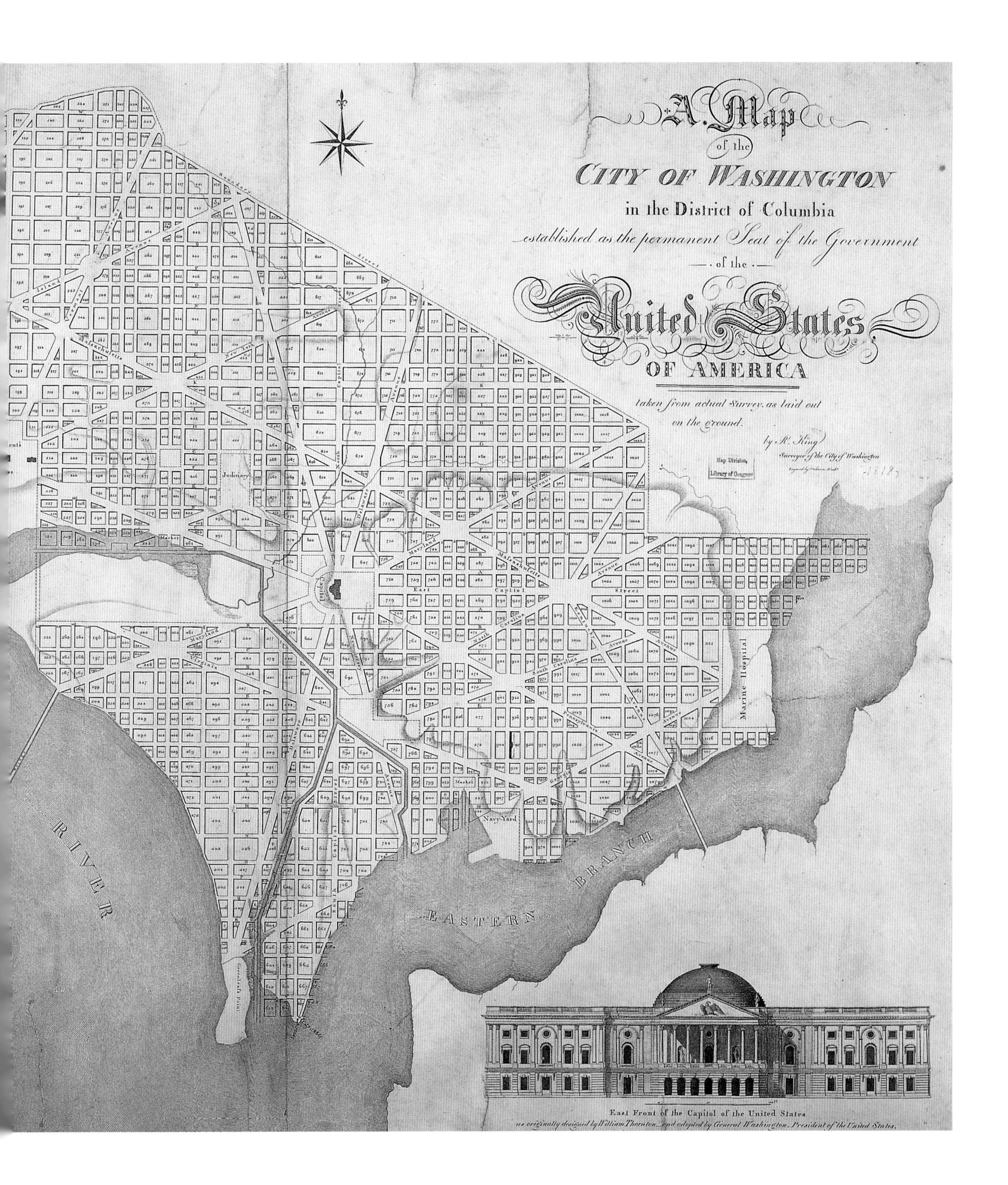
A Map
of the
CITY OF WASHINGTON
in the District of Columbia
established as the permanent Seat of the Government
of the
United States
OF AMERICA
taken from actual Survey, as laid out
on the ground.
by R. King
Surveyor of the City of Washington
Map Division,
Library of Congress
Judiciary
Marine Hospital
Navy Yard
RIVER
EASTERN BRANCH
East Front of the Capitol of the United States
as originally designed by William Thornton, and adopted by General Washington, President of the United States.

ALEXANDRIA CANAL LINKS TO GEORGETOWN

The Stone Plan

TITLE: Chart of the Head of Navigation of the Potomac River Shewing the Route of the Alexandria Canal...
DATE DEPICTED: 1838
DATE ISSUED: February 5, 1841
CARTOGRAPHER: William James Stone. Sr.; compiled from surveys of Lt. Col. Kearney, Major Turnbull U.S.T.E., W.M.C. Fairfax, M.C. Ewing, C.E.
PUBLISHER: U.S. Senate 1841, Senate doc.:178, 26th Congress, 1st session, Washington, D.C.
Lithograph map, scale [ca. 1:20,000] 47 x 91 cm
Library of Congress, Geography and Map Division, G3851 P55 1838 .S7

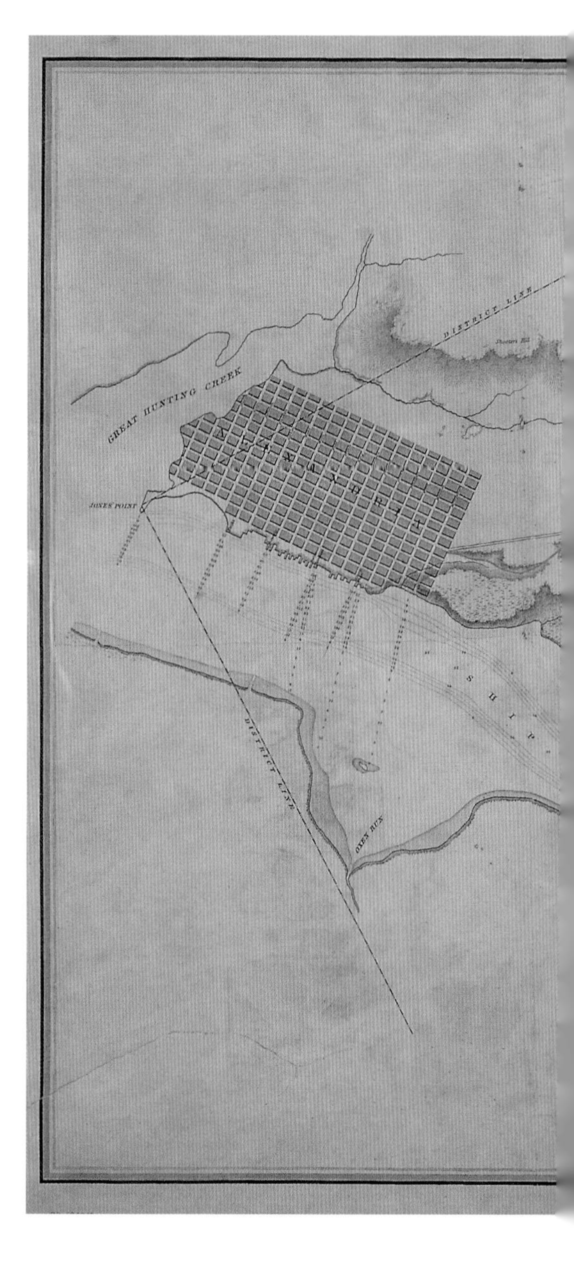

George Washington had prodded the Virginia Legislature as early as 1769 and again in 1772 to open a water route to Fort Cumberland. His explicit goal was to draw upon expanded western resources and induce trade at Virginia ports. A Virginia bill to raise funds through public subscription passed in 1775, and was adopted ten years later. Thus, the Patowmack Co. was set up with George Washington as its first president.

With the opening of the "Potomak Canal" at Great Falls by 1802, two aims were achieved: 1) the existing Potomac River channel was deepened, and 2) locks were constructed where feasible at perilous falls which had presented an impenetrable barrier to navigation. However, navigation was only practical during high water periods in spring and fall. The best channel (five were tested) was in Maryland and the joint venture between Maryland and Virginia ceased operations in 1828 when four-year old Chesapeake and Ohio Co. acquired Patowmack Co. franchise.

Georgetown was favored as the site of a canal that would connect to the federally funded Washington Canal. Thereby an extension of the canal through Washington City to Eastern Branch (Anacostia River) would be established. Ground was broken in 1828, bypassing the falls to access western markets and improve navigation. Alexandria citizens were unhappy, having subscribed $250,000 to C&O Canal to no advantage to themselves. They pressed for a connecting canal in Virginia.

In 1830 Congress, without granting funding, authorized formation of Alexandria Canal Co. to extend the C&O Canal by way of Aqueduct Bridge. Alexandria Canal originated at the Alexandria waterfront, traversing Virginia through Arlington to Aqueduct Bridge spanning the Potomac River, to meet the C&O Canal at Georgetown on the Maryland side. Georgetown citizens, feisty back then as today, organized a strong lobbying effort to initially defeat federal funding!

The project proved to be far more expensive than anticipated, requiring additional funds from Alexandria. After many delays, the Canal (drawn in 1838) opened at the end of 1843, six months after the opening of Aqueduct Bridge. Coal was the major product transported, along with a smattering of farm products. Erratic water supply from the C&O Canal, silting and breaks in the walls caused frequent problems. Poignantly, during the Civil War the bridge was drained, and used instead for military-conveyances purposes.

Congress had refused to assist with needed finances, while the Commonwealth of Virginia was constrained by the canal's location within the federal District of Columbia on the Virginia side. Incentive was mounting for retrocession of D.C. land within Alexandria County to Virginia. With retrocession accomplished in 1846, the Virginia Board of Public Works then provided some monetary aid. Nevertheless, Alexandria Canal Co. concluded operating in 1886. Some traces of the old Alexandria Canal can still be found in Virginia.

This map illustrates a nautical chart of the Potomac estuary within the District of Columbia. Relief is shown by hachures; depths, by contours and soundings. The magnetic meridian orienting north is drawn at lower right. A note states "The soundings are taken at ordinary low water. The flood of the tide is 3 1/2 feet. The soundings thus 2 3 4 5 6 shew the depth as it was in the year 1795 and the soundings on the dotted lines opposite to Alexandria are from a survey made in 1816 by Capt. J. T. Brooks. The hydrographic curves on this Chart have been laid down from near 7000 soundings the scale is too small to shew all the soundings. The soundings are in feet."

From the District Line, east to west, on

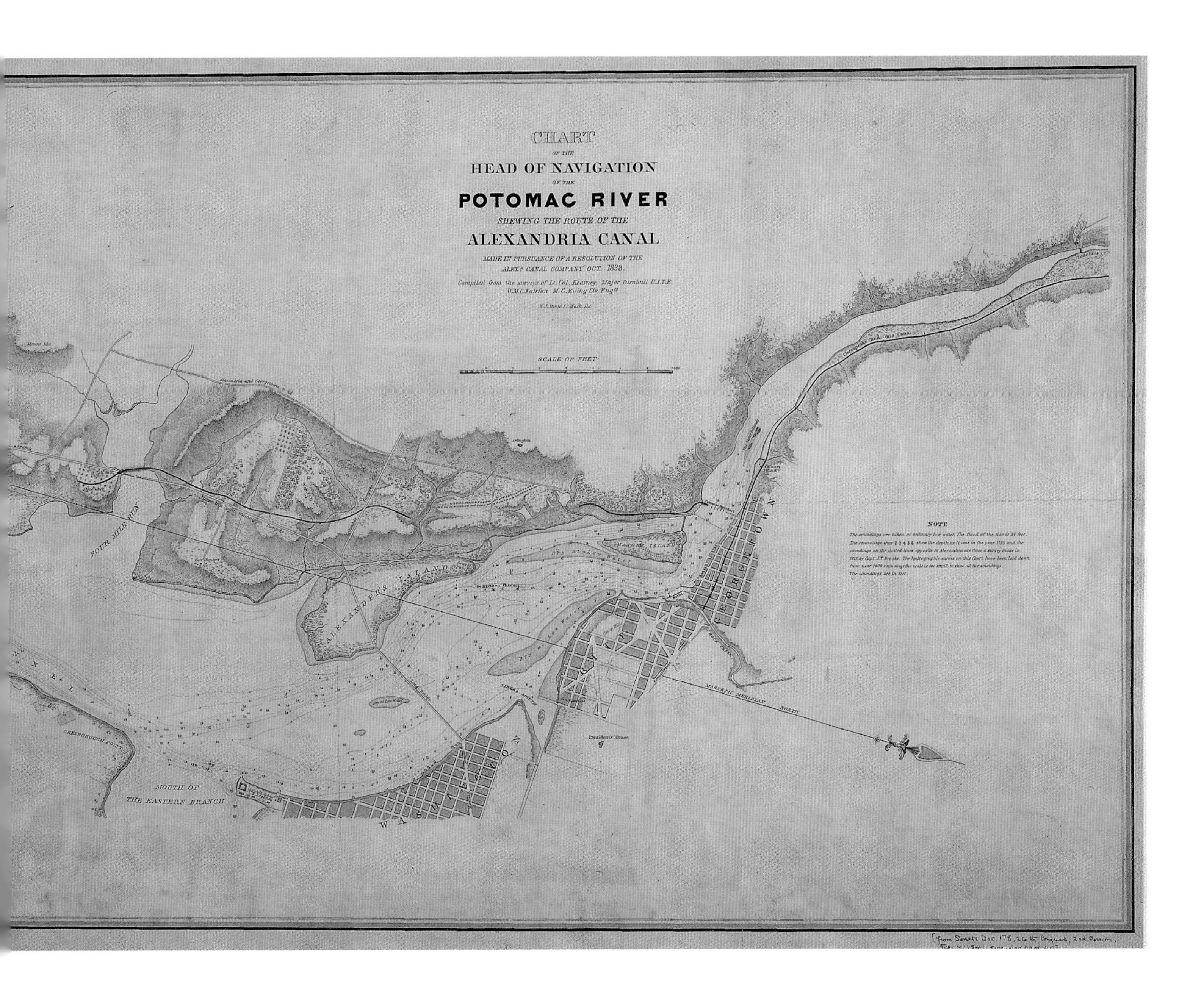

the Washington side of the Potomac, noteworthy sites are Oxen Run, Gresborough Point, U.S. Arsenal, General Van Ness', Camp Hill, and Cannon Foundry. The favored American grid street pattern is depicted at Alexandria, Georgetown and parts of Washington City.

Readers may not be conversant with markings along the river: Ship Channel, Four Mile Run, sections "Dry at Low Water," Long Bridge, Gravel Point on Alexander's Island, Mason's Island, Georgetown Channel, Potomac Aqueduct, and The Sister Islands. These small islands, now named "Three Sisters," were the center of a mid-1960s contentious bridge proposal and misguided scheme for a multi-loop interstate freeway system that would have destroyed neighborhoods and parts of The Mall and adjacent areas. The plan was routed by protests from a coalition of environmental activists, preservationists, and residents from both the east and west sides of the city.

From east to west in Virginia striking features are Jones' Point, Great Hunting Creek, Shooter's Hill, Mount Ida, Swann's, Alexander's, Toll Gate, and Gen. Hunter's. Alexandria and Washington Turnpike leads
to Long Bridge, while Alexandria and Georgetown Road leads to the Georgetown ferry near Potomac Aqueduct. Columbia Turnpike and Arlington are also named.

Alexandria Canal in Virginia is shown eight years before land was retroceded to Virginia. A solid black line indicates completed canal sections; a dotted line signifies unbuilt links. Land contours, orchards, woodlands, streams and farm roads along with other features highlight the sparse settlement in mid-nineteenth century. Today the roadway traverses the rolling hills and parkland adjacent to the Potomac, passing the site where the Alexandria Canal once carried the hope for a great commercial market.

FROM GARDENESQUE TO PICTURESQUE: THE MALL AND ENVIRONS

The Mills Plan and The Downing Plan

TITLE: PLAN of the MALL with the adjoining Streets and Avenues, the relative position of the Capitol, President's House, and other Public Buildings and particularly the improvements…
DATED DEPICTED: Future projection unrealized
DATED ISSUED: 1841
CARTIGRAPHER: Robert Mills
Manuscript, pen and ink, watercolor, I: 19 x 25 in, 2: 11 x 12 in
National Archives, Cartographic Unit, 1:RG77: CON 90-1 2:RG77:CON 90-2

TITLE: Plan showing proposed method of laying out the public grounds at Washington
DATE DEPICTED: Future projection
DATE ISSUED: 1851
CARTOGRAPHER: Andrew Jackson Downing
Manuscript, pen and ink, blue watercolor, and lead pencil, scale [ca. 1:1800], 112 x 175 cm
Library of Congress, Geography and Map, G3852.M3G45 1851 .D6 Vault

ROBERT MILLS: THE MALL

In the twenty years preceding the Civil War, construction of civic projects in central Washington contributed to the *city's* physical and cultural enhancement, as envisioned by its founders. Among the most significant were those by architect Robert Mills, including the 1836 Treasury Building at Pennsylvania Avenue and 15th Street and the 1830 General Post Office, later renamed Tariff Commission, at 7th, E and F Streets.

Mills also prepared designs for works on The Mall. In 1831 he was selected to devise improvements to Washington City Canal, which overflowed at high tide from Tiber Creek estuary at 17th Street, because of poor original construction. A growing interest in scientific research and instructive museum displays led to an 1842 invitation to produce a design for expansion of the Botanic Garden, to house specimen plants collected by Lieutenant Charles Wilkes from expeditions of 1838 to 1842.

A bequest to the United States by English scientist, James Smithson, established an institution to disseminate knowledge. At the request of Secretary of War, Poinsett, in 1841, Mills submitted a design for Smithsonian Institution.* Mills' English-Norman-architectural-style building included three fascinating unsolicited landscape concepts (one is shown). Eventually, however, Smithsonian "Castle" was designed by James Renwick in 1847. It is of interest that Mills' first drawing contained a "foot print" plan of his 1836 winning Washington Monument design.

Albeit contrary to L'Enfant's formal plan for The Mall, Mills' drawings remain a resplendent legacy of Washington's landscape heritage. His first plan included The Mall, Capitol Grounds, and Washington Monument Grounds west of 14th Street. Mills conceived two alternate solutions for Smithsonian Grounds between 7th and 12th Streets, inserting figurative symmetrical parterres in an informal field. These plans reflected an artistry incorporating both classical and romantic landscape themes. A virtual implied axis was projected between the Capitol and Washington Monument. Reservations surrounding Smithsonian Grounds revealed an affinity with late eighteenth-century French Romanticism, rather than the English Picturesque of A. J. Downing (p. 79), that was to become stylish within ten years.

In Mills' plan, foliage was densely clumped among meandering paths, architectural features, and clearly marked centers. In contrast, Smithsonian Grounds were imaginatively crafted in a novel, though inappropriate, landscape approach for the required refinement of The Mall. Pattern and color variation emitted pulsating vibrations and optical pleasure. We might well imagine this as a trendsetter design for less ceremonial locations today. Mills' plan retained north-south streets—7th, 9th, 12th, 14th—crossing The Mall in keeping with L'Enfant's plan. Canal Street, not named on other period maps, indicates that Mills probably envisioned a promenade. Years later, after the 1872 demolition of the Washington Canal, B Street (now Constitution Avenue) was created.

Multiple reasons have been advanced for Washington Monument's location off the intersecting axis between the White House and Capitol: poor soil conditions, a need for high ground to escape flooding, and Mills' aesthetic preferences. Whatever the reason, the shifted position adds an unexpected, skewed spatial result, affording splendid simultaneous views of several symbolic structures in sacred alignment. The Washington Monument, completed without the colonnade, was dedicated in 1885, affirming the humility, resolve, and dignity of the nation.

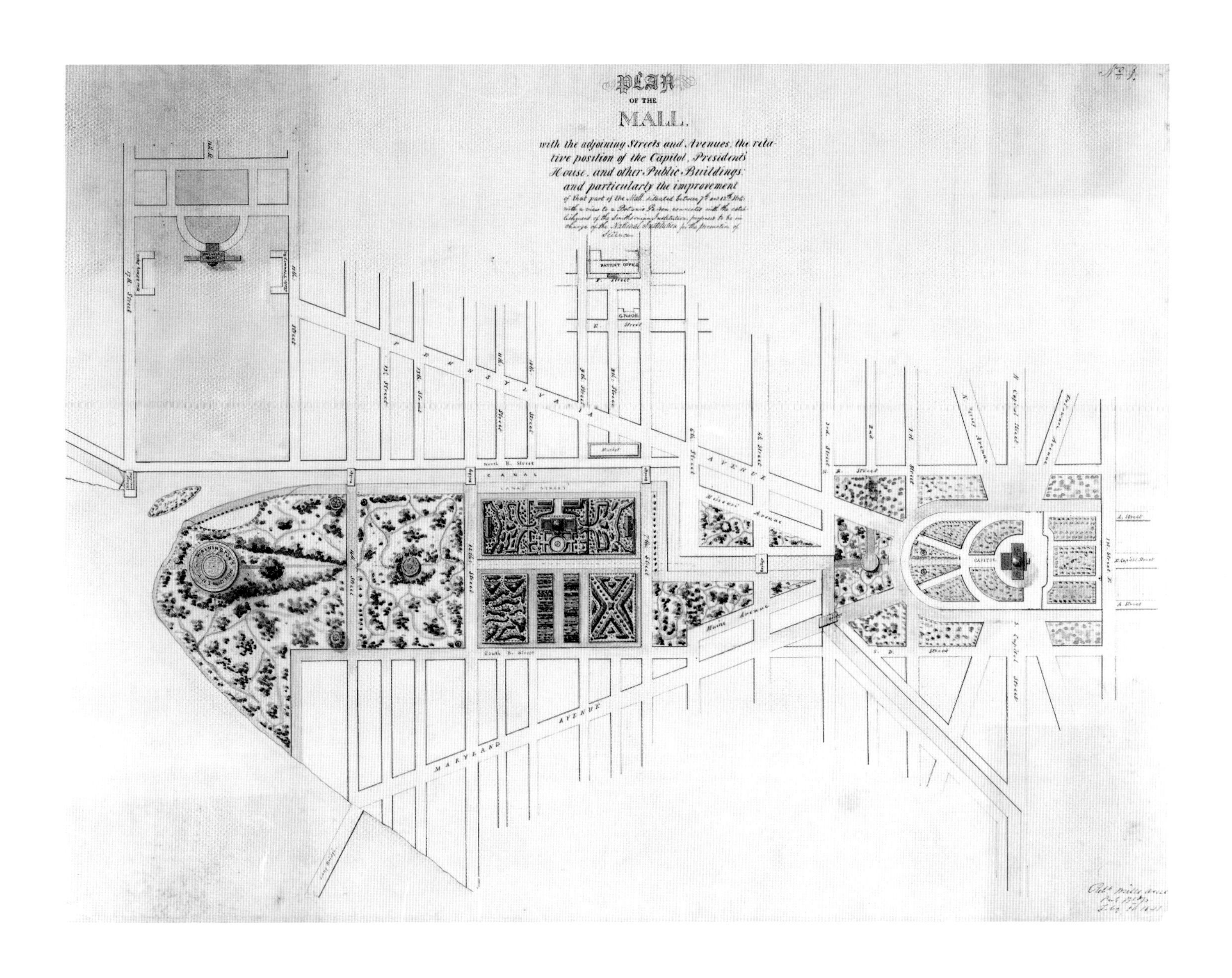
PLAN
OF THE
MALL.
with the adjoining Streets and Avenues; the relative position of the Capitol, Presidents House, and other Public Buildings;
and particularly the improvement
PATENT OFFICE
CANAL
CAPITOL
MARYLAND AVENUE
PENNSYLVANIA AVENUE

ANDREW JACKSON DOWNING: THE MALL

Changing sensibilities about landscape, popularized in Europe in the "Age of Enlightenment," spread across the Atlantic. America was coming into its own, philosophically, politically, and internationally. By the mid-nineteenth century, the country embraced the informal style of landscape design, yet with its own distinctive expression.

The "luminists" painted romantic wilderness scenes of unspoiled nature, suggesting a fundamental mainspring of fear combined with exalted beauty. The implication of an infinite expanse of untamed land extending westward reinforced a quintessential emphasis on nature. In the philosophical essays and poetry of the "transcendalists," Ralph Waldo Emerson, Henry David Thoreau and others, humankind was tied closely to nature, social issues, self-reliance and rugged individualism—values that were represented symbolically in the romantic landscape of the period.

Keen international interest in collecting and exchanging plant specimens for scientific purposes—a passion shared by Thomas Jefferson—endured during the nineteenth century. In 1842, the Botanic Gardens at the foot of the U.S. Capitol expanded to house its enlarged exotic plant and tree collections. Impetus to improve The Mall grounds came in 1846 when the Smithsonian Institution was established. Its first building was a turreted Norman-style castle designed by James Renwick.

In 1851, at the request of President Millard Fillmore, noted landscape gardener, Andrew Jackson Downing, submitted a plan to redesign The Mall and parks south and north of the President's House. Downing's name had been suggested to the president by Secretary of the Smithsonian, Joseph Henry; Mayor, Walter Lenox; and financier, William Corcoran. His practical experience had been largely designing rural estates in New York's Hudson Valley. Mount Auburn Cemetery near Boston influenced Downing's design for the new National Park on The Mall. (see also, Longstreth, R., ed. *The Mall in Washington, 1791-1992,* CASVA Symposium Papers XIV [Scott, Pamela and Therese O'Malley], Studies in the History of Art, No. 30, Washington National Gallery of Art, 1991.)

His landscape proposal, typified by sweeping romantic geometric order of a nineteenth-century French park, was a radical departure from L'Enfant's formalist symmetrical design for The Mall flanked by linear tree-lined avenues. He favored a naturalistic, picturesque setting with drives and paths curving through a botanical museum of living labeled specimen trees and shrubs. In a critique of L'Enfant's concept of straight streets and broad avenues, Downing signified the benefits of his own plan as providing a sense of relief from the urban grid, thereby acting as a complement, concealing many flaws.

Downing wrote down three objectives of his plan: form a national park as an ornamental compliment to the Capitol; influence the taste of the country by providing an example of a "natural style of landscape gardening"; and create a public museum with a collection of all trees and shrubs that would grow in Washington's climate, each labeled with its popular and scientific names.

North-south streets—7th, 9th, 12th, 14th—crossed The Mall, causing separate garden reservations, unified by lawn, serpentine paths and informal plantings. Ornamental wire suspension bridges, high enough for medium-size boats to pass beneath, and strong enough to support carriages, were proposed at Washington Canal. "Monument Park" surrounding the Washington Monument, designed by Robert Mills, would hold large-growth American trees clumped to open vistas to the Potomac.

"Evergreen Garden," densely planted with foreign and native species for winter enchantment between 12th and 14th Streets, would be a symmetrical maze-like oval parterre. The Smithsonian "Pleasure Grounds" between 7th and 12th Streets would offer great beauty and seclusion. Densely planted with rare trees and shrubs in the picturesque style, it would replicate how they were often found in nature.

Toward the east, "Fountain Park," between 7th and 3rd Streets, would boast two water features: a remarkable water display, also envisioned by L'Enfant, and an irregular-shaped pond (excavated soil used as fill along The Mall). A basin at the foot of the Capitol would furnish the fountain's water, while the pond would be supplied by the Canal or fountain's overflow. Also at the base of the Capitol was the Botanic Garden and three green houses where carriage ways would pass to exit through a new gateway (a balance to his classical marble arch gateway design at The Mall's west entrance).

Perpendicular to The Mall, just south of the President's House, Downing envisioned an open ground for parades, military reviews, pub-

lic festivities and other celebrations. A shaded forty-foot-wide circular carriage avenue of elm trees would surround the Parade, while ten-foot-wide footpaths would meander through lush plantings beyond.

With Downing's untimely, accidental death in 1852, his plan for The Mall lost its strongest advocate. The Civil War curtailed prospects for implementation. Moreover, with Congressional approval, railroad tracks crossed over the east end of The Mall, expanding trade and traffic. While elements of the plan were realized, each theme garden remained an independent entity, diminishing opportunities for a unified Mall until the turn of the century and the momentum of the Senate Park (McMillan) Commission Plan to restore the vision of the L'Enfant Plan.

At a time of increasing political turmoil and a desire by some to remove the federal capital to another more central location in the country, beautification of The Mall was thought to elevate the status of Washington and lend an aura of symbolic completion to the civic core. The social vision embodied in Downing's iconography was implicitly related to the agrarian ideal grounded in American democratic principles. These precepts, advocated by French physiocrats in eighteenth-century France, were part of our nation's collective memory. Indeed, Presidents Thomas Jefferson and George Washington embraced this utopian ideal for the private realm, as a representation of morality and equality. Their home plantations, styled as *ferme ornees,* ornamental farms, gave iconic meaning to these virtues, similar in manner to Downing's nineteenth-century picturesque plan.

Nevertheless, although a popular notion in its time, the plan was quite controversial among those predisposed to a more classical concept for the public realm. The plan by Andrew Jackson Downing for the nation's central Mall in Washington's public monumental core was in clear opposition to the philosophical imperatives embodied in the L'Enfant Plan.

* A black and white manuscript tracing of Downing's faded drawing "To accompany the Annual Report date Oct. 1867, of Br't. Brig. Gen'l. N. Michler, In charge of Public Buildings, Grounds, & Works" resides in the U.S. National Archives.

A WASHINGTON VIEW, PRE-CIVIL WAR

The E. Sachse View of the Capitol

TITLE: VIEW of WASHINGTON
DATE DEPICTED: 1852
DATE ISSUED: 1852
CARTOGRAPHER: Edward Sachse
PUBLISHER: E. Sachse & Co., Baltimore, MD
Colored lithograph drawn from nature on stone, 51 x 69.8 cm, 20 x27.5 in
Library of Congress, Prints and Photographs Division, Neg LC-UZC4-771; #PGA-D-Sachse,E. & Co.-View of Washington

Fascinating renderings of Washington, revealing a growing city framed by low hills, were created by Edward Sachse between 1850 and 1871. In this 1852 lithograph, the Capitol, and its dome by Charles Bulfinch as the focal point, was drawn at an oblique angle. Sited on 35 acres of land, Congress authorized expansion of new wings at each end. The Capitol, as pictured here, incorrectly anticipates the actual final design.

Compare the completed Capitol of 1871 (p. 92) with this early projected image. Except for the dome, the exterior central rotunda facade remained much the same after the additions. However, the imposing scale of the completed building—62,000 square feet—with its two new wings and dome prompted some members of Congress to suggest purchasing surrounding residential land.

This congenial view of Washington City Canal, passing along the north side of The Mall to join the Potomac in the distance, would provoke one to lament its 1872 demolition. The perspective renders the canal with equal importance to the broad diagonal sweep of Pennsylvania Avenue stretching northwest toward the President's House. Contrast Washington with Paris; consider the animation and delight induced by the ancient canals St. Martin and St. Denis. What splendid ambience the Washington Canal would have added to the city's contemporary urban walks.

On a far rise, the old Observatory appears to dialogue with the Washington Monument. (A new observatory was subsequently built at a Massachusetts Avenue site, where the vice presidential residence is also located.) Commanding a key position on The Mall, Washington Monument's image is deceiving. Shown as designed by Robert Mills and authorized by Congress in 1848, the colonnade at its base depicting Revolutionary War heroes was never built. The slender Monument was not completed until 1884. One of the tallest masonry structures—555 feet high—it is forty-five feet less than Mills proposed, and ten times higher than the width of its base. In the pre-Civil War years, construction halted at 156-feet high. When building resumed a decade after the War, the marble colors did not match. This graceful obelisk stands as a dynamic expression of nationhood.

Smithsonian Castle's deep pink stone occupies the middle-ground of The Mall. A 150-acre naturalistic landscape was proposed as a national park according to an 1851 plan by Andrew Jackson Downing. Its marshy soil would be prepared to accommodate a variety of trees and curving carriageways. South of The Mall the monumental scale of government buildings today belies this panorama of a once close-knit residential village at Independence and Maryland Avenues.

Although this picture presents a view of a compact city, accounts of the period tell a different story. This was a city designed for the future, a future yet unfulfilled. One day it might become a grand world capital; now, it was unattractive, sprawling incohesively with few permanent residents. Streets ended suddenly—usually unpaved, muddy and squalid. The city canal was a noxious eyesore.

Urban fabric north of Pennsylvania Avenue was comprised of small-scale commercial and residential structures of requisite rosy brick, interspersed with stately Neoclassical ivory civic buildings of marble and limestone, including the Treasury Department, Patent Office, Post Office, and City Hall. Towering above was the majestic Capitol, enduring as a compelling metaphor for American democracy.

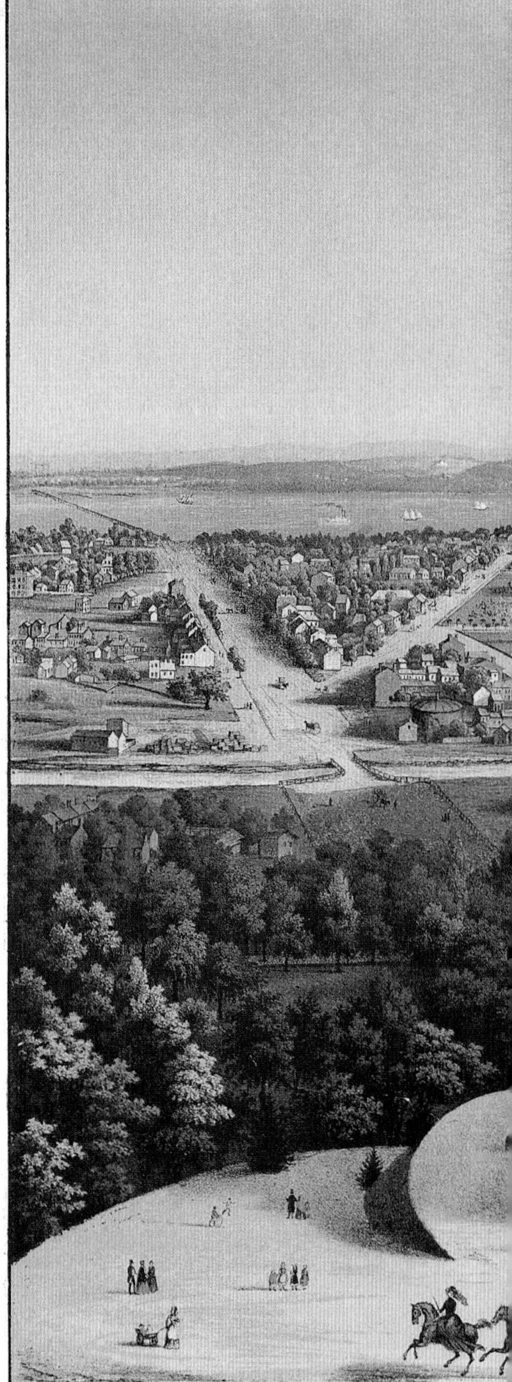

Lith. and print. in colors by E. Sachse & Comp
VIEW OF WASHINGTON.
Published and sold by E. Sachse & Co. Baltimore Md.
North Liberty Street.

PROMOTING WASHINGTON: MAP WITH GAZETTE; AMENITIES, MARRED BY RACIAL PROFILE CHART

The Colton Atlas

TITLE: Georgetown and the City of Washington, the Capital of the United States of America
DATE DEPICTED: 1855
DATE ISSUED: 1855
CARTOGRAPHER: Unknown
PUBLISHER: G. W. & C. B. Colton & Co., New York
Colored engraving, 14 x 17 in
Courtesy, The Albert H. Small Washington Collection, 226 .MP.NC.L.F.
(Also in Iris Miller Collection; and Library of Congress, Geography and Map Division, G3850.1855 .J12)

ESSAY BY ROBERT L. MILLER

"There is a map of Washington accurately laid down; and taking that map with him in his journeyings a man may lose himself in the streets, not as one loses himself in London between Shoreditch and Russell Square, but as one does so in the deserts of the Holy Land, between Emmaus and Arimathea. In the first place no one knows where the places are, or is sure of their existence, and then between their presumed localities the country is wild, trackless, unbridged, uninhabited and desolate... If you are a sportsman, you will desire to shoot snipe within sight of the President's house."—Anthony Trollope[1]

"As in 1800 and 1850, so in 1860, the same rude colony was camped in the same forest, with the same unfinished Greek temples for work-rooms, and sloughs for roads."—Henry Adams[2]

If Trollope was bemused by a semi-wild Washington, other European visitors of the 1850s were as dismissive as Adams. Undoubtedly maps reinforced such attitudes. Every plan of Washington displays a certain gap between medium and message, but surely the disconnect peaked in the 1850s. Elegant, stylish, technically powerful, the era's commercial maps blended existing conditions and future projections, accurate minutiae and romantic misinformation, to describe a place at best unfinished and at worst raw, stalled and stagnant.

At the start of the decade banker William Wilson Corcoran and Smithsonian Institution Secretary Joseph Henry recruited landscape designer Andrew Jackson Downing to redesign The Mall as a romantic park and arboretum—a scheme that limped forward following Downing's 1852 death in a steamboat explosion. The plan barely surfaced on the Keily (1851, not shown)[3] and Colton (1855) maps, however, which generally showed strikingly little development in four years. Nor did the Boschke (1857, 1861) survey maps (p. 85-86), the first Washington maps to delineate all built-up areas, reveal much progress beyond the separate villages that defined the capital in 1850.

Most federal buildings, too, while nearer reality than most streets, were still under construction. The Capitol's new House and Senate wings reflected an expanding Union. More often, however, piecemeal construction mirrored Congress's wavering interest in the federal city. The privately sponsored Washington Monument was faring even worse, and would remain an unfinished stump for three decades.

As historian Margaret Leech wrote of the city in 1860: "These were the sights of the Federal Metropolis—six scattered buildings, a few dubious statues and one-third of an obelisk — and, barring an inspection of the government greenhouses, or a drive to the Navy Yard, the Arsenal or the Observatory, there was nothing more to be seen within the city limits."[4]

Undaunted—or perhaps encouraged—by this sparse content, Keily and Colton, both map-makers, set out to embellish their subject. In the spirit of contemporary publishers of city views (and contemporary producers of the first world's fairs), their approach, like their street plans, applied artistic license and looked to the future.

Consequently, the Washington Monument according to Robert Mills' 1836 design materialized on both Keily and Colton maps, complete with peristyle. Vignettes of a finished Treasury, Patent Office, and Capitol appeared on one or both maps as well. So did population charts, tables of building elevations and tables of avenue widths, and labels bristling with swash capitals. Numbered blocks, both real and projected, filled out the color-tinted, heavily outlined areas of the wards, the dominant units of government here as in most American cities.

The Colton map added a typeset page that described both landmarks and other geographic data, from "Astronomical Position" to "Productive Industry." Elsewhere, however, the writer editorialized freely, if only to pad out the material ("The mechanical operations of [the Navy Yard] are various and extensive, and the skill of the workmen and the excellence of the materials employed have been satisfactorily tested on every sea...") The Washington Monument, again, was described as complete with rotunda, statues, and even "the tomb of the great chief."

Both cartographers, however, presented accurate population figures—apparently a first in Washington map-making. Both Colton's and Keily's tables of U.S. Census data carefully separated "White Inhabitants," "Colored Inhabitants" or "Freed Colored," and "Slaves" of this still thoroughly Southern city. Keily's map tables, taken from 1850 only, included statistics by wards for "male" and "female," while Colton's data, on his supplementary information sheet, presented comparative census data from 1800 to 1850 in ten-

year increments. Colton added totals of the "insane," "blind," and "deaf and dumb," and helpfully noted that comparisons with earlier censuses should be adjusted for the 1846 retrocession of trans-Potomac District of Columbia to Virginia.

If today Colton and Keily's efforts seem more charming than dishonest, it is at least partly due to the sheer beauty of the metal engraving. That this art peaked in 1850s America is confirmed by U.S. paper money, which, like our stock certificates and diplomas, would remain firmly rooted for 150 years in the era's aesthetic: a late, overripe Greek Revival classicism, just turning to the florid Neo-Baroque of Second Empire Paris.

While often associated with the agrarian Old South—columned plantations, perfumed letters in copperplate handwriting—this antebellum moment was also one of precise, elegant technology: clipper ships, the yacht America, the American 4-4-0 locomotive, and cast iron buildings. Like the great expositions' crystal palaces, crisp iron and glass frames filled with bulbous housewares, it was a time of contradiction and transition. So it was for Washington. If these maps gave the capital a completeness it mostly lacked, they accurately reflected its evanescent Southern charm, and the thin, steely tension of a nation both on the brink of international power, and about to come apart at the seams.

Endnotes:

[1] North America, Anthony Trollope, J.B. Lippincott, Philadelphia, 1862 [First printing by Chapman And Hall, London, 1862. Lippincott is the first American edition; second is Harper & Bros., New York, 1863. Out of print. Quoted Leech, op.cit., p. 10.]

[2] The Education of Henry Adams, Henry Adams, Edited with an introduction and notes by Ernest Samuels, Houghton Mifflin Company, Boston, 1974, [First commercial edition, Massachusetts Historical Society, 1918], Ch.VII, p. 99

[3] "Map of the City of Washington, D.C., established as the Permanent Seat of the Government of the United States of America." James Keily, Surveyor. [Lloyd Van Derveer: Camden, N.J. 1851.] Library of Congress, Geography and Map Division. G3850 1851 .K4 Vault

[4] Reveille in Washington, Margaret Leech, Carroll & Graf, New York, 1986, [First edition Harper & Row, New York, 1941], Ch. 1, p.7

EXPANDING URBAN DENSITY AMID PROSPERITY

The Boschke Map

TITLE: Map of Washington City, District of Columbia, Seat of the Federal Government…
DATE DEPICTED: 1857
DATE ISSUED: 1857
CARTOGRAPHER: A. Boschke, C.E.
PUBLISHER: J. Bien, New York
Colored lithograph, 61.5 x 58.5 in
Courtesy, The Albert H. Small Washington Collection, 295.MP.NC.E.F.
(Also in Library of Congress, Geography and Map Division, G3850 1857 .B61)

BEYOND THE CENTRAL CORE: A BROADER PROSPECT

The Boschke Map

TITLE: Topographical Map of the District of Columbia
DATE DEPICTED: Surveyed in the Years 1856, 57, 58 & 59
DATE ISSUED: 1861
CARTOGRAPHER: A. Boschke
ENGRAVER: D. McClelland
PUBLISHER: Washington: D. McClelland, Blanchard & Mohun, Washington, D.C.
Engraved map, scale [1:15,840], 99 x 103 cm
Library of Congress, Geography and Map Division, G3850 1861 .B6 Vault

ESSAY BY ROBERT L. MILLER

Boschke's maps brought to pre-Civil War Washington a new standard of accuracy and density of information. He carried out his surveys in 1856-1859 and published his "Map of Washington City" in New York in 1857 as a colored lithograph roughly five-feet square, including an ornate border with 20 conventional vignettes of local sites. A second, more detailed map, showing the entire District of Columbia in a two-sheet, forty-inch square format, was engraved by Washington's David McClelland four years later.

Boschke copied earlier mapmakers in drawing all of Washington's many platted but unbuilt streets with line weights that "shadow" each block, making them appear slightly three-dimensional. In 1857, however, he became the first to show every building, public and private—clustered settlements centered on the President's House, Pennsylvania Avenue, the Capitol, Southwest's "Island," and the Navy Yard—a pattern that would soon vanish in the explosive growth that would follow the Civil War.

The 1861 District of Columbia map depicted this same realistic pattern of built up areas, at smaller scale but still highly detailed, as the center of a vast, irregular patchwork of woods, pasture, fields and water. Relatively featureless and without scale when viewed from a distance, it looks to modern eyes like some huge region in a satellite photo, while Washington and Georgetown resemble some equally huge megalopolis. On close inspection, however, Boschke's scale registers, as we begin to recognize individual buildings. This was a "partial cadastral" map; that is, the surveyor had started to indicate individual properties and some owners' names, with accurate boundary lines and tree lines. Contours and hatching showed the hills rising steadily

MAP
of
WASHINGTON CITY
DISTRICT OF COLUMBIA
SEAT OF THE FEDERAL GOVERNMENT.
Respectfully dedicated to the
SENATE AND THE HOUSE
OF REPRESENTATIVES
of the
UNITED STATES OF NORTH AMERICA.
Surveyed and published by
A. BOSCHKE C.E.
1857.
GEORGETOWN
POTOMAC RIVER
EASTERN BRANCH

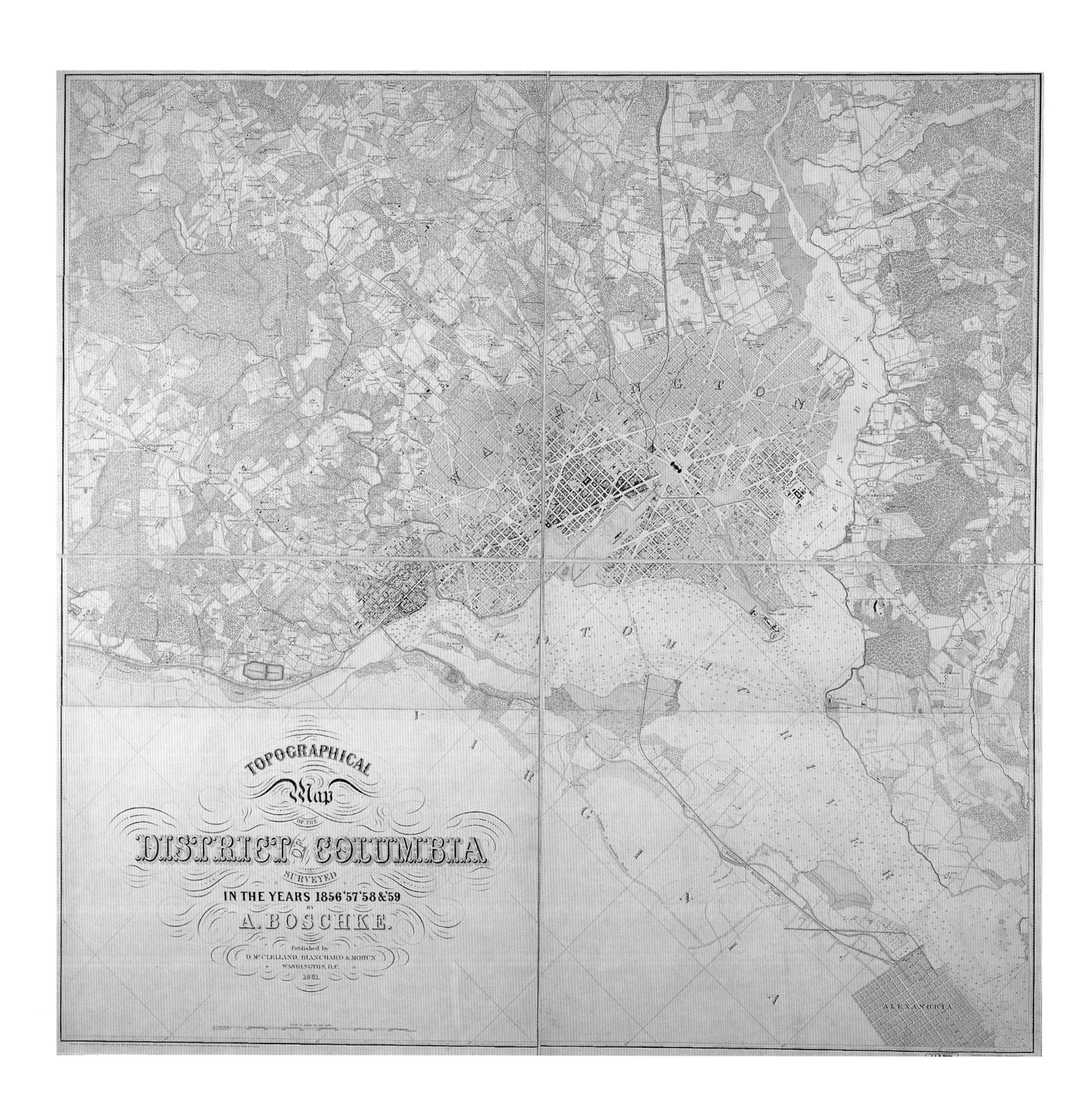
TOPOGRAPHICAL
Map
OF THE
DISTRICT OF COLUMBIA
SURVEYED
IN THE YEARS 1856'57'58&'59
BY
A. BOSCHKE.
Published by
D. McCLELLAND, BLANCHARD & MOHUN
WASHINGTON, D.C.
1861.
WASHINGTON
POTOMAC RIVER
VIRGINIA
ALEXANDRIA

from the city's alluvial plain; soundings outlined the depths of the Potomac and Eastern Branch (Anacostia River).

Perhaps most significant for the future, this 1861 map accurately traced the lines of regional roads and railroads, predictors of future growth and today's commercial and commuter arteries. On close study, too, the topographical data gave information of more immediate importance to Boschke's contemporaries—high ground suitable for forts and batteries that might be needed to defend the city.

In 1860-1861 "might be" was surely the operative phrase. As Boschke worked, war seemed more or less likely from day to day. On one hand, a compromise or negotiated dissolution of the Union might preserve peace. On the other, if Maryland and Virginia joined in secession, Washington might—as one Southerner suggested from the floor of Walter's new senate chamber—make a logical capital for the future Confederacy. Following the fall of Fort Sumter, Lincoln's first commanding general, the aged Mexican War hero Winfield Scott, privately feared that the District was indefensible, and made plans to turn the Treasury into a bunker where the President and Cabinet could hold off attack. Tense days followed during which the city was guarded only by local militias of questionable loyalty; while New England troops slowly made their way to Washington, by rail and steamboat, skirting the hostile city of Baltimore.

When these soldiers arrived many bivouacked in the halls of the Capitol. One regiment slept under the stars in the Rotunda below Walter's uncompleted dome. Meanwhile, the big cast iron dome segments waiting to be put in place had been arranged around the building as a kind of instant armor. In a similar mood of panic and improvisation, the War Department seized the plates for Boschke's new District map, leaving only a few proof impressions in private hands.

Soon construction of a ring of forts was underway on the high ground bordering the District Line. Undoubtedly Boschke's 1861 map played a role in changing the image of the District of Columbia, from wild and indistinguishable piece of Maryland to urban stronghold. Forty years later, the preserved circle of Civil War forts—and the image of a unified District—would form the basis for the park system's twenty-eight-mile Fort Drive that, no less than the rediscovery of L'Enfant, is a legacy of the Senate Park (McMillan) Commission Plan.

CONFIDENCE IN DESTINY: "TEN-MILE SQUARE"

The Arnold Map

TITLE: Topographical map of the Original District of Columbia and Environs, Showing the Fortifications around the City of Washington
DATE DEPICTED: 1862
DATE ISSUED: 1862
CARTOGRAPHER: E. G. Arnold, C.E.
PUBLISHER: G. Woolworth Colton, New York
Color Engraving, 31 x 34 in
Courtesy, The Albert H. Small Washington Collection, 009.MP.NC.E.F.
(Also in Library of Congress, Geography and Map Division, G3851.S5 1862 .A71 Vault)

STRATEGIC BATTLEFIELD POSITIONS

The Bruff Map

TITLE: Army Map of the Seat of the War in Virginia, Showing the Battle Fields, Fortifications, etc. on & near the Potomac River
DATE DEPICTED: 1861
DATE ISSUED: 1861
CARTOGRAPHER: J. G. Bruff
PUBLISHER: John Disturnell, New York, and Hudson Taylor, Washington, D.C.
Colored map, 29 x 27.5 in
Courtesy, The Albert H. Small Washington Collection, 315. MP.NC.E.F

At the commencement of the Civil War, Washington was a small town, relatively undeveloped, with a population of just over 60,000. Few streets were paved. An open sewer carried trash. Washington City Canal was noxious and stagnant. Sanitary regulations were poor. The meager police and fire departments were inadequate.

Within months after the War began, temporary residents from northern states, thousands of troops and other supporters of the war effort, increased the population to over 100,000. Temporary buildings were erected. To accommodate the sick and wounded, seventy hospitals sprung up in many former public and private facilities, in use until the War's end. The ranks of government employees increased from 3,466 to 7,184.

In addition to troop arrivals and departures, the city became a haven for former slaves who had escaped or had been abandoned by their masters. Nicknamed "contrabands," they came full of hope for freedom and a new life, as the District of Columbia banned slavery on April 16, 1862.

A limited number of city improvements took place during this period, including enlarging the Capitol and completing the dome, rebuilding Long Bridge across the Potomac River at 14th Street, and developing the Washington Aqueduct to bring water from the Potomac River at Great Falls to the city. The first streetcar line opened with horse-drawn cars going from the Navy Yard to Georgetown in 1862.

The Aqueduct Bridge, connecting Georgetown with Virginia, was used as a roadway—planks were laid for use by day and removed at night to stem a Confederate attack. Loss of some early battles by Union forces left Washington in a state of confusion. Dreary and dusty, with an endless sound of army wagons and ambulances, the city was a depressing scene. Looking from the Capitol toward the Potomac River, one saw a panorama of Union regiments grouped as tent cities.

A circle of sixty-eight forts and batteries in a thirty-seven-mile ring (including on the Virginia side of the Potomac) were built to protect the city under the direction of the Secretary of War, Edwin Stanton, following the Union defeat at Bull Run (Manassas, Virginia) in 1861. The first areas to be safeguarded were the heights at Arlington, bridges, and ridges around the Navy Yard and Capitol. Fortified connections were spaced systematically at 800 to 1,000 yards.

In 1901 the Senate Park Commission, and subsequently, the National Capital Park and Planning Commission in 1932, proposed a Fort Drive. This wide, 28-mile long parkway, within the District of Columbia's outer row of hills, would link old Civil War forts suitable for conversion to parks. From southeast to northwest the old forts included Fort Greble, Fort Carroll, Battery Ricketts, Fort Stanton, Fort Wagner, Fort Baker, Fort Davis*, Fort Dupont*, Fort Chaplin, A Battery, Fort Mahan, Fort Bunker Hill, Fort Totten, Fort Slocum, Fort Stevens*, Fort De Russye, Fort Reno, Fort Bayard, Battery Kemble, Battery Vermont*, and Battery Parrott. Fort McPherson in Arlington National Cemetery, was preserved for its historic value. Defense positions, except batteries, for later consideration as parks consisted of Fort Snyder, Fort Meigs,

COMPREHENSIVE REGIONAL WAR MAP
The United States War Department Map

TITLE: Extract of Military Map of N.E. Virginia Showing Forts and Roads
DATE DEPICTED: 1865
DATE ISSUED: 1958 facsimile of 1865 map
CARTOGRAPHER:
PUBLISHER: United States War Department, Engineer Bureau, Kensington, MD
Photo-processed facsimile map, scale [1:63,360], 60 x 42 cm
Library of Congress, Geography and Map Division, G3851.S5 1865 .U51 1958

MILITARY PERSPECTIVE FROM THE NORTH
The Prang View

TITLE: Birds-Eye View of the Seat of War, Arranged After the Latest Surveys, showing views of & their R.R. Connections & the Genereal Surface of the Country. . .
DATE DEPICTED: ca. 1863
DATE ISSUED: ca. 1863
CARTOGRAPHER: Unknown
PUBLISHER: L. Prang & Co., Boston
Colored lithograph, 13.5 x 19.25 in
Courtesy, The Albert H. Small Washington Collection, 049.MP.NC.L.F.

Fort Lincoln, Fort Saratoga, Fort Stemmer, Fort Kearny, Fort Reno, and Fort Gaines. (Those forts marked with an asterisk [*] are now owned by the Federal Government and are under the jurisdiction of the National Park Sevice.)

The popular Arnold Map located the circle of forts, marking their high elevations surrounding the central city. The "ten mile square" was drawn, recalling Andrew Ellicott's 1793 and 1794 boundary plans. Although the District's Virginia portion had been re-ceded in 1846, Alexandria remained a major port for Union troops. Only the original City of Washington was color-coded on the map by ward, and railroads and roadway connections were indicated. Border flourishes, Astronomical Positions and Population Charts from 1800 to 1860 revealed racial statistics for "whites, free colored, slaves," as in other recent maps (p. 82).

Lloyd's Military Map of 1861 (not shown, by Viele and Hastings), published under the auspices of the American Geographical and Statistical Society, assembled several maps on one sheet to present a broad War perspective.

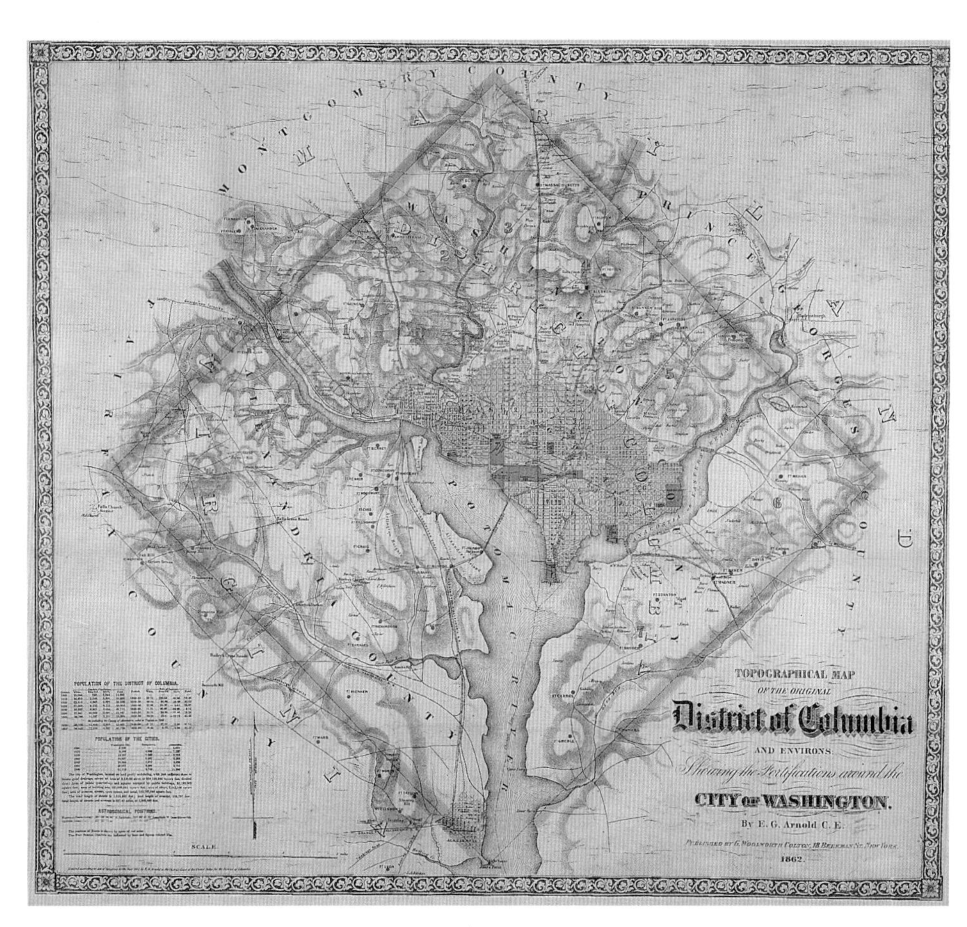

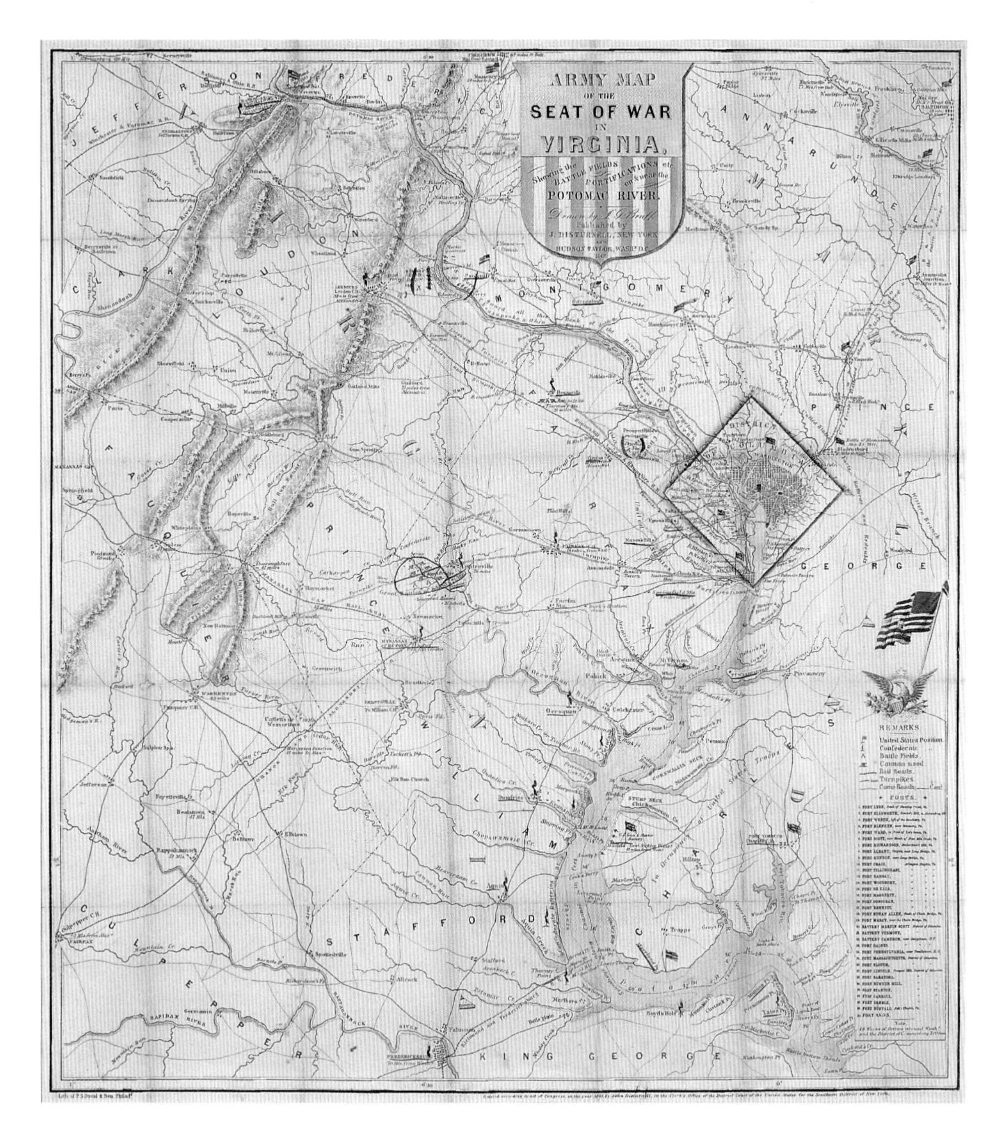

The focal point was two central images, the United States map with a line drawn between north and south, and the District of Columbia bordered by four colorful symbolic figures. Washington was also depicted on a regional map which indicated railroads, rivers and water bodies, and the Appalachian Mountain range—similar to the depiction on the 1775 Fry and Jefferson Map (p. 26). This was an interesting example of the plethora of Civil War maps created during the war years.

The 1861 regional Bruff Map depicted conditions early in the Civil War, listing 34 forts. Red, white and blue flags identified strategic positions and towns controlled by the both the Union and Confederate forces. Capitalizing on symbolic imagery and colors, this map concentrated upon the Potomac River and Virginia as the "Seat of War," a play on words relevant to early District maps as the "Seat of Government." Also indicated were battlefields, cannons, railroads, turnpikes, common roads and canals. While expository in design, the message projected was unity.

The U.S. War Department's Engineer Bureau produced a number of facsimile maps in 1865 with minor notation variations. A map note entitled "Extract of Military Map of N.E. Virginia" stated "The Coast Survey Maps were used in the Compilation, North of the Potomac, outside of the District of Columbia." Drawn in black and white, this information-packed government document showed forts, railroads, roads, bridges, watercourses, vegetation imprints, relief by hachure and depths by soundings. Union roads were marked by heavy black continuous lines. Of special interest, rural householders were named.

Prang's engaging "Birds-Eye View" was an unusual portrayal of the Chesapeake Bay region looking from north to south with Baltimore and Ft. McHenry at lower left. Cities and villages were drawn in three dimensions, highlighting the pivotal position of Washington, and nearby towns of Georgetown and Alexandria. From this angle, the strategic location of Ft. Washington at a bend in the Potomac facing Mount Vernon becomes apparent.

We may use Prang's map to follow the route of British troops during an earlier war—the War of 1812. Americans anticipated a frontal attack on Washington from the Potomac River. Imagine the surprise when the British arrived in the evening of August 24, 1814 from a northern overland course, led by General Robert Ross and Admiral Cockburn. They took a more direct route, sailing up the Patuxent River, proceeding on land to Upper Marlboro and Bladensburg, before advancing from behind to Washington. Flames lit the night sky as the President's House, the Capitol, and other civic and private buildings burned. Fortunately, many valuable documents had been removed from Washington, including the Constitution and the Declaration of Independence.

Civil War maps evoked a spirit of nationalism. The War itself transformed Washington, both physically and emotionally. War's emotional trauma, and the country's subsequent physical and social reconstruction, compelled Americans to seek a new identity. Within a decade, Washington, formerly a small backwater town, would become a cosmopolitan city, a consequence of demographic changes, population growth, new construction and infrastructure, and the city's enhanced self-image.

Extract of
Military Map of N.E.Virginia
Showing Forts and Roads
Engineer Bureau
War Dept
1865.
Scale one inch to the mile
Note.
The Coast Survey Maps were used in the Compilation North of the Potomac outside of the Dist. of Columbia.

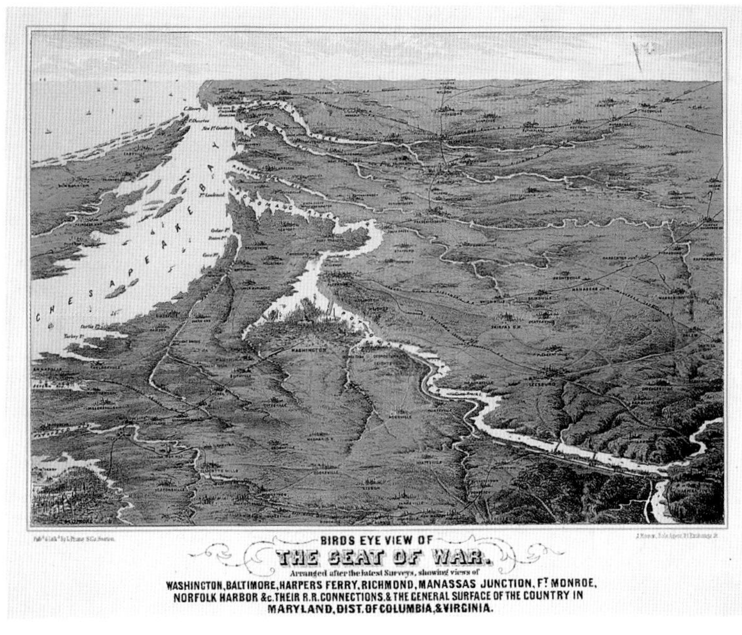
BIRDS EYE VIEW OF
THE SEAT OF WAR.
Arranged after the latest Surveys, showing views of
WASHINGTON, BALTIMORE, HARPERS FERRY, RICHMOND, MANASSAS JUNCTION, F.T MONROE,
NORFOLK HARBOR &c. THEIR R.R. CONNECTIONS, & THE GENERAL SURFACE OF THE COUNTRY IN
MARYLAND, DIST. OF COLUMBIA, & VIRGINIA.

A WASHINGTON VIEW, POST-CIVIL WAR

The E. Sachse View of the Capitol, 1871

TITLE: View of Washington City
DATE DEPICTED: 1871
DATE ISSUED: 1871
CARTOGRAPHER: Edward Sachse
PUBLISHER: E. Sachse & Co., Baltimore, MD
Chromolithograph, 17.75 x 26.75 in
Courtesy, The Albert H. Small Washington Collection, 002.PR.BV.E.F.
(Also in Library of Congress, Prints and Photographs Division, Neg. LC-USZ C4 770, #PGA-D-Sachse, E, & Co.-View of Washington City)

The gleaming white marble-clad cast-iron dome of the Capitol conveys the essence of the noble experiment in democratic government. Adorned with Neoclassical ornamentation, the building presents a paradigm for the ideals of the nation's founders.

Published shortly before his death in 1873, we see here Edward Sachse's final view of the Capitol with its east front additions and new dome—the symbolic focal point of Washington—by Thomas U. Walter. In this west-oriented view, exaggerated for artistic effect, the imposing scale of the building—62,000 square feet—highlights the grandeur of its new soaring dome. Some members of Congress had urged the acquisition of surrounding homes and additional Capitol Hill land to create a more regal setting.

In contrast to Sachse's 1852 view (p. 80), the city's character had changed considerably by 1871. In this view, the Capitol and urban fabric are better integrated, expressing faith in the future. Once again, Sachse worked with varying degrees of accuracy. The unfinished Washington Monument, in fact built 370 feet east and 123 feet south of the designated point on the L'Enfant Plan, was still shown according to Robert Mills' original design with the colonnade at the base. The Mall itself was depicted as a picturesque landscape with fountain, curving lanes, and densely planted trees. By 1869, only the north end of the Treasury Building on Pennsylvania Avenue was occupied—although Sachse portrayed a completed structure.

Immediately following the Civil War, construction began on several new public buildings, with occasional departures from the classical style. Mirroring population growth, the city had filled in between the Botanic Garden and its greenhouses at the foot of the Capitol and the distant heights of Georgetown. New government programs, federal bureaucracy, increasing industrialization and commerce thrust Washington into a position of immense importance. Society was transforming from an agricultural rural base to urbanization. New residential neighborhoods began to develop as scores of European immigrants were drawn to Washington along with vast numbers of Americans—both whites and newly freed African-Americans.

Undoubtedly, the city was becoming more attractive and pleasant, albeit with pockets of great poverty and blight. Appalling conditions and rampant crime south of Pennsylvania Avenue (now called Federal Triangle) provoked the nickname "Murder Bay." People lived here in crowded shanties. Filth from Center Market and stagnant water from the canal fostered an unhealthy, shocking situation just in sight of the Capitol.

Within a year, under the extraordinary leadership and authority of Alexander Robey Shepherd, the Washington City Canal would become an underground culvert, converted to B Street (now Constitution Avenue) at ground level. Intolerable conditions of roads, utilities and other infrastructure would be upgraded with a comprehensive plan for improvements.

IEW OF WASHINGTON CITY.

SANITATION, HEALTH AND WELFARE

The Gedney Statistical Maps

TITLE: Exhibit Chart showing Streets & Avenues of the Cities of Washington and Georgetown (water mains), improved under the Board of Public Works, D.C.
DATE DEPICTED: 1873
DATE ISSUED: November 1, 1873
CARTOGRAPHER: J. F. Gedney
PUBLISHER: Board of Public Works, Washington, D.C.
Colored map, 42 x 52 cm
Library of Congress, Geography and Map Division, 1:G3851 .N42 1873 .E9, 2: G3851 .N44 1873.E9

REMARKABLE IMPROVEMENTS IN CITY ENHANCEMENTS

The Greene Statistical Maps

TITLE: City of Washington, Statistical Maps
No. 1. Valuation of Real Property—1878-79
No. 2. Street Pavements—1882
No. 3. Shade Trees—1880
No. 4. Gas Lamps—1880
DATE DEPICTED: 1878-1882
DATE ISSUED: July 1, 1880
CARTOGRAPHER: Lieut. Francis Vinton Greene
PUBLILSHED: Washington, D.C.
Colored maps (from set of 12), scale [ca.1:12,000], 53 x 75 cm
Courtesy, The Albert H. Small Washington Collection, 116 MP.ND.E.F.
(Also in Library of Congress, Geography and Map Division, G3851 .F8 s12 .G7)

The post-Civil War image of Washington summoned most unfavorable observations, especially from European journalists and diplomats familiar with the elegant capitals of the continent. War had interfered with the city's intent to upgrade infrastructure and its commitment to build an improved market by 1862. Congressional impatience with basic conditions prompted a movement to relocate the federal city to the center of the country, perhaps at St. Louis.

Annual reports by the federal Corps of Engineers, Officer in charge of Public Buildings, Grounds, and Works, also called attention to disastrous physical conditions, recommending improvements to both places under federal control and those that were not federally supervised. Corps-controlled federal sites included most parks and public reservations, all avenues and many streets, the aqueduct water system, and much of The Mall. A critical 1867 report by Major Nathaniel Michler faulted the deplorable conditions of the Washington City Canal, and the locations and offensive state of the markets and their environs.

Michler advised channeling the unsightly, unsanitary Washington City Canal entirely underground, in a manner similar to that of the culvert under the Botanic Garden at the foot of the Capitol. He was appalled by the Mount Vernon Square Market (at 8th and K Streets, New York and Massachusetts Avenues, N.W.), located on a principal L'Enfant axis and designated public reservation. He was equally outraged by Center Market located at nearby Pennsylvania Avenue.

Of the avenues and streets, Michler criticized the inferior, unpaved, deeply rutted, and often impassable roadways, and noted the necessity for sewers for drainage. He stressed that the nation's capital must remedy these problems. A capital is expected to facilitate the comings and goings of its residents and visitors, and to offer splendor appropriate to its status. He urged that Andrew Jackson Downing's Comprehensive Plan for The Mall be carried out in each direction beyond the Smithsonian Institution Grounds. In order that north-south street traffic might conveniently pass across The Mall, he devised a special gate. Moreover, Michler advocated the display of water and statuary in public reservations, in keeping with L'Enfant's seminal concept.

Michler's report set the stage for congressional action and the entrance of an ambitious

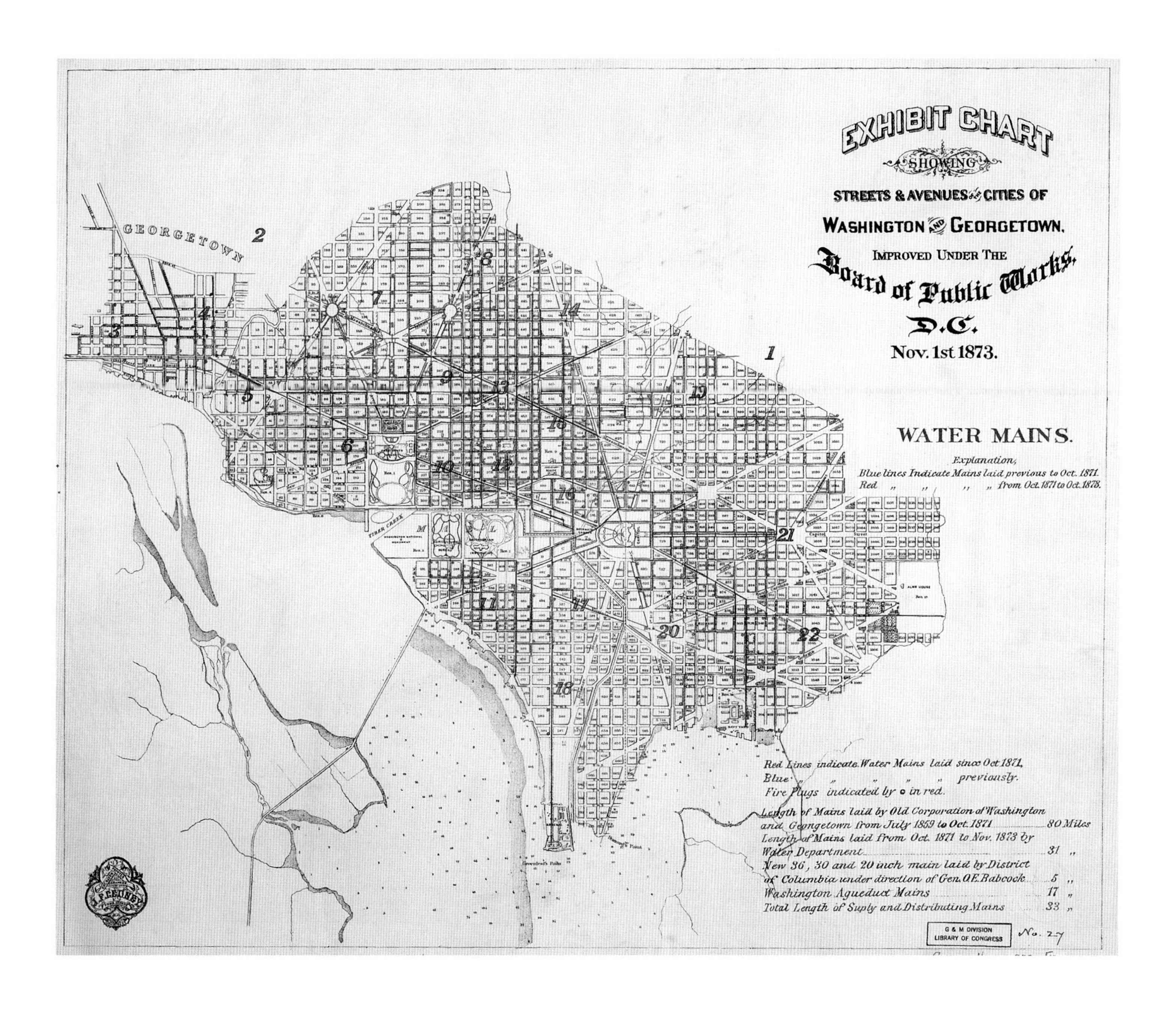
EXHIBIT CHART
SHOWING
STREETS & AVENUES of the CITIES OF
WASHINGTON AND GEORGETOWN,
IMPROVED UNDER THE
Board of Public Works,
D.C.
Nov. 1st 1873.
WATER MAINS.
Explanation,
Blue lines Indicate Mains laid previous to Oct. 1871.
Red „ „ „ „ from Oct. 1871 to Oct. 1873.
GEORGETOWN
TIBER CREEK
Red Lines indicate Water Mains laid since Oct 1871,
Blue „ „ „ „ „ previously.
Fire Plugs indicated by o in red.
Length of Mains laid by Old Corporation of Washington and Georgetown from July 1859 to Oct. 1871 ... 80 Miles
Length of Mains laid from Oct. 1871 to Nov. 1873 by Water Department ... 31 „
New 36, 30 and 20 inch main laid by District of Columbia under direction of Gen. O.E. Babcock ... 5 „
Washington Aqueduct Mains ... 17 „
Total Length of Suply and Distributing Mains ... 33 „
G & M DIVISION
LIBRARY OF CONGRESS
No. 27

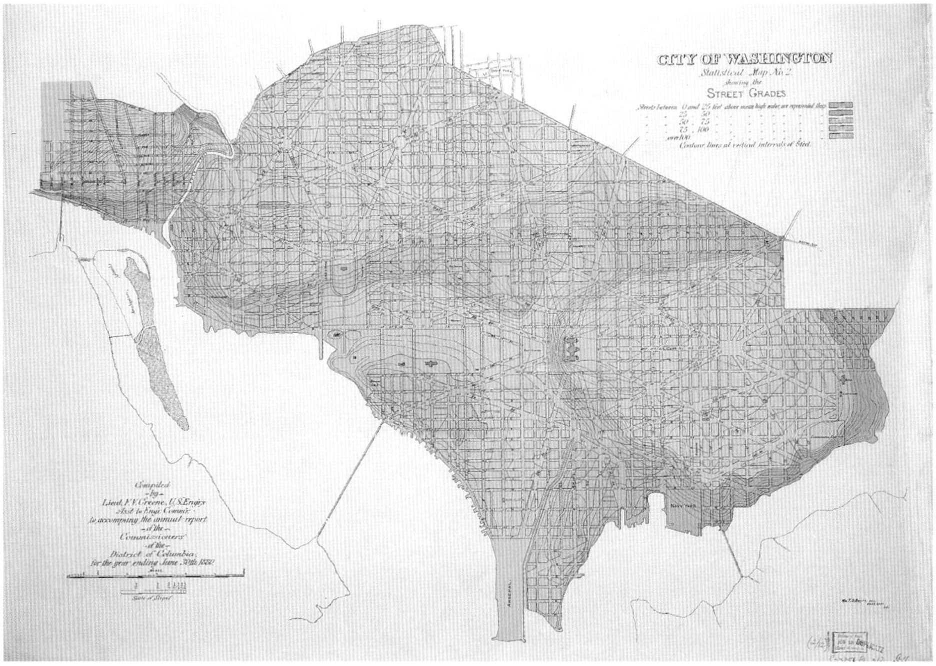

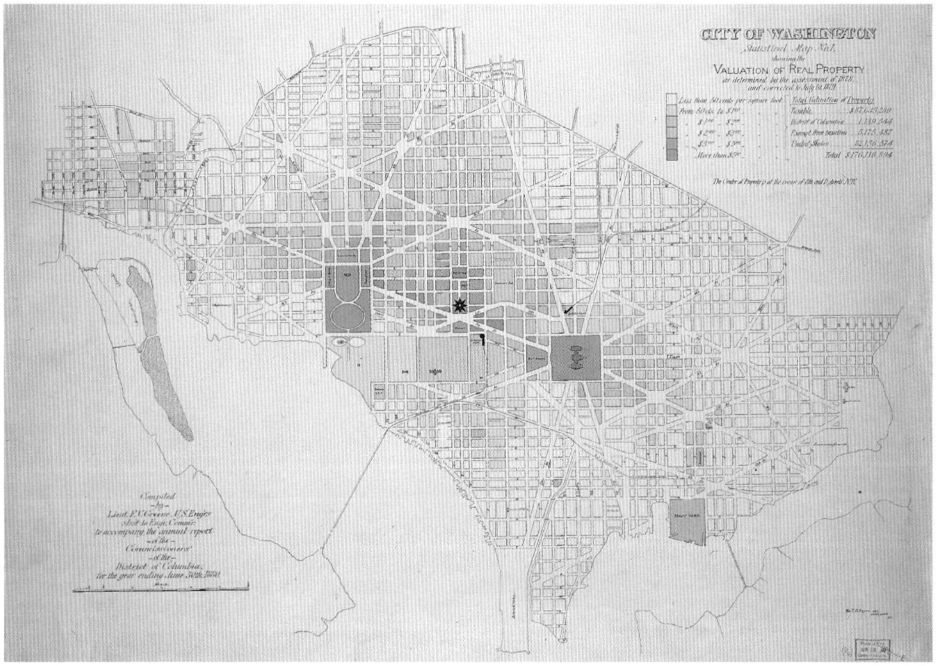

young man, Alexander Robey Shepherd, later called "Boss" Shepherd, who would carry out a massive public works agenda. An appropriation of $500,000 enabled construction of a new State, War, and Navy Building (now Old Executive Office Building) on Pennsylvania Avenue near the White House. Congress created a Territorial Government in 1871 with a governor, legislature, and five-member Board of Public Works. Three quarters of a century after Washington became the nation's capital, Shepherd, a businessman involved in construction and land development, served as head of the Board of Public Works, and later briefly as the Territorial Governor. People still speak of the great public improvements program that he oversaw.

In three years, more than two-thirds of the 300 miles of streets and avenues were paved with wood, concrete, or macadam. Soon Washington had more paved streets than most American cities. Improved topographical grading and roadway paving expedited the creation of a far-reaching "street-rail" system. An extensive sewer, water main and gas main system was laid, plus 128 miles of sidewalks, and 3,000 gas lamps. Washington City Canal was filled and Tiber Creek was channeled underground.

Many parks were fashionably landscaped with graceful trees, elaborate fountains, fences, lodges for a park concierge, and meandering walkways in the prevailing romantic landscape style (p. 77) of Downing's era. Small triangles and squares, formerly trash heaps, became beautifully planted reservations scattered throughout the city. No longer were farm animals fanning out through streets and parks as in a third-world country. Nearly 60,000 trees (some still flourishing today) were planted along street curbs. This image of sylvan magnificence led many to proclaim Washington the "City of Trees." At the arrival of the twenty-first century, the Garden Club of America received an endowment from the Casey Foundation to restore Washington's forested canopy of trees. Aerial photography and mapping will enable documentation of current conditions and replacement.

Colorful statistical maps and reports documented these assets, along with real estate values. Gedney's "Exhibit Charts" showed improvements made under the Board of Public Works, such as gas (not shown) and water mains for "Cities of Washington and Georgetown," designated red and blue (red—water mains laid from October 1871 to 1873; blue—laid previously). Fire plugs were indicated by a red circle. Gedney's map noted that the "Old Corporation of Washington" had laid 80 miles of water

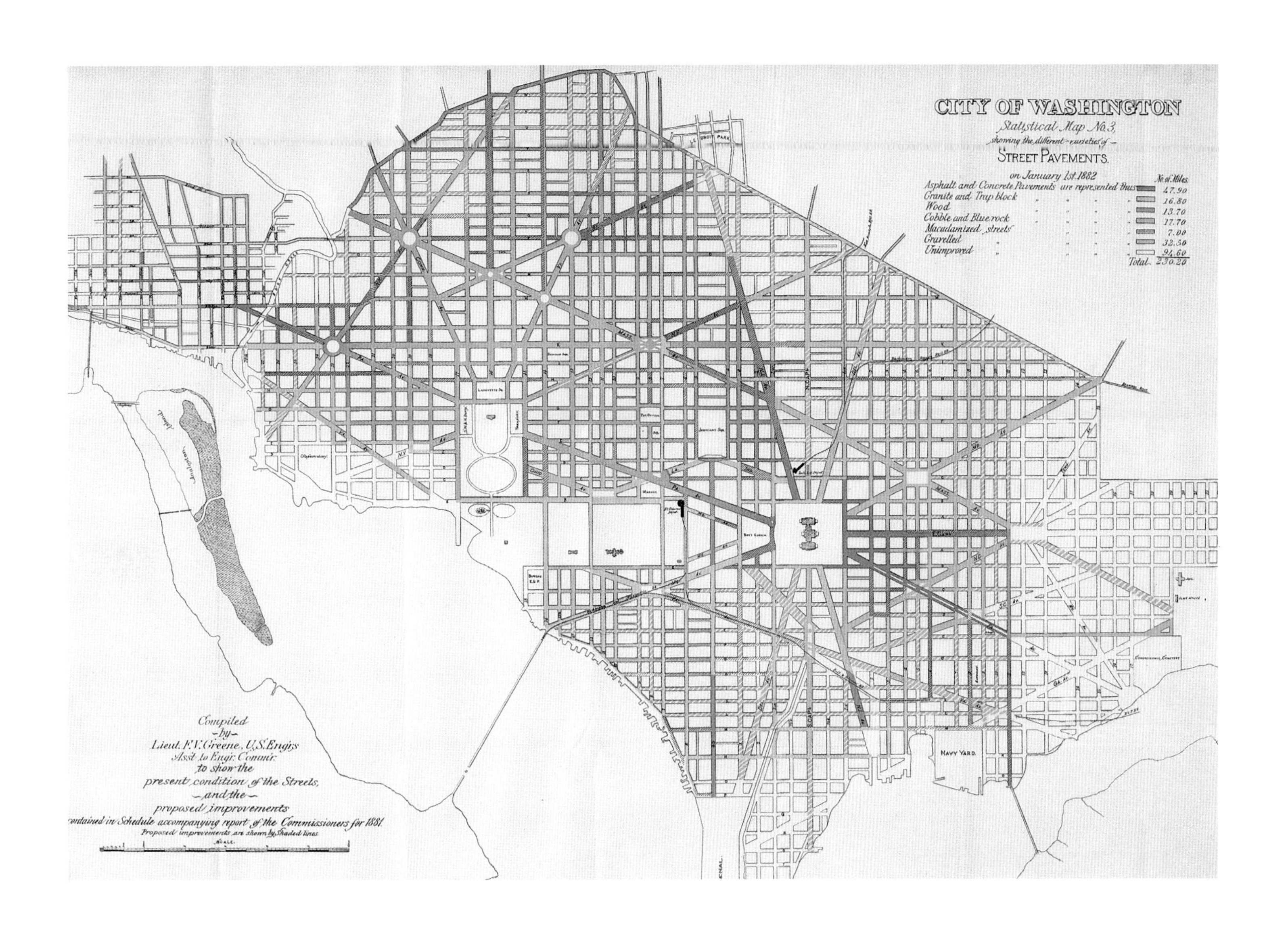
CITY OF WASHINGTON
Statistical Map No. 3,
showing the different varieties of
STREET PAVEMENTS.
on January 1st 1882
No. of Miles.
Asphalt and Concrete Pavements are represented thus 47.90
Granite and Trap block 16.80
Wood 13.70
Cobble and Blue rock 17.70
Macadamized streets 7.00
Gravelled 32.50
Unimproved 94.60
Total 230.20
NAVY YARD
Compiled
by
Lieut. F. V. Greene, U.S. Engrs
Ass't to Engr. Commr.
to show the
present condition of the Streets,
and the
proposed improvements
contained in Schedule accompanying report of the Commissioners for 1881.
Proposed improvements are shown by Shaded lines.
SCALE.

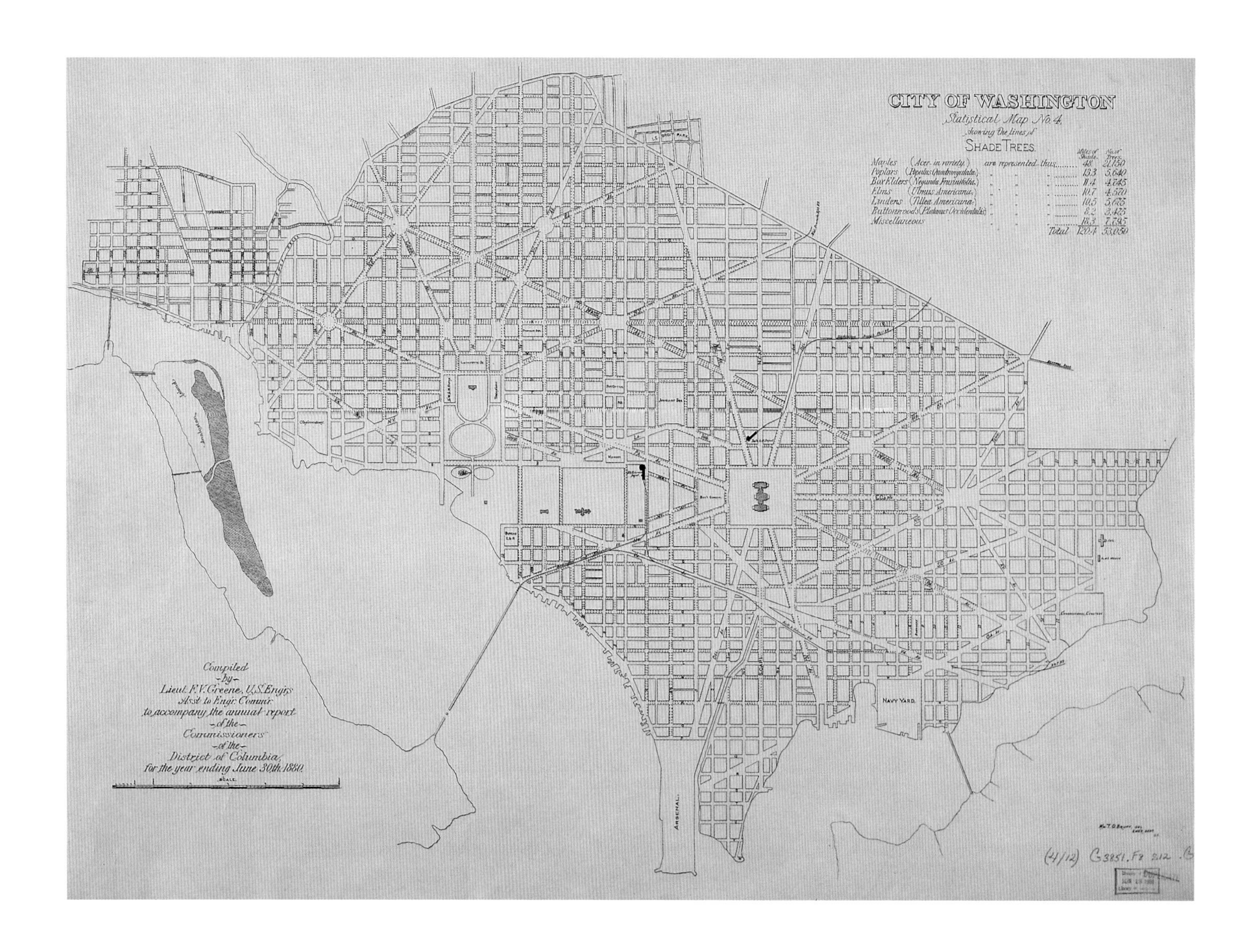

mains between 1859 and 1871, while more than 31 miles had been laid in the two subsequent years. Previous to 1871, there were 69 miles of gas mains, while 39 additional miles were laid during the next two years. Both maps noted public reservations by name and number and water depths by soundings.

Five of twelve "Statistical Maps" compiled by Lieutenant Greene "to accompany the annual report of the Commissioners of the District of Columbia" synthesized the types of improvements initiated by "Boss" Shepherd that contributed to an up-to-date praiseworthy image. Map No. 1, in pastel tones, indicated "Valuation of Real Property" per square foot. Map No. 3 presented a surprising range of "Street Pavements" noted by "No. of Miles"—47.9 asphalt and concrete, 16.8 granite and trap block, 13.7 wood, 17.7 cobble and blue rock, 7 macadamized, 32.5 graveled, and 94.6 unimproved. Colored dots filled Maps No. 4 and 5, to represent the abundant "lines of Shade Trees" and "location of Gas Lamps." These amenities allowed people to develop a taste for beauty and perceive the elegant gardenesque quality that now characterized the city. Map No. 2, Greene's "Street Grades," rendered in soft pastel shades, showed contours at 5-foot vertical intervals. Map No. 12 (not shown) portrayed a "Schedule of Street Sweeping," showing the number of streets swept daily, twice per week, once per week (the largest, by far), once in two weeks, and wooden pavement at once per month. The flurry of improvements encouraged the production of other statistical maps.

Entwistle's map (not shown) is of interest for its diagram showing "Divisions of the City," its color-coded "Street Rail Roads," its pictograph symbols of churches, hotels, markets, "Places of Amusement," approximately 11 Public Schools, and named public reservations.[1] The Peters Co. "Conventioneering guide" (p. 102) is presented in a similar fashion.

Not unlike today, these necessary yet huge civic engineering and sanitary works eclipsed cost

projections! A depression in 1873 exacerbated financial problems. Funding, in part through a bond issue and assessments on private property, could not prevent accusations of corruption and wrong-doing, nor muster support for Shepherd. In 1874, after only three years, the Territorial Government was abolished. A temporary form of government harkened back to the District's first form of rule—three commissioners appointed by the President and approved by the Senate. By 1878, a new government was established through the "Organic Act," a municipal corporation somewhat similar to the present form. Thus, modern Washington began, energized by passage through an astonishing era.

[Two important companies began producing maps relevant to the discussion above (not shown), which continue to be widely used and updated: Baist Atlas first published in 1903 (11th edition was 1960) and Sanborn Maps begun in 1888. The Baist Atlas was preceded by the G. M. Hopkins Company Maps in 1878 and 1892. Related to infrastructure, the Sanborn Maps pertained to fire insurance, while the Baist Atlas was made for real estate purposes. Consequently, Baist shows ownership of land tracts, often with names of owners, and block and lot numbers, thus facilitating tracing property ownership, historical and legal research, real estate and city planning needs. Smaller in scale than the Sanborn Maps, the Baist maps offer larger geographic coverage of neighborhoods. As a fire insurance document, the Sanborn Maps show fire hydrants, water mains, fire walls, roof composition, and construction materials for buildings, such as wood and brick. These maps can be found in the Library of Congress, D.C. Office of the Surveyor, D.C. Library, and in private offices.]

[1] "Entwistle's Handy Map of Washington and Vicinity: Showing Public Buildings, Churches, Hotels, Places of Amusement and lines of Street Rail Roads." Published in 1876 by J.C. Entwistle. Washington, D.C. In the collection of the Library of Congress, Geography and Map Division, G3850 1876 .E5.

CHARTING LAND RECLAMATION AND HARBOR POTENTIAL

The Hilgard Engineering Map

TITLE: Washington and Georgetown Harbors, District of Columbia
DATE DEPICTED: 1887
DATE ISSUED: 1882
CARTOGRAPHER: J. E. Hilgard
PUBLILSHER: D.C. Government
Engraving, 18 x 28.5 in
Courtesy, The Albert H. Small Washington Collection, 135.MP.ND.E.F

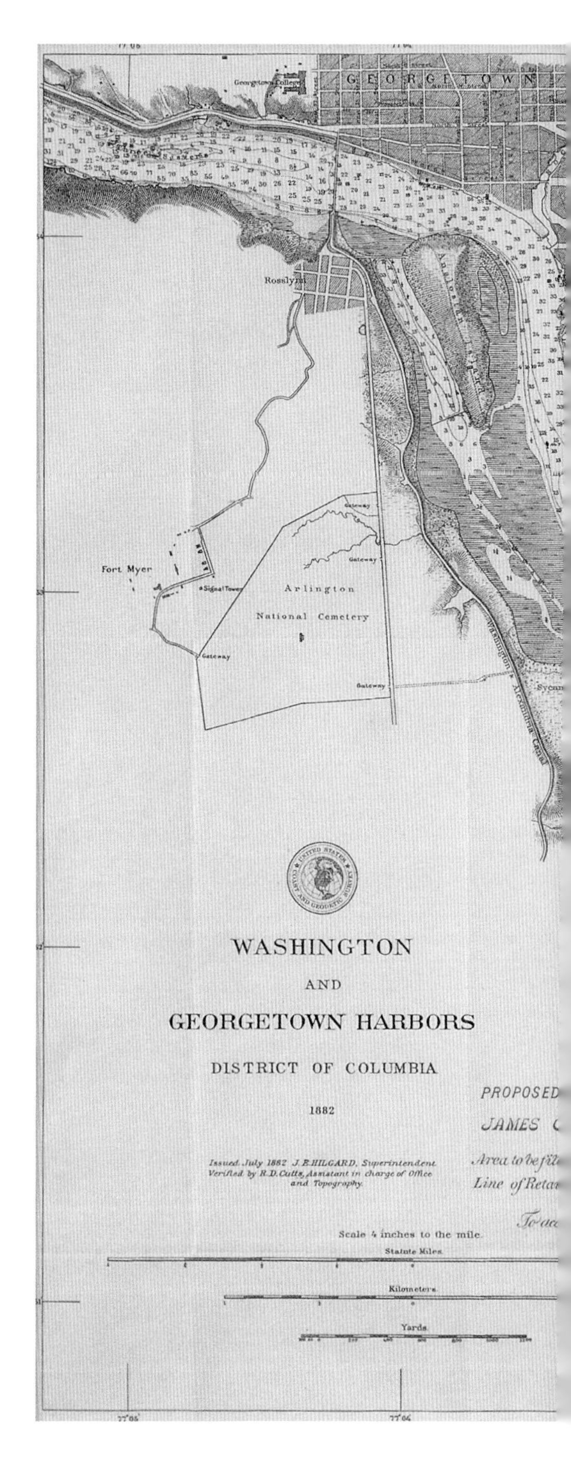

A proliferation of railroads and failure of the C&O Canal to advance Washington as a major port city resulted in little attention being paid to navigation on the Potomac River. Rapid physical growth commensurate with splendid civic and residential architecture rekindled a desire for expanded commercial opportunities. In 1881, a Senate committee investigation led to Congressional appropriation of funds to revive and improve navigation and reclaim the tidal flats.

Over a twenty-year period, 600 acres were reclaimed through a process of dredging the river channel to increase its depth. Dredgings were dumped in the adjacent marshlands above high tide and flood levels. The deeper river channel enabled water craft to reach wharves and docks. Essentially, between the newly created West and East Potomac Parks, the Tidal Basin functioned as a tidal reservoir. Its automatic water gate allowed water to fill up during high tide when the gates were closed. During low tide the gates opened between the Basin and channel to release excess water and debris.

Before the Potomac Flats landfill project was undertaken, the Potomac River bank stretched quite close to the Washington Monument's south side. Reclamation both altered aspects of L'Enfant's plan for The Mall, and expedited a "kite-shaped" symmetrical design that would be posed by the 1901 Senate Park Commission Plan.

Hilgard's map, sweeping across the southern portion of Washington and Georgetown, emphasizing the rivers, was rife with numbers—tides, contours, and soundings for rivers and landfill areas. A "Sewer Canal" connected the old Washington City Canal (Tiber Creek) under the proposed reclamation area with the Potomac River. A "Proposed Improvement of James Creek Canal to accompany Report of December 16th 1887" by Lieutenant Colonel Peter C. Hains was applied in red pen. To the southeast, the grid pattern mews of Congressional Cemetery and the U.S. Jail can be seen alongside the Eastern Branch, or Anacostia, its recent name.

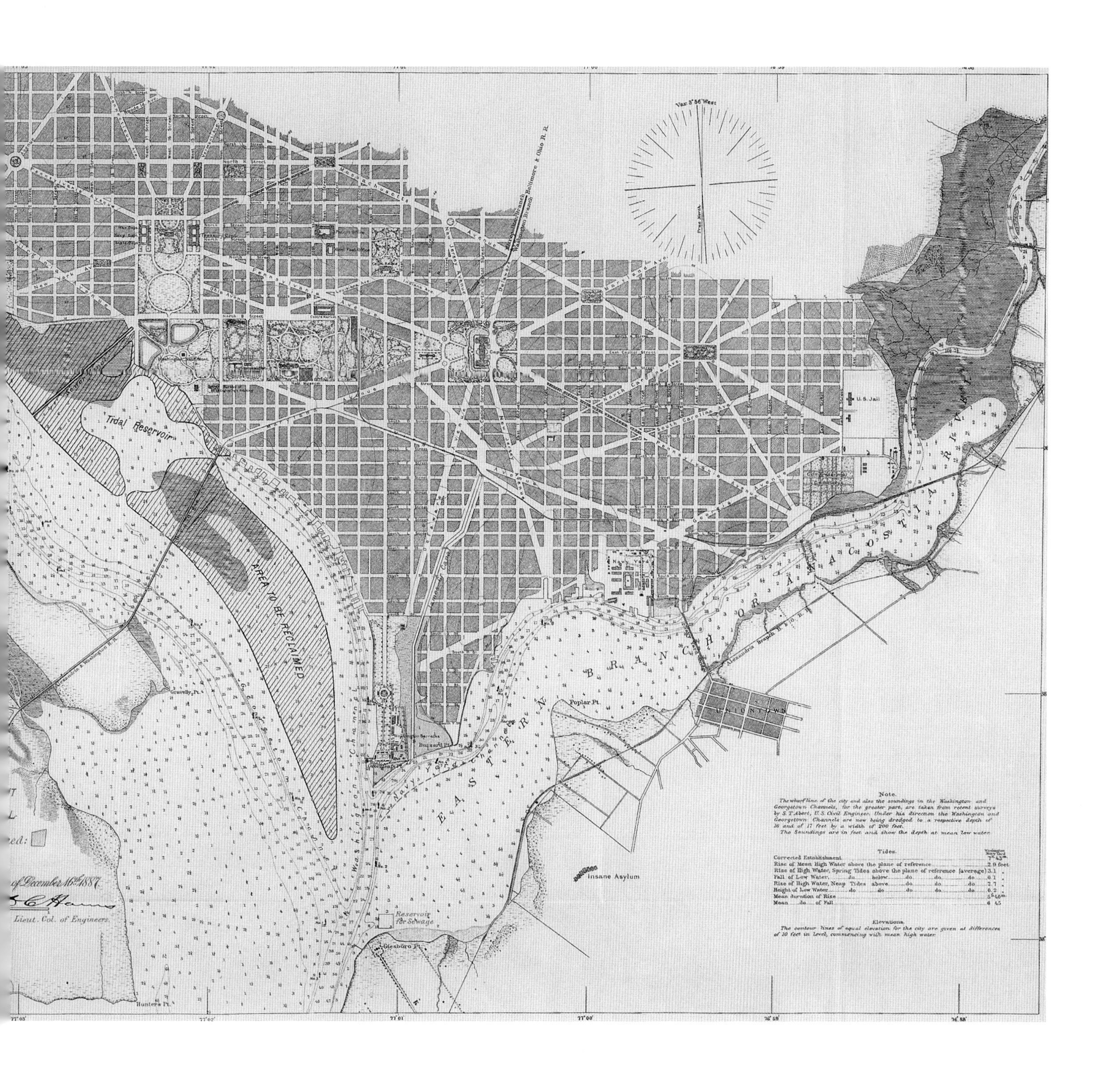
Var. 3°56' West
True North
Tidal Reservoirs
AREA TO BE RECLAIMED
U.S. Jail
Navy Yard
EASTERN BRANCH OR ANACOSTIA RIVER
Washington Channel
Navy Yard Channel
UNIONTOWN
Poplar Pt.
Buzzard Pt.
Gravelly Pt.
Insane Asylum
Reservoir for Sewage
Giesboro Pt.
Hunters Pt.
Alexandria & Washington R.R.
Alexandria Branch B. & O. R.R.
Baltimore & Ohio R.R.
of December 16th 1887.
Lieut. Col. of Engineers.
Note.
The wharf line of the city and also the soundings in the Washington and Georgetown Channels, for the greater part, are taken from recent surveys by S. T. Abert, U.S. Civil Engineer. Under his direction the Washington and Georgetown Channels are now being dredged to a respective depth of 16 and of 17 feet by a width of 200 feet.
The Soundings are in feet and show the depth at mean low water.
Tides.
Washington Navy Yard
Corrected Establishment 7h 43m
Rise of Mean High Water above the plane of reference 2.9 feet
Rise of High Water, Spring Tides above the plane of reference (average) 3.1
Fall of Low Water, do below do do do 0.1
Rise of High Water, Neap Tides above do do do 2.7
Height of Low Water do do do do do 0.2
Mean duration of Rise 5h 40m
Mean do of Fall 6 45
Elevations.
The contour lines of equal elevation for the city are given at differences of 10 feet in level, commencing with mean high water.

CONVENTIONEERING GUIDE, "IN-TOWN" SUBURBS

The Peters Co. Map for International Christian Endeavor

TITLE: Fifteenth International Christian Endeavor Convention
DATE DEPICTED: July 8-13, 1896
DATE ISSUED: 1896
CARTOGRAPHER: Unknown
PUBLISHER: Washington: The Committee, C1896 (Norris Peters, Co., Washington)
Colored map, scale [ca. 1:17,000,], 45 x 58 cm
Library of Congress, Geography and Map Division, G3850 1896 .U5

Inevitably, tourist and conference maps exhibit an eye-catching flamboyance. And, the Fifteenth International Christian Endeavor Convention's Official Map of Washington, prepared by the Committee of '96, was no exception.

A powerful graphic design prominently portrayed the city's division into four quadrants intersecting at the Capitol. Bold north-south and east-west lines distinguish neighborhood sectors that had been envisioned since the inception of the L'Enfant Plan.

The map makes clear the systematic order of streets fanning out from the Capitol. Those streets aligned in a north-south direction were named alphabetically; while those aligned in an east-west direction were given numerical names. This map was useful to Washington's many visitors, and even to its residents, who may not have perceived the street identification concept.

Referencing the iconography of L'Enfant's plan, equal radial distances were drawn, centering upon another branch of government, the president. With the Ellipse and White House as the emblematic center, the radials unfolded in reasoned, cosmic-like symbolism.

A principal feature was the "Location of Street Railways and Transfer Stations." This form of public transportation was relatively new. Wharves for four different means of water transport were noted—an urban design component frequently stated on the wish-lists of many twentieth- and twenty-first-century city planners.

A large portion of the sheet was filled with religious edifices and hotels. Also listed on the map were depots, public buildings, educational institutions, foreign legations, hospitals, monuments, statues, parks, squares, and cemeteries. Additional points of interest were mentioned on the verso. For tourists seeking recreation or a quiet respite from conference activities, public parks and reservations were delineated in green. Few maps of this period depicted Rock Creek from its mouth passing through the National Zoological Park and Rock Creek Park into upper northwest.

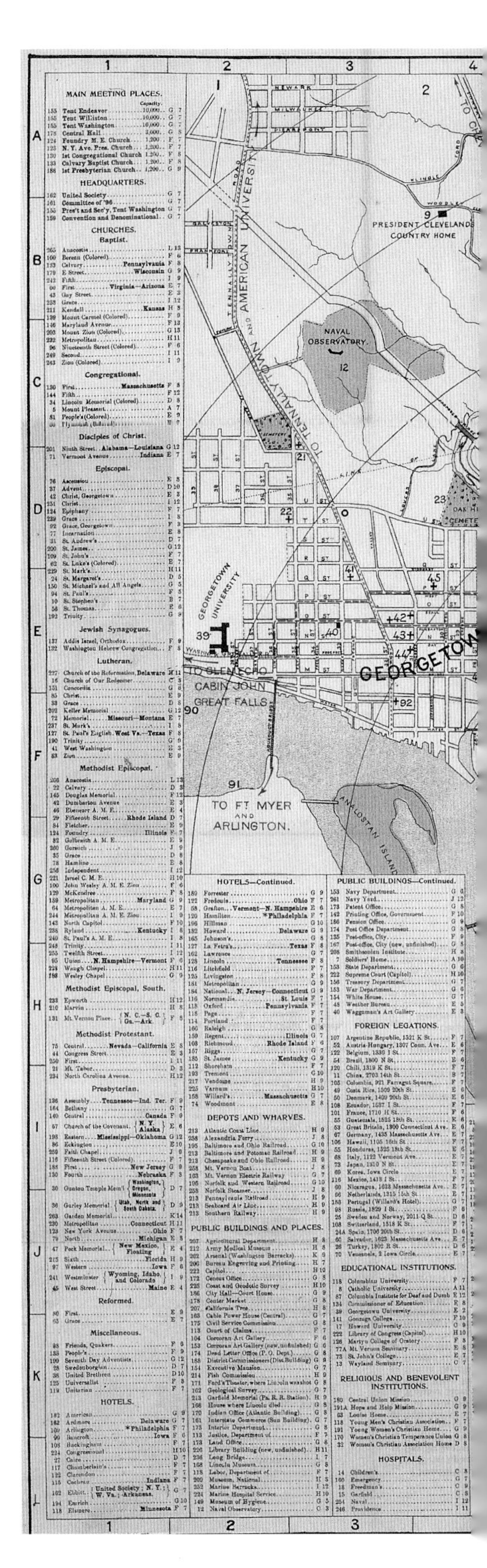

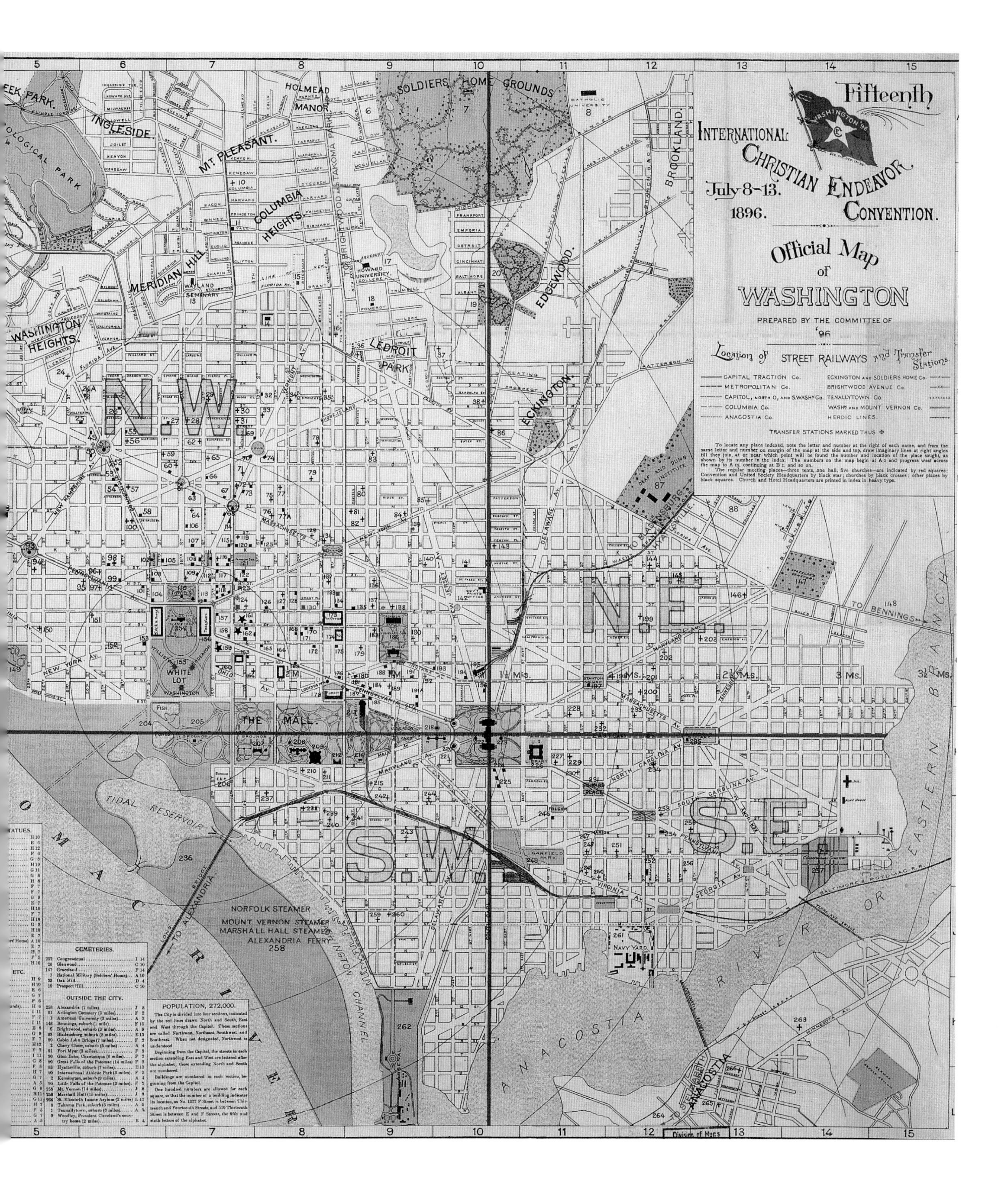
Fifteenth
International Christian Endeavor Convention.
July 8–13. 1896.
Official Map of WASHINGTON
PREPARED BY THE COMMITTEE OF '96
Location of STREET RAILWAYS and Transfer Stations
CAPITAL TRACTION Co.
METROPOLITAN Co.
CAPITOL, NORTH O, AND S.WASHN Co.
COLUMBIA Co.
ANACOSTIA Co.
ECKINGTON AND SOLDIERS HOME Co.
BRIGHTWOOD AVENUE Co.
TENALLYTOWN Co.
WASHN AND MOUNT VERNON Co.
HERDIC LINES.
TRANSFER STATIONS MARKED THUS
N.W.
N.E.
S.W.
S.E.
THE MALL.
WHITE LOT
TIDAL RESERVOIR
ANACOSTIA
EASTERN BRANCH OR ANACOSTIA RIVER
POTOMAC RIVER
SOLDIERS HOME GROUNDS
MT. PLEASANT.
COLUMBIA HEIGHTS.
MERIDIAN HILL
INGLESIDE.
HOLMEAD MANOR
LEDROIT PARK
ECKINGTON.
EDGEWOOD.
BROOKLAND.
WASHINGTON HEIGHTS.
NAVY YARD.
GARFIELD PARK
TO ALEXANDRIA.
TO BENNINGS
NORFOLK STEAMER
MOUNT VERNON STEAMER
MARSHALL HALL STEAMER
ALEXANDRIA FERRY
CEMETERIES.
OUTSIDE THE CITY.
POPULATION, 272,000.
Division of Mazes

SEGREGATION AS A FACTOR OF CITY SCHOOLS
The Rogers Statistical Map

TITLE: School Buildings
DATE DEPICTED: 1925 - 1926
DATE ISSUED: 1926
CARTOGRAPHER: William J. Rogers
PUBLISHER: Wm. J. Rogers, Washington, D.C.
Colored lithograph, 69 x 84 cm
Library of Congress, Geography and Map Division, G3851 .E68 1926 .S3

ESSAY BY CHARLENE DREW JARVIS AND IRIS MILLER

Discovering this 1926 statistical map, in the Library of Congress catalogue and then in person, was startling. In this world, there are some things that one may know exists which one has not experienced personally. Why then, should it seem so arrogant, indeed, that a city had the audacity to produce a map with information about school buildings, "White or Colored," by race?

Today, when we think about officially sanctioned segregated schools, we may wonder, how was it possible? For those growing up in the north, schools were segregated only to the extent that people lived in homogeneous, or segregated, neighborhoods. Many northerners, perhaps most, experienced education in integrated classrooms.

Washington, D.C. was a "Southern City." Yet, it was, and still is, the capital of the United States of America. In 1960, 100 years after the Civil War and even today, one can meet people of many races who have grown up here and have attended schools segregated by race without a choice. Factors such as skills, intelligence, location of residence were not a consideration. In truth, southern schools in America were segregated until the Supreme Court decision eliminated segregation in schools.

The segregation of the public school system in the District of Columbia had paradoxical consequences for "colored" students. Despite the inequity of resources between the Division II (colored) and the Division I (white) schools, "colored" schools had some of the most highly educated principals, teachers, and staff in the entire school system. Many had earned their degrees at the finest northeastern and mid-western colleges and universities. Unable to find positions in their fields in a job market that was unwelcoming, many African-American scholars taught in the Division II schools. Students, therefore, sat at the feet of, and were inspired by, teachers of their own race who expanded their horizons and stimulated their intellectual curiosity. Most importantly, those teachers had an expectation that their students would achieve. Those teachers also embodied the middle-class ethic, demanding courtesy, controlled behavior, and accountability.

Furthermore, because the neighborhoods in which schools were located were also segregated, and because suburban flight had not started, teachers and their students lived in the same neighborhoods. The teacher had moral authority in the community.

Despite having to ride past neighborhood schools to go to segregated schools, despite unequal resources, despite lowered expectations from the external community, many students in the District's Division II schools prospered intellectually and culturally.

Then, the very long struggle for integration in the public school system was won in 1954 in the Supreme Court in Brown vs. Board of Education. The administrative offices of Division I and Division II schools disappeared. "Negro" students were introduced into formerly white schools. The environment for students was uncomfortable at best and hostile or demeaning or violent at worst. The system of "tracking" students into college and non-college tracks was introduced and often reflected the lowered expectations that white principals, teachers and staff had for their new charges. For students who remained at "Negro" schools, many of their teachers and staff pursued job opportunities in their fields that had long been denied to them. And thus, the "brain drain of integration" occurred from the former Division II schools.

Evidence of the ability of students educated in Division II schools in the District of Columbia schools abounds: Hugh Price, Executive Director of the National Urban League; Colonel Frederick Drew Gregory, Commander of a shuttle mission to space; Wesley Williams, partner at the law firm of Covington & Burling; Carol Thompson Cole, head of the White House Task Force on the District of Columbia; Charlene Drew Jarvis, scientist, elected official, President of Southeastern University; and Paul Reason, Commander of the Naval Fleet for the entire mid-Atlantic region.

The 1926 map of "colored" and "white" schools in the District of Columbia reveals the geography of segregation by race. What it does not reveal is the paradoxical academic and cultural strengths of schools that were separate and ostensibly "unequal."

Given this history, in looking back to earlier maps perhaps it should not be surprising, therefore, to discover maps from the 1840s and 1850s with "Population Charts" noting statistics (p. 82) by race. We note that during this period, the country experienced a strong "movement for popular education" and free public schools. By mid-century the District of Columbia was a

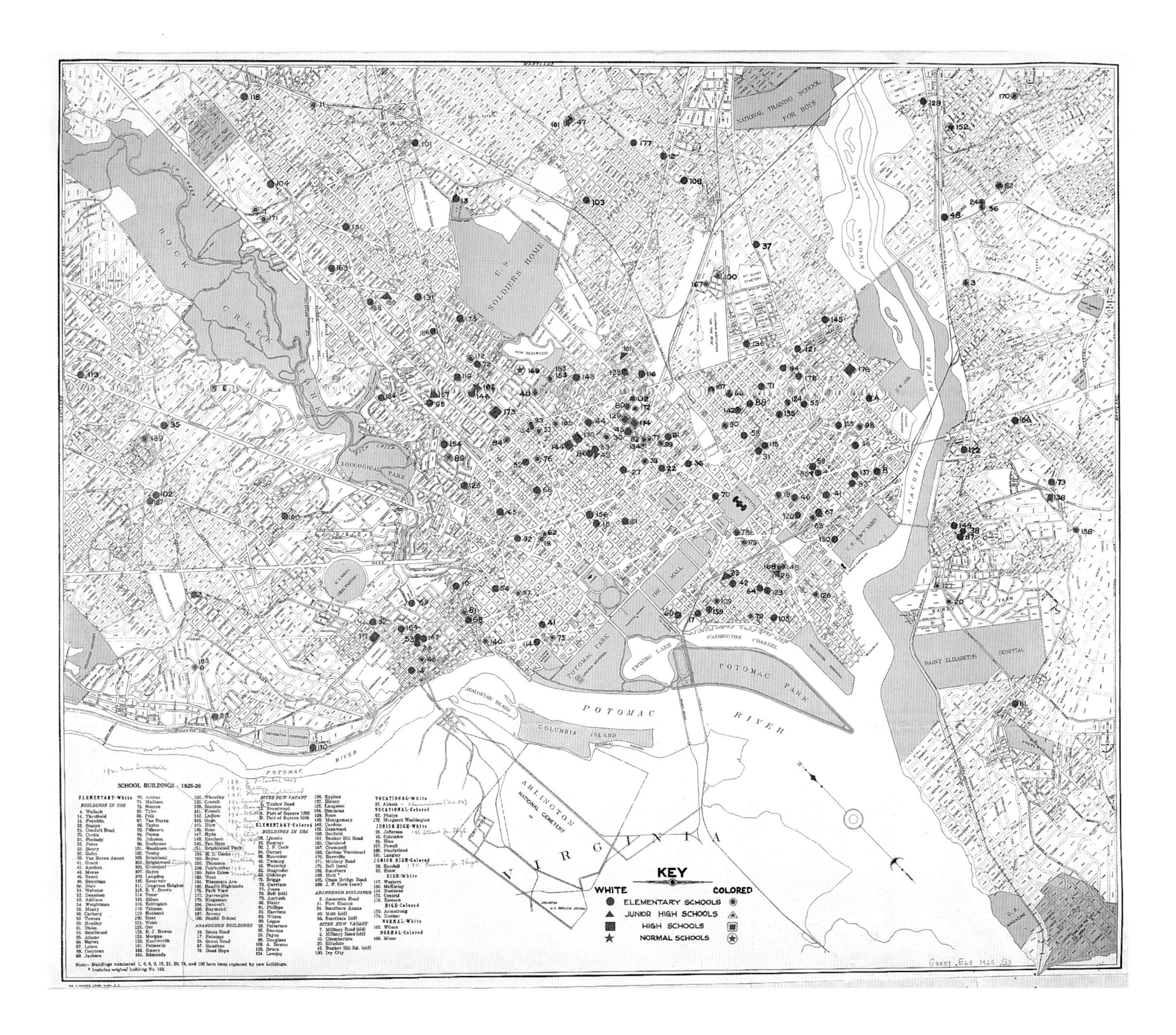

leader in the public education movement. Some sources suggested approximately 23 schools existed, while the 1876 Entwistle Map indicates 11 public schools.

Likewise, public libraries and museums were promoted. Increasing emphasis was placed upon building facilities for extending scientific knowledge. Society in general, and all politicians in particular, could agree upon these educational objectives. However, in the rhetoric, the use of the term "public" remained unclear.

Rodgers' Map is useful to the extent that it displayed the full breadth of the District to its boundaries, designating parklands, military facilities, and reservoirs—and numbering blocks (squares) according to the original system. Dotted lines showed projected streets not yet laid in neighborhoods of low population density. Schools were listed according to "Buildings in Use" and "Buildings Abandoned." Pencil notes have been added. A key, with symbols in red, separated elementary, junior high, high, and normal schools by "Whites" and "Colored." The map indicates that the greatest population density of young families was clustered in the original L'Enfant city, downtown and east and south of the Mall, and that the races were scattered rather evenly throughout except for Anacostia, which was primarily white and not yet settled. This map is a statement about the age of schools still in existence after 75 years. Static in time, it is a reflection of population density, but not a predictor. To uncover this astonishing map—of 1926—nevertheless makes one wonder about the implications and damage imposed upon the city's collective memory.

ESSENCE OF A LATE-NINETEENTH-CENTURY AMERICAN CITY

The A. Sachse View and Vignettes

TITLE: Bird's- Eye View
DATE DEPICTED: 1883-1884
DATE ISSUED: 1884
CARTOGRAPHER: A. Sachse
PUBLISHER: A. Sachse and Co., Baltimore, MD
Colored panorama, not to scale, 108 x 165 cm
Library of Congress, Geography and Map Division, G385 .A3 1884 .S3 Vault, Oversize

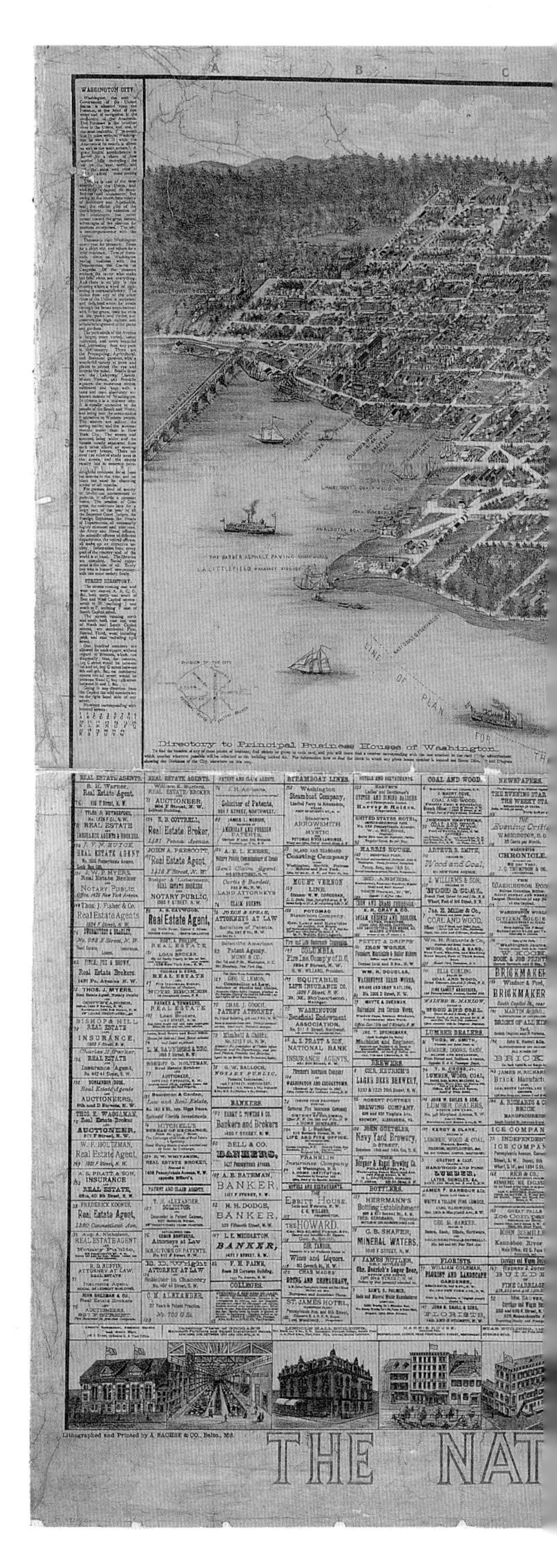

Washington grew substantially from 1880 to 1920, becoming a sophisticated city and tourist attraction. Yet the scale of the inner city remained low and sprawling, changing only incrementally.

With the increasing wealth of its citizens, many old squat houses were replaced by larger, attractive structures, even mansions, with remarkable architectural ornament. Northwest Washington offered fashionable neighborhoods for government officials and foreign embassies. Georgetown, at the western edge, had been incorporated into the city since the mid-1870s when Congress altered the city's governmental organization. Large vacant tracts on Capitol Hill, to the east, and precincts south of The Mall had largely been filled in.

In 1884, outward growth was occurring at the edges of the city, in the surrounding hills, as "in-town suburban" neighborhoods became sites for new development and land speculation. The major diagonal avenues of the L'Enfant Plan, named for the states, were extended to peripheral areas providing access to these new neighborhoods.

Drainage, grading and road paving was now improved thanks to the Board of Public Works and the brief, albeit forceful leadership of "Boss" Shepherd (1871-1873). Public reservations, squares and circles at intersections of radial avenues, were newly designed and landscaped by the Corps of Engineers, with equestrian statues as focal points to honor Civil War heroes.

This rare, three-plate view by Adolph Sachse, of the Baltimore firm which provided earlier Washington views by Edward Sachse (pp. 80 and 92), offered perhaps the best view of Washington at the end of the nineteenth century. Produced in a manner similar to an advertisement, vignettes across the bottom of the sheet show a curious mix of buildings—a restaurant, furniture and carpet house, coal and wood dealers, iron foundry, insurance company, hotels, two newspapers, real estate and law offices, and Heurich's Brewery.

Advertisements in both lower corners comprised a "Directory to Principal Business Houses of Washington" and "Public Buildings, Parks, Hotels." A newspaper style column at the upper left, "Washington City," included text and "Street Directory."

Ships and sailing craft were pictured actively operating on the Potomac River, but a dotted line foretold of the designated the proposed reclamation area of the Potomac Flats. Red brick was the predominant building material of choice, almost obscuring the white marble and limestone of the symbolic civic structures. Railroads occupied an imposing place on The Mall, while curving roadways had been installed according to the Downing Plan (p. 77).

The perspective, with its picture plane up-tilted and distorted by the artist, permits improved scrutiny of the city plan and its individual buildings and spaces. The city as garden is apparent. Diagonal avenues bisect the grid of streets intersecting to form small parks. The public reservations conform to a late-nineteenth-century landscape design tradition of Paris, by Adolphe Alphand, and Germany, by Peter J. Lenne. The parks project a classical-romantic composite of geometric symmetry, with curvilinear paths laid out adjacent to linear tree-lined allées and streets, completing the urban-garden image.

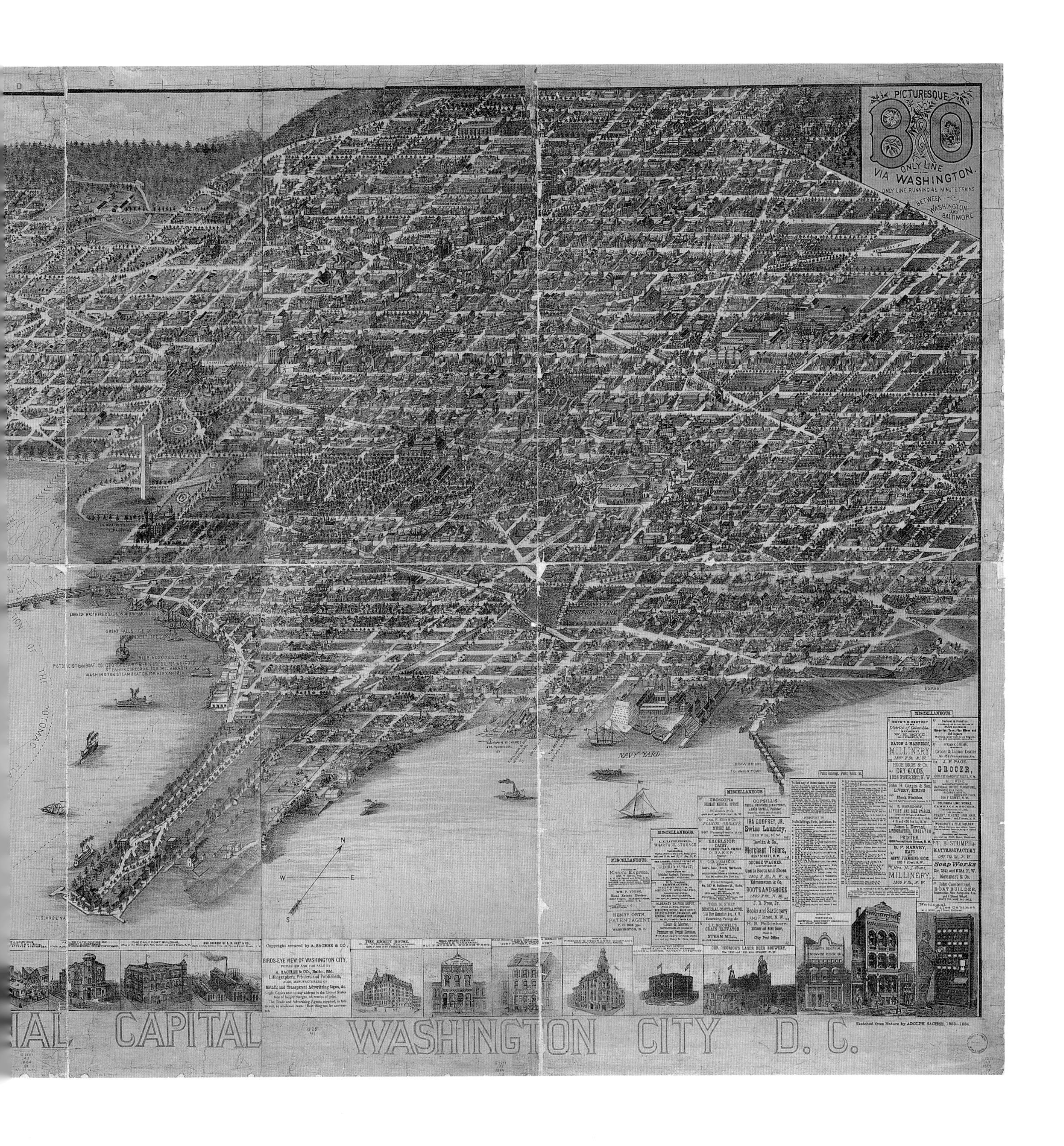
PICTURESQUE
B&O
ONLY LINE
VIA WASHINGTON.
ONLY LINE RUNNING 45 MINUTE TRAINS
BETWEEN WASHINGTON AND BALTIMORE
NAVY YARD
THE POTOMAC
MISCELLANEOUS.
MILLINERY
DRY GOODS
GROCER
BOOTS AND SHOES
Swiss Laundry
Merchant Tailors
Soap Works
BIRDS-EYE VIEW OF WASHINGTON CITY,
A. SACHSE & CO., Balto., Md.
IAL CAPITAL
WASHINGTON CITY D. C.
Sketched from Nature by ADOLPH SACHSE, 1883–1884.

MERGING URBAN FANTASY AND ASPIRATION

The Hoen View

TITLE: Bird's-Eye View of the National Capital Including the Site of the Proposed World's Exposition of 1892 and the Permanent Exposition of the Three Americas
DATE DEPICTED: 1892
DATE ISSUED: 1888
CARTOGRAPHER: A. Hoen, lithographer, A. Hoen & Co., Baltimore, MD
PUBLISHER: E. Kurtz Johnson
Colored lithograph, 26.75 x 36 in
Courtesy, The Albert H. Small Washington Collection, 286. PR.BV.E.F.
(Also in Library of Congress, Georgraphy and Map Division, G3851.A35 1888 .J6 Vault)

The World's Columbian Exposition of 1893, in Chicago, profoundly transformed the fashion and destiny of urban design in the United States for generations. The dignity and beauty of their transitory city by the water inspired Chicago citizens and in turn the designers who would create and re-create cities worldwide in the spirit of the City Beautiful Movement.

The 400-year anniversary of Christopher Columbus's discovery of America inspired several cities to bid for the distinction to host a grand celebration. Each hoped to receive the guaranteed federal funding for the exposition, which was ultimately won by Chicago.

With the support of business leaders, Daniel Burnham masterminded the Chicago Exposition. He displayed a keen understanding of the ordering power of Neoclassicism, bringing considerable organizational savvy. He mobilized experts to collaborate on specific tasks: Charles McKim, Frederick Law Olmsted, Jr., Augustus St. Gaudens, Richard Morris Hunt, and Charles B. Atwood—the first three men later became part of Burnham's team in the 1901-1902 redesign of Washington, D.C. (p. 116). The experience became precedent for his and Edward H. Bennett's 1909 Plan of Chicago, and their prolific urban designs for many cities.

In Washington, community leaders had devised strategies to win the World's Fair for the federal city. To enhance their position, they proposed to combine the event with a continuing exposition—Permanent Exposition of the Three Americas. The reclaimed tidal Potomac Flats near four lakes and the Washington Monument were envisioned to house three permanent buildings, one temporary exposition building, and the National Zoological Garden. Notwithstanding advantages offered by the federal site, Congress ultimately named Chicago as its choice.

This misleading birds-eye view exaggerates the Capitol, Washington Monument, and other buildings to tout the city's advantages and the proposed exposition site's preeminent location within the national capital. Sprawling red brick background buildings present a contrasting image of a cohesive community and a seductive quality of life.

LAKE
LAKE
LAKE
LAKE
SITE FOR STATUE OF COLUMBUS
SITE FOR NATIONAL ZOOLOGICAL GARDEN
ILDING FOR
SITION OF 1892
A C
R I V E R
INTERIOR DEPT.
WASHINGTON MONUMENT
POST OFFICE DEPT
PENSION BUILDING
DEPT OF JUSTICE
AGRICULTURAL DEPT.
BUREAU OF ENGRAVING
SMITHSONIAN INSTITUTION
NATIONAL MUSEUM
BOTANICAL GARDEN
CAPITOL
LITH. BY A. HOEN & CO. BALTIMORE
EYE VIEW OF THE NATIONAL CAPITAL
INCLUDING THE SITE OF THE PROPOSED
TION OF 1892 AND PERMANENT EXPOSITION OF THE THREE AMERICAS.

GARDEN CITY ARTICULATED

The Du Bois Tourist Guide Map

TITLE: The Altograph of Washington City or Stranger's Guide
DATE DEPICTED: 1892
DATE ISSUED: 1892
CARTOGRAPHER: James T. Du Bois
PUBLISHER: Norris Peters Co., Photo - Litho., Washington, D.C.
Colored lithograph map, 27.75 x 37.75 in, 63,5 x 88.8 cm
Courtesy, The Albert H. Small Washington Collection, 299. M.P.ND.E.F.
(Also in Library of Congress, Georgraphy and Map Division, G3850 1892 .D8)

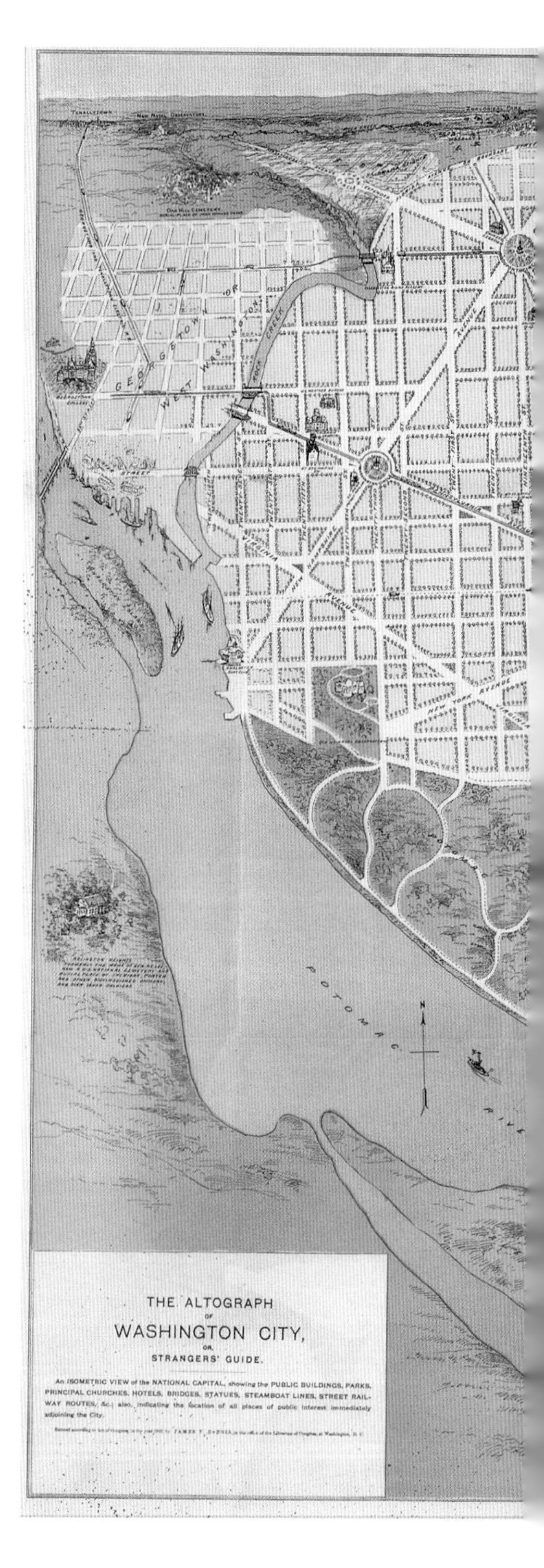

A unique American map for its time, this "altograph" superimposes three-dimensional isometric views of structures, parks, and riverboats on a plan view. Its clarity is disarming. Contrasting the soft pastel colors of pink, green and blue as ground, the figural objects and landscape are projected in black and white. Public buildings, hotels, and churches are clustered in the central core. Bridges, statues, steamboat lines, street railway routes, parks and city blocks edged by continuous rows of trees extend toward the city edge. In the upper right corner is a list for tourists, or "Strangers," of almost two dozen places "To Go."

Compared with three-dimensional views of the 1890s, the quintessential difference in this arresting drawing is the focus upon the abundance of parks spreading across the city, signifying city as garden. The repetitive series of small parks—nineteenth-century romantic, geometric configurations with statues of heroes occupying center—confers L'Enfant's vision of an ordered sequence of neighborhoods within a unified entity. The larger parks and public spaces address the national civic realm, a place for all, residents and visitors alike.

The Potomac Flats invoke the legacy of landscape woven through the city inherent in L'Enfant's plan. This reclaimed land along the Potomac River in the 1880s was reworked for more than two decades. The goals were to improve navigation while simultaneously alleviating the flooding of marshlands near the President's House, The Mall and along the river.

This imaginative altograph portrays an inclusive urban community—a flourishing downtown, honorific civic development, and parks distributed among the city's neighborhoods. Attention is drawn to selected principal buildings. The Library of Congress next to the Capitol is here depicted as completed. Other government buildings included State, War and Navy (now the Old Executive Office Building) west of the White House, and the Pension Building (now the National Building Museum) north of Judiciary Square. The "new" Post Office on Pennsylvania Avenue was depicted although it was not yet built.

The map confirms that Washington had become a major city with a burgeoning population. Street railways and rail lines crisscross the city. Universities and cemeteries are shown on the outskirts, along with a New Naval Observatory in upper northwest. The onset of the "in-town suburbs," or "Edge City," is just birthing on the fringe. The implication here is that a cohesive downtown will coexist with expansion beyond the present boundaries, heralding the promise of the future.

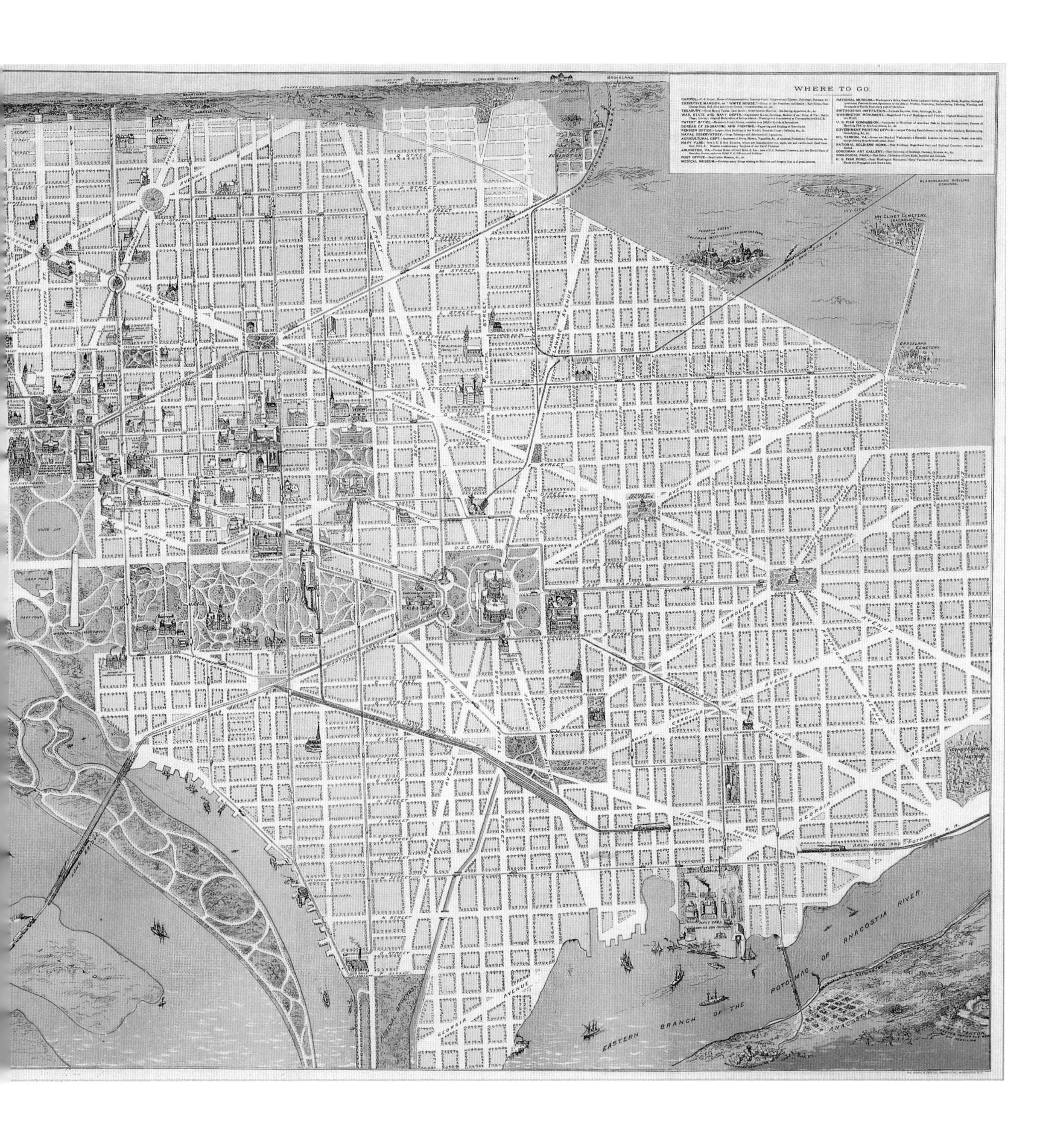
WHERE TO GO.
CAPITOL.
EXECUTIVE MANSION, or "WHITE HOUSE."
TREASURY.
WAR, STATE AND NAVY DEPTS.
PATENT OFFICE.
BUREAU OF ENGRAVING AND PRINTING.
PENSION OFFICE.
NAVAL OBSERVATORY.
AGRICULTURAL DEPT.
NAVY YARD.
ARLINGTON, VA.
POST OFFICE.
MEDICAL MUSEUM.
NATIONAL MUSEUM.
SMITHSONIAN INSTITUTION.
WASHINGTON MONUMENT.
U. S. FISH COMMISSION.
GOVERNMENT PRINTING OFFICE.
MT. VERNON, VA.
NATIONAL SOLDIERS' HOME.
CORCORAN ART GALLERY.
ZOOLOGICAL PARK.
U. S. FISH POND.
GLENWOOD CEMETERY
BROOKLAND
CATHOLIC UNIVERSITY
BLADENSBURG DUELLING GROUNDS.
IVY CITY
MT OLIVET CEMETERY (CATHOLIC)
"KENDALL GREEN" OR COLUMBIA INSTITUTION FOR DEAF AND DUMB
U.S. CAPITOL
EASTERN BRANCH OF THE POTOMAC OR ANACOSTIA RIVER
BALTIMORE AND POTOMAC R.R.
ANACOSTIA

NOSTALGIC SOCIAL VISION

The Currier and Ives View

TITLE: The City of Washington, Birds-Eye View from the Potomac, Looking North
DATE DEPICTED: 1892, after 1880 view
DATE ISSUED: 1892
CARTOGRAPHER: Currier & Ives (after earlier 1880 drawing by C.R. Parsons)
PUBLISHER: Currier & Ives, New York
Colored lithograph, 20.5 x 33 in
Courtesy The Albert H. Small Washington Collection, 286.PR.B.,E.F.
(Also in Library of Congress, Prints and Photograph Division,
Neg.LC.USZC4-843, #PGA-D-Currier & Ives, City of Washington)

Comparison of a selection of late-nineteenth-century views suggests innocence inevitably linked to optimism—essentially, an embellished graphic style exemplifying nostalgia, or a sense of security. Nearly thirty years after the Civil War, Washington had become a respectable city. It had been transformed from the squalid town of filth and unpaved roads to one of great beauty. Dominated by spacious tree-lined streets, parks and fine architecture, visitors would write enthusiastically of its merits. Implicit in its heroic plan was an underpinning for dignified public buildings that were emerging in the central area.

This lithograph by Currier and Ives personifies an aura of sentimentality. Its rendering in soft earth tones is punctuated by the symbolic presence of numerous white civic buildings. Densely packed dwellings are depicted in a virtual cartoon-like fashion, masking the poor condition of many buildings in the foreground. Although the publication date was 1892, the image is essentially the same as that drawn by C. R. Parsons in 1880. Selected revisions include the completed Washington Monument of 1884, but without the corresponding lengthening of its reflection in the river. New buildings of the time are missing and the fish ponds are here still shown as they had previously existed.

The numerous boats of many types suggest a nineteenth-century aspect of mystery and danger—a connection with the distant world. Along Massachusetts Avenue streets are defined by a continuous wall of dwellings opening upon public reservations with commemorative statues. The scale and density metaphorically implies a sophisticated European city.

Public transportation enabled growth beyond the original boundary of the City of Washington. Horse-drawn car lines followed by electric car lines and a commuter railroad fostered real estate speculation and development of new neighborhoods. The 1890s saw a corresponding proliferation of promotional maps of subdivisions and suburban communities such as Chevy Chase, Oak View, Brookland, Eckington, Kalorama, 16th Street, and Wesley Heights.

As the twentieth century drew near, optimism was tinged with an undercurrent of dissatisfaction regarding urban conditions. The centennial of the transfer of the "Seat of Government" to Washington would be celebrated in 1900. Pressure was mounting to improve the public realm and to return to the classical language of urbanism and architecture as embodied in the L'Enfant Plan. Government engineers and influential architects prepared designs for the enhancement of public space, particularly for The Mall, the Potomac Flats and Federal Triangle. Among those (not shown) were concepts by Theodore Bingham, Samuel Parsons, Jr., Glenn Brown, Cass Gilbert, and Paul J. Pelz. These proposals inspired extensive dialogue regarding appropriate treatment of The Mall area. Early in 1901, the Senate Park (McMillan) Commission would be formed to masterfully resolve the challenge.

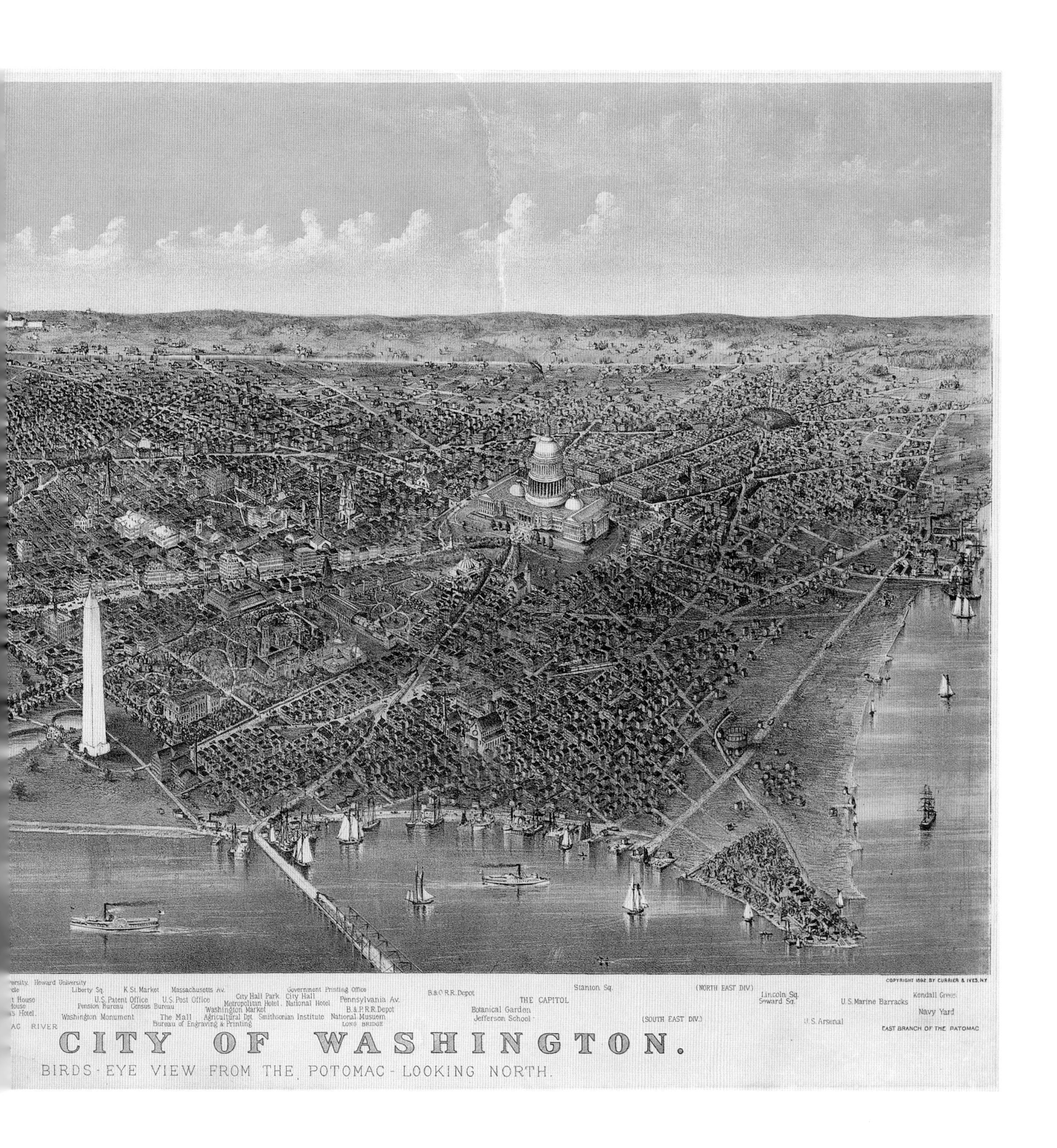
Howard University
Liberty Sq.
K St. Market
Massachusetts Av.
Government Printing Office
B.&O.R.R. Depot
Stanton Sq.
(NORTH EAST DIV.)
COPYRIGHT 1892 BY CURRIER & IVES, N.Y.
U.S. Patent Office
U.S. Post Office
City Hall Park
City Hall
Metropolitan Hotel
National Hotel
Pennsylvania Av.
THE CAPITOL
Lincoln Sq.
Seward Sq.
U.S. Marine Barracks
Kendall Green
Pension Bureau
Census Bureau
Washington Market
B.&P.R.R. Depot
Botanical Garden
Navy Yard
Washington Monument
The Mall
Agricultural Dpt.
Smithsonian Institute
National Musuem
Jefferson School
(SOUTH EAST DIV.)
U.S. Arsenal
Bureau of Engraving & Printing
LONG BRIDGE
EAST BRANCH OF THE PATOMAC
RIVER
CITY OF WASHINGTON.
BIRDS-EYE VIEW FROM THE POTOMAC-LOOKING NORTH.

LOST PERIOD IN TIME: URBAN SCALE BEFORE MODERNISM

The Olsen Axonometric Perspective

TITLE: Washington, the Beautiful Capital of the Nation
DATE DEPICTED: ca. 1921
DATE ISSUED: ca. 1923
CARTOGRAPHER: William Olsen
PUBLISHER: A. B. Graham Co., Washington, D.C.
Lithograph, 49 x 75 cm
Library of Congress, Geography and Map Division, G3851.A3 1923 .O4

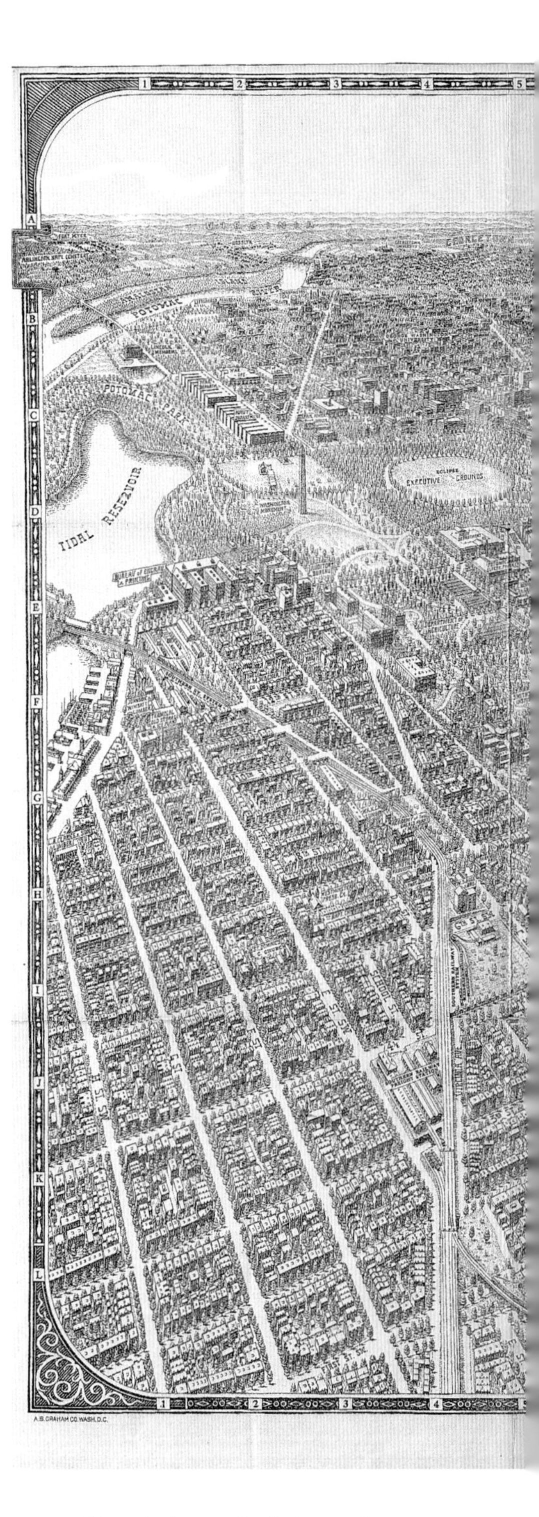

Town planning will vary in any epoch according to current fashion and local circumstances. Many towns and neighborhoods, initially admirably planned, have been allowed to develop without guidance, often with negative results. External demands for change in Washington have met resistance or approval according to diverse imperatives. Fortunately, substantial preservation of L'Enfant's plan has conserved both its aesthetic ideology and functional reasoning.

In order to protect the federal city's dignified urban fabric, two commissions were formed: the U. S. Commission of Fine Arts, in 1910; and the National Capital Park and Planning Commission (later known as the National Capital Planning Commission), in 1924. However, these review commissions did not always shield the community from forces of the zeitgeist, that is, the prevailing attitudes favoring modern monumentality in structures and public space.

Plausibly, this is a decisive pictorial view of Washington before its seminal metamorphosis in scale and building type. At the lower left, the southwest section is seen before the onset of urban renewal's scandalous modern panacea that was to destroy the character and residential quality of this waterfront neighborhood. (See text p. 144 and visual results post-1970, p. 156 Passonneau Map.) By chance, because economic forces have claimed hold upon the city's northwest and suburbs, fewer changes have occurred on the eastern side, except near the Capitol.

This post-World War I view depicted a number of recent public buildings, including the huge red brick Pension Building with its gigantic glass-roofed atrium, the Neoclassical Library of Congress, House of Representatives and Senate Office Buildings, and Union Station. Although the railroad tracks had been removed from The Mall, "temporary" wartime office buildings beyond the Washington Monument yet remain.

At upper right, the "New [McMillan] Reservoir," Pumping Station, and Filtration Plant, designated an historic landmark in 1991, were drawn as completed in 1905. For years, the sand filtration plant's towers and underground vaults stood unused behind a chain-link fence, as newer methods had replaced the old system. In 2000 a city-community process evolved in anticipation of reaching consensus between preservation and adaptive uses to bring new life to the 25-acre abandoned site.

Ushering in a new era for control of the Potomac River and Potomac Flats, in-filled land created East and West Potomac Parks, Washington Channel and Tidal Basin (Reservoir). More than 600 acres were reclaimed by river dredging and filling marshlands above flood and high tide levels. The result was improved navigation and recreation facilities. The Tidal Basin was devised with automatic gates and openings through which river water could enter at high tide. At low tide the gates allowed water to flow outward, thus inhibiting waste from entering.

Pictorially, this image describes a rational pattern for growth and development in Washington's remarkable evolution. For the most part, the viewer accurately sees the buildings and spaces, although the up-tilted picture plane concealed hilly topography and bowl-like features at the old city edge. As early twentieth-century designers were shaping plans to implement the Senate Park Commission recommendations for The Mall and adjacent areas (p. 116), Washington would soon be forever transformed. Views such as this expose the consequences of our past indifference and what has been lost.

WASHINGTON
MARYLAND AVE
EAST CAPITOL

NOBLE PARADIGM FOR THE TWENTIETH CENTURY

The Senate Park Commission, Hoppin Aerial Perspective

TITLE: View of General Plan
DATE DEPICTED: Future projection
DATE ISSUED: 1902
CARTOGRAPHER: Francis L.V. Hoppin
PUBLISHER: Senate Park Commission, 1902
Watercolor on paper, 33.75 x 71.5 in, 86.3 x 181.6 cm
U. S. Commission of Fine Arts

REAFFIRMING CLASSICAL URBANISM

The Senate Park Commission, The Mall Plan

TITLE: The Mall Plan
DATE DEPICTED: Future projection
DATE ISSUED: 1902
CARTOGRAPHER: Unknown
PUBLISHER: Senate Park Commission, 1902
Watercolor on paper, 45.5 x110 in, 115.7 x 279.7 cm
U. S. Commission on Fine Arts

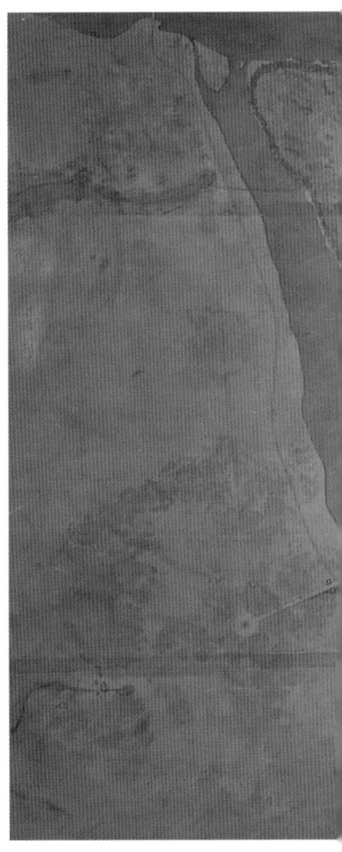

One world's fair profoundly changed the course of American urban history. The ideological underpinnings of the Chicago World's Columbian Exposition of 1893 sparked the City Beautiful Movement that dominated American urban design during the next half century. The men and women involved in the 1890s reform movement were often the same leaders in city beautification. The result was a reassessment of the country's cityscapes and urban parks, as radical Neoclassical landscape interventions penetrated the American polis.

Influenced by recent European models, particularly works by Adolphe Alphand and Baron Georges-Eugene Haussmann in Paris, proponents of the turn-of-the-century new urbanism vocabulary embarked on a fresh course. Communities would soon be enriched by massive redesign of urban environments—buildings and landscape. In the United States, Washington, D.C. became the first and most impressive paradigm, the quintessential city as a work of civic art.

Philosophically, aesthetic improvement of entire physical environments is the harbinger of increased urban identity and civic consciousness for large segments of the population. As cities were becoming heavily industrialized, urban reform offered psychological benefits and improved quality of life.

Despite obvious European connections—Americans studying at the Beaux Arts in France and the Garden City Movement in England—the City Beautiful Movement was a distinctive American urban phenomenon. It straddled ties to European roots and precedents, concurrent with a need to seek an American cultural ideal. Affirming the spirit of the L'Enfant Plan, the

Senate Park Commission Plan, 1901-1902, (also known as McMillan Commission) struck a compromise between prevailing American attitudes toward glorification of nature, inherent in Downing's plan (p. 79), and the Baroque formality of L'Enfant's populist celebratory design.

The Improvement of the Park System of the District of Columbia, Senate Report No. 166, 1902, was proposed by Daniel H. Burnham, Charles F. McKim, Frederick Law Olmsted, Jr., Augustus Saint-Gaudens, and Charles Moore (the secretary and administrative assistant, and previously secretary and architecture advisor, to the proposal's sponsor, Senator McMillan). Fortuitously, this dramatic far-reaching civic proposal was followed by work of these and other designers, whose prolific projects across the nation during the twentieth century's first three decades transformed urban composition. It is no wonder that this beautification movement, with its sense of ritual and logic, achieved global influence, reaching as far as New Delhi, India; Canberra, Australia; and Baguio, Philippines.

During the summer of 1901, the Senate Park Commission team traveled through European cities and gardens, absorbing their lessons in spatial design and details of civic art. They concentrated on France and England, touring Paris, Versailles, Vaux-le-Vicomte, Fontainbleau, Compiegne, London, Hampton Court, Oxford and Hatfield House. They were also inspired by Rome, Venice, Vienna, Schonbrunn, Berlin, and Budapest. Noting the differences between designed landscapes and piecemeal urban development collaged over time, they integrated ideas with reality. What emerged was a noble assemblage—a symbolic vision of what a city might become.

Implementation of a modified design for civic and park amenities generated by the Senate Park Commission took almost 35 years. The report consisted of a printed document and an exhibition of models, drawings and paintings. After a number of iterations and compromises, the document pragmatically integrated democratic and romantic values with elitist views. A synthesis of original imperatives anticipated visual drama and clarity in a preserved ceremonial core, and a city rich in public parkland.

Senate Report No. 166, included two sections: I. Report of the Senate Committee on the District of Columbia to the full Senate (14 pages); and II., Report of the Park Commission (100 pages plus appendix). Photographic images and drawings were distributed throughout the text. Section I. set out the problems and process followed in establishing the Commission. It highlighted the need for a comprehensive plan, referencing the original city plan, with assessment of the Commission plans.

Section II. reviewed Washington's entire system of parks, open space, and opportunities for preserving existing and potential wilderness and park areas, and for providing additional recreational areas. A critical consideration was the central core—the public buildings, monuments, memorials, reservations and landscape settings worthy of a federal capital. The text consisted of analytical discussions of each issue, complimented by examples, references, photographs, and sketches.

In essence, the report championed the following themes: provide enclosure for The Mall and U.S. Capitol; develop a unified government and museum precinct validating civic Neoclassical iconography; create a "Federal Triangle" north of The Mall between Constitution Avenue, 15th to 6th Streets, and a widened Pennsylvania Avenue; eliminate residential and commercial uses from this precinct; remove all railroad tracks and Pennsylvania Railroad from The Mall to Capitol Hill; situate a new Union Station; reject any future proposal to build a roadway underpass below The Mall (as in New York's Central Park). Ambitious plans were devised for The Mall system, Washington Monument Grounds, Lincoln Memorial, Memorial Bridge, Analostan (Teddy Roosevelt) Island, and Arlington National Cemetery.

Admittedly, the report contained several radical departures from the seminal plan of L'Enfant. Reclamation of the Potomac Flats, begun in the 1880s, extended the Monument Grounds, one mile east to west, for a total of 2 1/2-miles. The Mall's central "Grand Avenue" was deleted, as were dramatic vistas along the greenswards perpendicular to the White House and Capitol at the Potomac River. Closing these axes with commemorative memorials (Lincoln and later Jefferson) provided an animated paradigm of classical landscape symmetry, but shattered the evocation of timelessness and tranquility inherent in an infinite vista.

Resolution of a unified composition of open space and public buildings with the Washington Monument as nexus of The Mall, Capitol, and White House axes was vital. To reconcile the Monument's adjusted location from its intended cross-axial site on L'Enfant's plan, The Mall's path was redrawn at 3-degrees southward, creating a serendipitous arrangement of reciprocal views among civic structures. The resulting irregularity in point of convergence was an ingenious tactic, while The Mall's novel course is hardly discernable except with a tape measure.

Essential to the bold 1902 plan was synthesizing "simplicity with dignity" and the evolving development in urban life of a civic-spirited image a national capital should project. Countless dynamic ideas included the need for beautiful fountains, multi-functional recreational facilities, and public baths. Consideration of climate was critical. Formal public reservations serving neighborhoods and conservation of natural landscapes were advocated. Monumental gateways were suggested at ridges on North Capitol and 16th Streets, beyond which a formal boulevard, Savannah Parkway, was proposed at the Soldiers Home near the McMillan Reservoir. The Senate Park Commission Report set a coherent agenda for comprehensive urban planning and design—landscape art to rival all European capitals.

The impact of the proposal was strengthened by the remarkable presentation of drawings and models. The watercolor images remain among the most exquisite pieces of art related to urbanism, communicating a visible manifestation of the grandeur of the new plan for the federal city.

INVESTIGATING ALTERNATES: THE MALL AND FEDERAL TRIANGLE
THE NATIONAL CAPITAL PARK AND PLANNING COMMISSION SCHEME

TITLE: The Mall
DATE DEPICTED: Prior to 1901 & Built or Authorized after 1901
DATE ISSUED: 1928-1932
CARTOGRAPHER: Unknown
PUBLISHER: National Capital Park & Planning Commission
Photo process base map, applied elements, ink and watercolor added, 26 x 43 in
National Archives, Cartographic Unit, RG 328 New File 1.00 (05.20) 25909

EXTENDING THE MALL AXIS EASTWARD
THE NATIONAL CAPITAL PARK AND PLANNING COMMISSION SCHEME

TITLE: The Mall
DATE DEPICTED: Future projection
DATE ISSUED: 1929
CARTOGRAPHER: Unknown
PUBLISHER: National Capital Park & Planning Commission
Photo process base map assembled from pieces, 14 x 44 in
National Archives, Cartographic Unit, RG 328 New File 1.00 (05.20) 25910

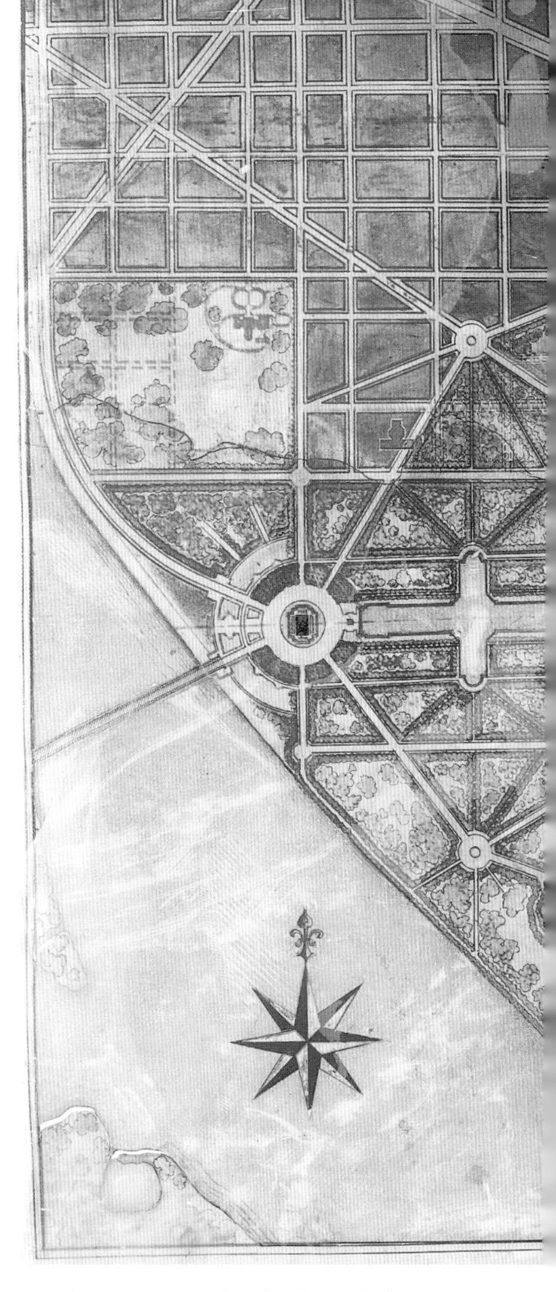

Reading newspaper and Congressional accounts of animosity and wrangling in the years following the Senate Park Commission report seems pointedly contemporary. From the beginning, the 1902 report engendered controversy about removing trees from The Mall, a 1908 proposal for placement of Grant's Statue at the Capitol's base and moving the Botanic Garden and greenhouses to 1st and 2nd Streets, S.W., the latter not accomplished until 1935.

Controversy surrounded the May 17, 1910 authorization of the Commission of Fine Arts to grant review of the city's federal buildings and improvements of public grounds. Three of its first seven appointed members had been with the Senate Park Commission: Chairman Daniel Burnham, Frederick Law Olmsted, Jr., and Charles Moore. Undoubtedly, this board was an instrumental arm supporting the legacy of the 1902 proposal. The 1926 establishment of what became the National Capital Planning Commission was another principal oversight board that consistently affirmed the City Beautiful ideology.

This map represents the struggle to carry forward the Commission's legacy. Albeit frequently misrepresented as the Senate Park Commission Plan, it is instructive to study its total composition and individual sites—The Mall, Federal Triangle, the Capitol and selected nearby sites—and to scrutinize its title. First, the title: the map is not a plan of 1901. It is a study of "Buildings Erected Prior to" and "Projects Built or Authorized After 1901."

The map is one of a series of studies made by the National Capital Park and Planning Commission (now the National Capital Planning Commission) between 1928 and into the early 1940s. Using overlay techniques,

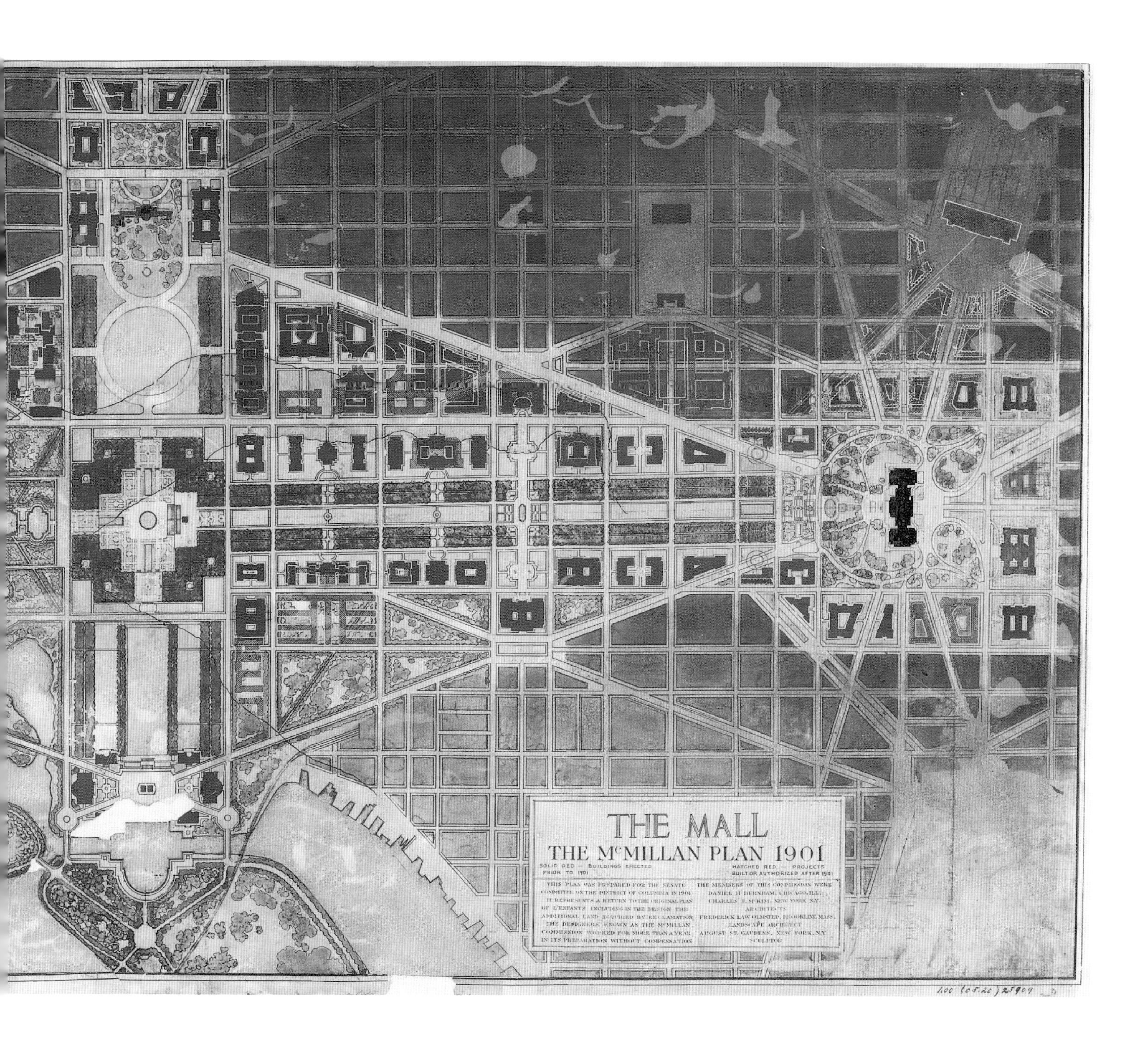
THE MALL
THE McMILLAN PLAN 1901
SOLID RED — BUILDINGS ERECTED PRIOR TO 1901
HATCHED RED — PROJECTS BUILT OR AUTHORIZED AFTER 1901
THIS PLAN WAS PREPARED FOR THE SENATE COMMITTEE ON THE DISTRICT OF COLUMBIA IN 1901
IT REPRESENTS A RETURN TO THE ORIGINAL PLAN OF L'ENFANT'S INCLUDING IN THE DESIGN THE ADDITIONAL LAND ACQUIRED BY RECLAMATION
THE DESIGNERS KNOWN AS THE McMILLAN COMMISSION WORKED FOR MORE THAN A YEAR IN ITS PREPARATION WITHOUT COMPENSATION
THE MEMBERS OF THIS COMMISSION WERE
DANIEL H BURNHAM, CHICAGO, ILL.
CHARLES F. McKIM, NEW YORK N.Y.
ARCHITECTS
FREDERICK LAW OLMSTED, BROOKLINE MASS.
LANDSCAPE ARCHITECT
AUGUST ST. GAUDENS, NEW YORK, N.Y
SCULPTOR

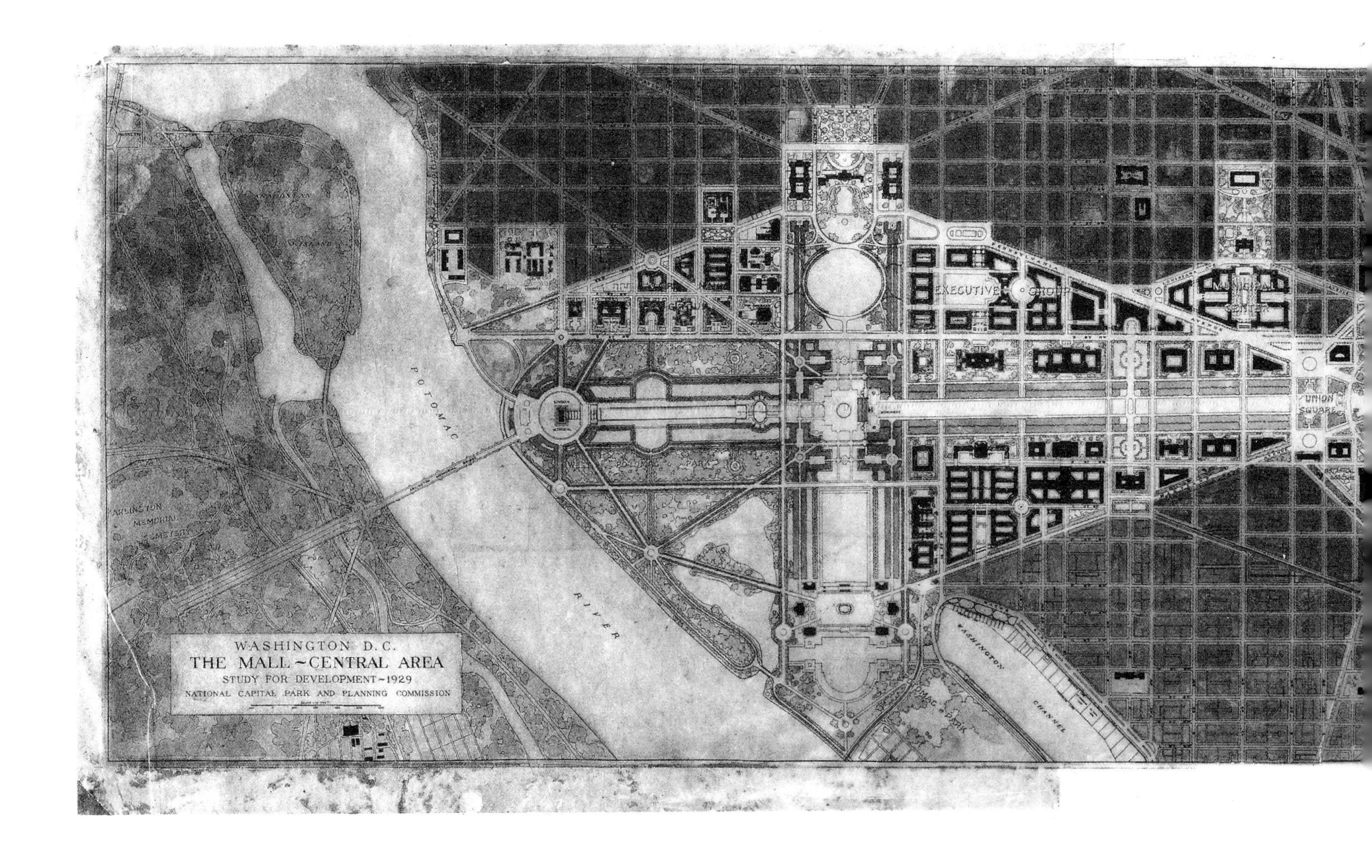

design and mapping exercises explored ideas for new buildings, street closings, and changes in open space and urban connections. Many drawings from this period can be found in the U.S. National Archives, with reproductions in numerous books and journals.

When the Federal Triangle was completed in the 1940s, except for the federal Post Office on Pennsylvania Avenue, its buildings followed mandated guidelines for consistency in architectural style, massing, cornice lines, and use of materials on facades and roofs. Perceived as a pedestrian-unfriendly precinct, the complex was a barrier between The Mall and the commercial city north of Pennsylvania Avenue, a zone lacking in shops and restaurants. Moreover, a questionable precedent of closing streets was set.

As built, The Mall and the abutting Washington Monument Reservation reveal major variations from the 1902 design. The Monument Grounds, proposed as a formal symmetrical rectangle, remain an open green to this day. Curiously, halfway between the Capitol and White House at 8th Street along The Mall, an elaborate cross-axial formal landscape motif was also never realized. Further, the proposed idiosyncratic long basin between the Monument Grounds and Lincoln Memorial, designed much like Versailles (1807 Plan, p. 18), was simplified.

Among the first of a series of memorials and buildings to test a commitment to the Neoclassical style was Union Station, built in 1907, and the adjacent City Post Office erected soon thereafter, both buildings designed by Daniel Burnham. Determining a suitable site and design for the Lincoln Memorial, authorized by legislation in 1908, became one of many hotly debated issues. Between July 1911 and February 1912 "swampy" Potomac Park was selected over several other sites, and Henry Bacon was named architect. His design resembled the proposal set forth by the Senate Park Commission. After much Congressional maneuvering, the project was approved in 1913, halted during World War I, and dedicated on May 30, 1922. With the construction of the majestic Lincoln Memorial, The Mall axis and preservation of the formality of the L'Enfant Plan was assured.

In 1938, after bitter arguments and redesign, the Senate approved construction and funding for the Jefferson Memorial on reclaimed

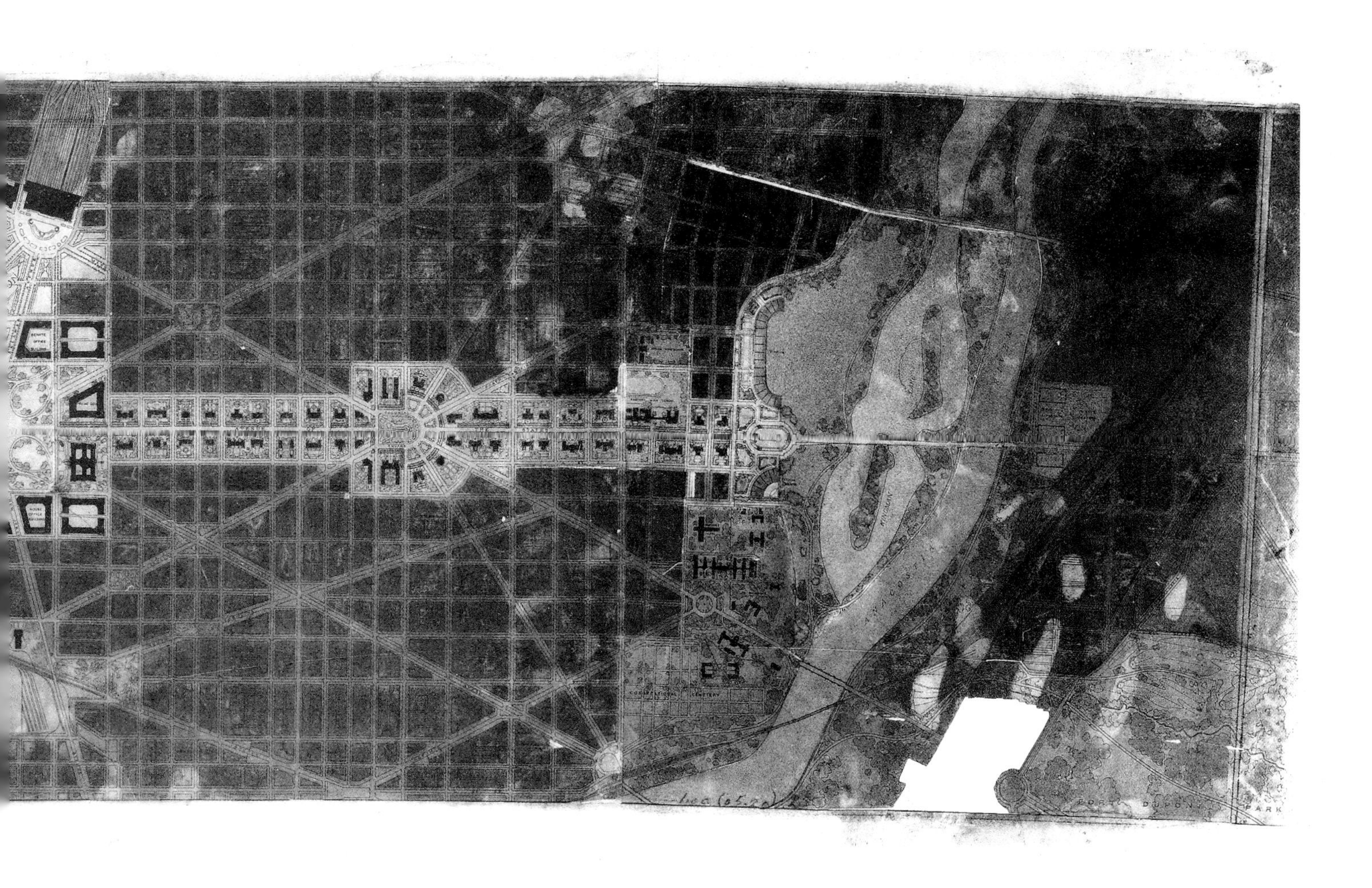

land on axis with the White House. The powerful design of architect, John Russell Pope, also emanated from a 1902 scheme.

With these memorials in place, an extraordinary Neo-baroque landscape unfolded in the federal ceremonial core. A continuum in a time-motion sequence of harmonious spaces and structures evoked a sense of parallax, whereby the viewer's perception was determined by a processional movement through space toward and around objects, foreshortening and elongating views. Symmetry was offset by hints of disorder, which heightened the dramatic visual effect. The influence of Andre LeNotre's landscapes and the French hunting forest is much in evidence in the drawings, signifying a place of delight and refuge.

During the first half of the twentieth century, the city was rife with new public and semi-public buildings forming an enclosure around The Mall and President's Park, conforming with the Senate Park Commission's fundamental public space concept. Along 17th Street and Constitution Avenues, commencing with Paul Cret and Albert Kelsey's Pan American Union (now the Organization of American States), buildings with a penchant for classical iconography were erected; similarly, at Lafayette Square and the Capitol. Unifying additions to The Mall included the S.I. Freer Gallery of Art by Charles Platt, the National Gallery of Art by John Russell Pope, the S.I. National Museum of Natural History by J.D. Hornblower (Hornblower and Marshall), and later the S.I. National Museum of American History, by Walker O. Cain (McKim, Mead and White). McKim, Mead and White's 1931 Watergate and 1932 Arlington Memorial Bridge, which brought the Lincoln Memorial and Custis-Lee Mansion into alignment, integrated the ensemble of public and symbolic elements.

A succession of unrealized studies (also by the National Capital Park and Planning Commission) redirected focus toward an "East Mall" extension from the Capitol to the Anacostia River. Paradoxically, this less affluent section of town might have enjoyed an aesthetic amenity of which future generations can only dream. Devising a traditional French boulevard enclosed by Neoclassical buildings and embellished by a park at Lincoln Square, this was to have been a realization of L'Enfant's vision.

ADVANCING PARKS IMPROVEMENTS AND ROADWAYS SYSTEM

The Senate Park Commission, Park Plan #287

TITLE: Map of the District of Columbia…Commission on the Improvement of the Park System
DATE DEPICTED: Future projection
DATE ISSUED: 1901-1902
CARTOGRAPHER: R. A. Outhet, for Senate Park Commission:
D.M Burnham, C.F. McKim, A. St. Gaudens, F.L. Olmsted, Jr.
PUBLISHER: A. Hoen and Co., Baltimore, MD
Colored lithograph, scale [1:24,000], 68 x 68 cm, 27.5 x 27.25 in
Library of Congress, Geography and Map Division, G3851.G5 1901 .09

PARKS NETWORK FOR COLLECTIVE SOCIAL GOOD

The Senate Park Commission, Park Plan #289

TITLE: Map of the District of Columbia…Commission on the Improvement of the Park System
DATE DEPICTED: Future projection
DATE ISSUED: 1901-1902
CARTOGRAPHER: R. A. Outhet, for Senate Park Commission:
D.M Burnham, C.F. McKim, A. St. Gaudens, F.L. Olmsted, Jr.
PUBLISHER: A. Hoen and Co., Baltimore, MD
Colored lithograph, scale [1:24,000], 68 x 68 cm, 27.5 27.25 in
Library of Congress, Geography and Map Division, G3851.G5 1901 .091

Among a city's most valuable assets are its parks. This was an underlying aspect of the Senate Park Commission's report. As large and small oases of beauty, retreat and recreation, parks enhance the public realm and elevate the quality of life. The interaction between the urban street (as outer space) and the natural landscape (as inner space) induces an exhilarating sensory experience.

The idea of park—an open space, town square, or large piece of land for pleasurable pursuits—is ancient. But in the eighteenth and nineteenth centuries it was transformed to serve an urban industrialized society. The phenomenon of urban landscapes developed to improve a community's physical condition and offer health benefits of light, fresh air, and psychological advantage. Parks induced serenity, laughter, dazzling ephemeral experiences. "A place to go," a place to meet friends, a place of hope—it seemed simple, a matter of civility. Yet many communities, even today, are devoid of this necessary amenity.

The Senate Park Commission's contribution to Washington greatly exceeded The Mall design alone, for which it is best known. The Commission stressed that a "living" city must have open space scattered throughout to serve the populations of its neighborhoods, along with public parks in the central core to benefit residents and tourists. Following their travels in Europe, the four commissioners were anxious to bring their theories and vision to reality. As advocates of the sublime beauty of parks, such as those of Paris and London, they were inspired by refined views of nature's organic picturesque settings and the more formal landscape interventions.

Regardless of real estate values and land costs, European cities reserved land for parks from which people daily derived great pleasure. In Washington, "seat of government," the Commission advocated that citizens had a right to no less. Understanding this need, the Commission selected land to be "wisely" added to the park system. Some locations, it was reasoned, might bring less financial return for private uses and, therefore would be appropriately procured for public purposes. Exhaustive study was initiated to justify a sound basis for these recommendations. Maps were studied, experts consulted, and proposals delineated.

Encircling Washington at the turn of the century, woodlands were still quite natural. Nevertheless, development and roads were encroaching. Before it "will be forever too late,"

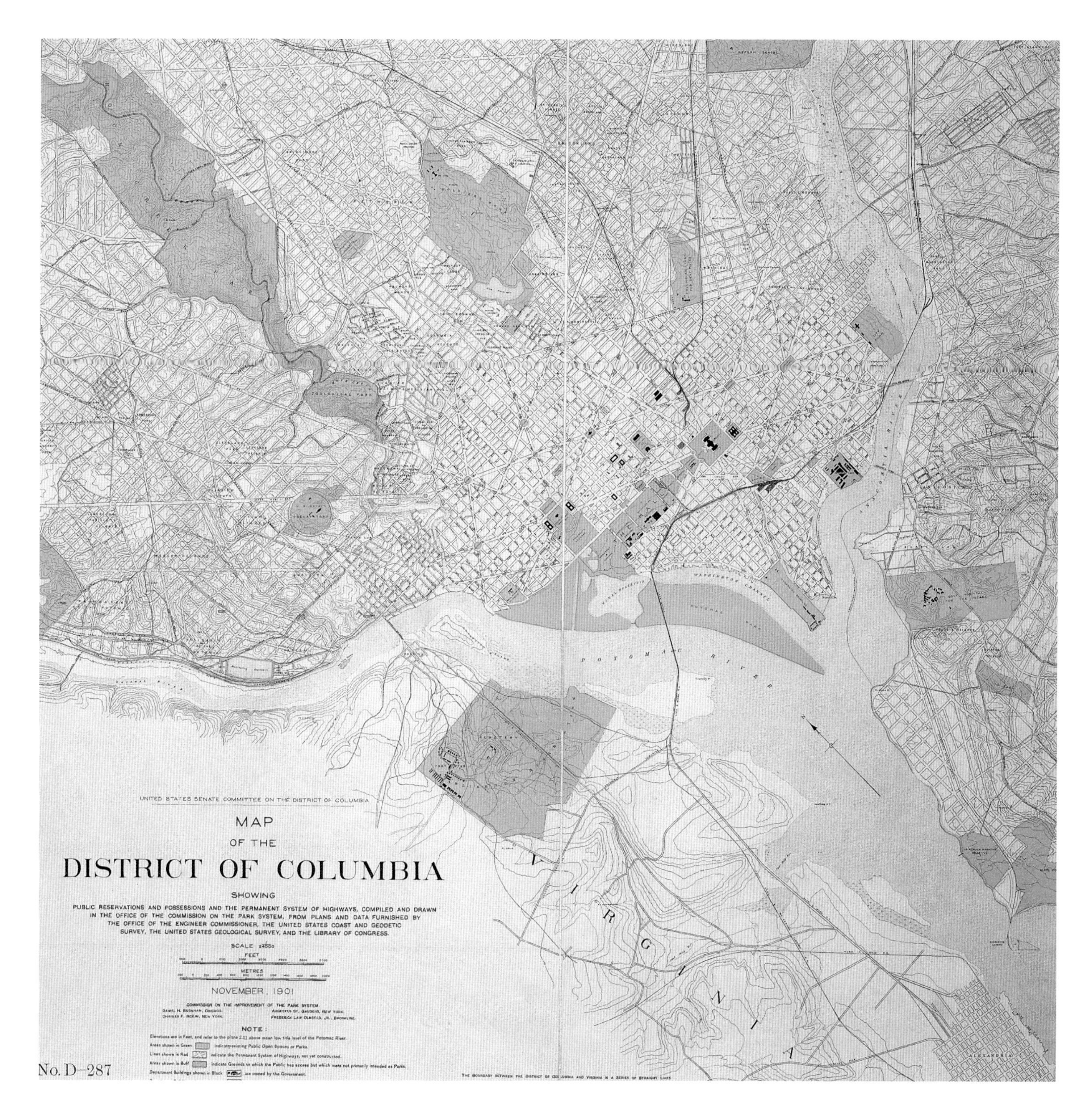

the commissioners urged that park space be secured and natural beauty be preserved. "Fortunately," they noted, "the larger areas necessary for an adequate park system have either been acquired or are awaiting reclamation." Pointing out areas of natural beauty, they focussed on conservation of a chain of unique abandoned Civil War forts (p. 91) on impressive summits, wild and picturesque spots along the rivers, and many smaller tracts to extend the park system in keeping with L'Enfant's tenets while correlating to the rapid pace of development.

Terrain, climate, oppressive summer heat and humidity required these imperatives: maintain shade trees, create micro-climates, conserve hilltops with breezes and valleys with cooler settled air, and be indulgent with water in fountains, basins, springs and brooks. During the first half of the twentieth century before the era of air-conditioning and when cities were relatively safe, on hot summer evenings entire Washingtonian families picnicked and slept in the parks.

Compared with other large American and European cities, Washington lacked an optimal

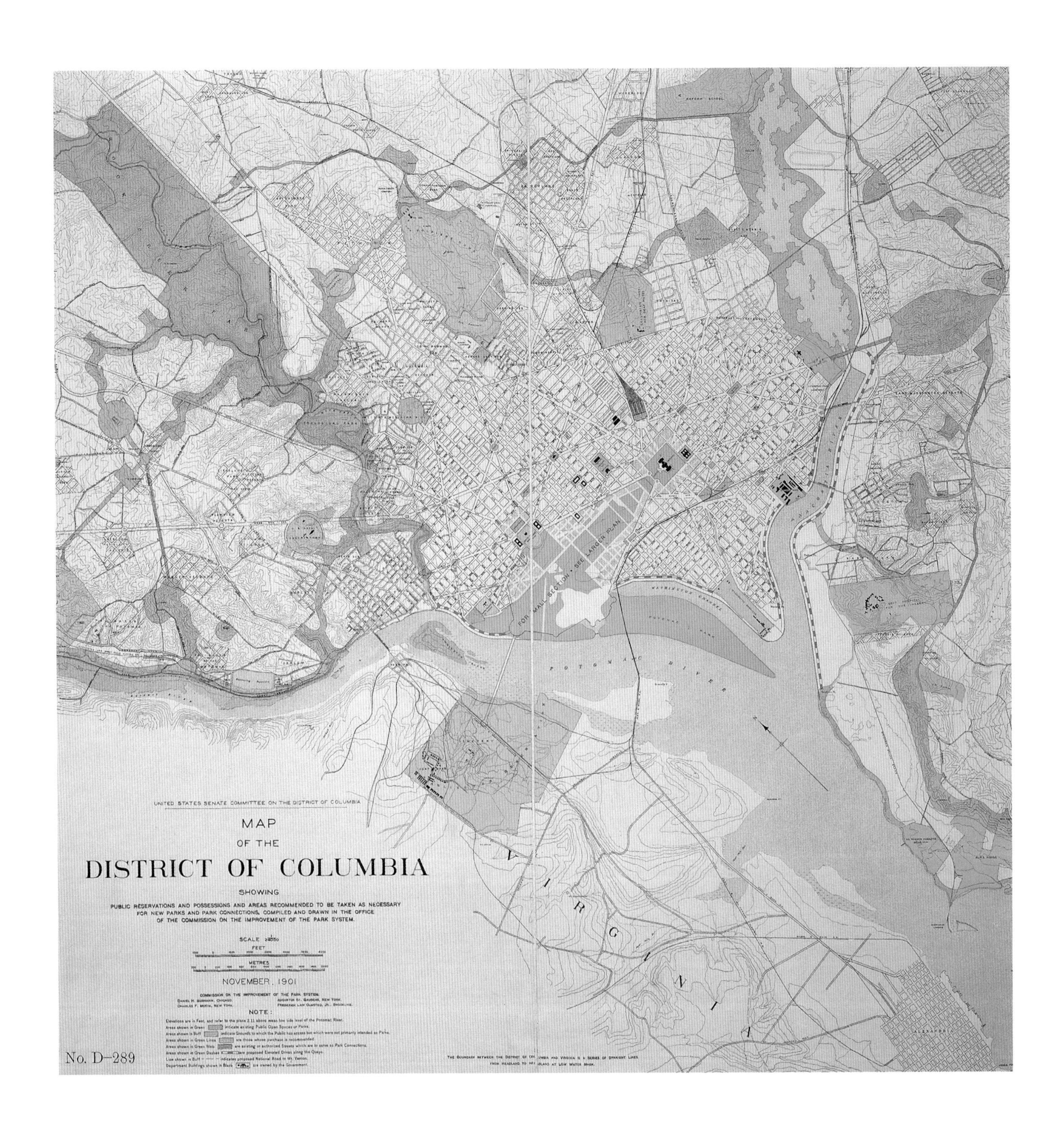
UNITED STATES SENATE COMMITTEE ON THE DISTRICT OF COLUMBIA
MAP
OF THE
DISTRICT OF COLUMBIA
SHOWING
PUBLIC RESERVATIONS AND POSSESSIONS AND AREAS RECOMMENDED TO BE TAKEN AS NECESSARY FOR NEW PARKS AND PARK CONNECTIONS, COMPILED AND DRAWN IN THE OFFICE OF THE COMMISSION ON THE IMPROVEMENT OF THE PARK SYSTEM.
SCALE 1/24000
FEET
METRES
NOVEMBER, 1901
COMMISSION ON THE IMPROVEMENT OF THE PARK SYSTEM.
DANIEL H. BURNHAM, CHICAGO.
CHARLES F. MCKIM, NEW YORK.
AUGUSTUS ST. GAUDENS, NEW YORK.
FREDERICK LAW OLMSTED, JR., BROOKLINE.
NOTE:
Elevations are in Feet, and refer to the plane 2.11 above mean low tide level of the Potomac River.
Areas shown in Green indicate existing Public Open Spaces or Parks.
Areas shown in Buff indicate Grounds to which the Public has access but which were not primarily intended as Parks.
Areas shown in Green Lines are those whose purchase is recommended.
Areas shown in Green Web are existing or authorized Streets which are to serve as Park Connections.
Areas shown in Green Dashes are proposed Elevated Drives along the Quays.
Line shown in Buff indicates proposed National Road to Mt. Vernon.
Department Buildings shown in Black are owned by the Government.
No. D–289
FOR MALL SECTION • SEE LARGER PLAN
POTOMAC RIVER
VIRGINIA

number of parks in outlying areas. But residents and the multitudes of visitors who came here annually made liberal use of existing parks, mostly in the civic center.

An extraordinary level of detail was contained in the Senate Park Commission report, maps and sketches about enhanced park development. Within the old city limits, the Commission noted that the existing 275 reservations were sufficient. For the remaining four-fifths of outlying areas, it proposed a record 1,000 to 1,200 parks. To correct the gap between reality and the ideal, 17 small parcels of 364 acres total were designated for parks. These and others were to be selected by an "authorized board". They also called for public reservations near schools with space suitable for playgrounds.

In essence, the report criticized a lack of variety in the 301 evenly distributed squares and circles within the city. Seldom were historical sculptures (typically Civil War generals on horses) integrated into the design; nor were shade trees, benches, and fences. Greater individuality in park design was necessary to extend possibilities for recreational programs for various age groups in scattered locations.

A labyrinth of ideas for playgrounds and large park uses was proposed. Occasional phrases reflected attitudes of an all-male design team of the period: shallow wading pools for small children "in view of sheltered seats for mothers and nurses," and larger areas for tag "for schoolboys," organized team sports and outdoor gymnasia devices. Other items included places for passive recreation, provision for crowds at bandstands, night lighting, fireworks displays at national celebrations, evergreen parks for winter effects, basins with brilliant aquatic flowers, and refreshing "living" water displays.

In the interest of social and environmental relevance, the Commission discussed larger parks and the means by which they were connected. Contemplating Potomac Parkway (p. 128) between Lincoln Memorial and Rock Creek and Georgetown harbor quay, a promenade and park drive with frequent arched openings at high elevations anticipated concurrent industrial and commercial uses, as in European river cities. Auspiciously, the first segment of the riverfront was ultimately developed as a recreational area, although the two-tier configuration remains only as a drawing.

Two radically different alternatives were studied for Rock Creek and its adjacent parkway (p. 28), an "Open Valley" and a "Closed Valley" Plan. Because of "economy, convenience, and beauty," the Open Valley Plan to maintain the creek bed and hollow was recommended for adoption. This proposal was finally realized in 1936.

Sweeping views along Maryland's Potomac River escarpment, where "land falls rapidly … [disclosing the] perfection of sunset views ... [must be] preserved from the intrusion of future buildings by the acquisition of the slope." The Commission envisioned the parkway continuing to the District line where parkland would be developed. Preserving the C&O Canal was critical for promenade and canoeing, likewise Potomac Drive as a pleasure drive near water's edge. The region's rare grandeur and most picturesque scenery between Cabin John and Great Falls was "one of the greatest cataracts of our Atlantic water shed which should be preserved on both sides as a national park and connected to the city with a continuous river drive." (See C&O Canal, p. 70)

Among the parks contemplated on the west side of the city were Battery Kemble Park (Fort Kemble and Battery Parrott), a Georgetown Parkway to the Naval Observatory, and Piney Branch. Among those to the east were Lincoln Avenue Valley and Eckington Parkway, Patterson Park and Gaulladet. The mountain-like character and high elevation of Mount Hamilton Park(now in the National Arboretum) near Bladensburg Road was to be preserved and accessed by a "winding drive to the summit." Savannah Street was to be widened to form a parkway boulevard near the Soldiers Home and neighboring universities.

Crucial to this visionary concept was the acquisition of land that "can be made an important supplement to the park system" at the new (McMillan) reservoir near Howard University. Similarly, the riverbanks at the Anacostia River shed were to be connected with an informal parkway and improved waterfront conditions. Reclaimed lands would offer recreational uses of numerous types in all seasons. River water would be purified. At the turn of the twenty-first century, these park areas again dominate city planning discussions regarding their recreational potential.

Underlying the motives of the Senate Park Commission was a faith in Washington as an urban paradigm of democracy, a model that must offer cultural parity and a noble environmental aesthetic to affirm a high quality of life for the American public. Then as now, the eyes of the citizens focused on Washington, a vision penetrating the collective consciousness of a nation.

At the conclusion of the written Senate Park Commission document, three large colored fold-out maps (two are shown here) highlighted the philosophical concepts and actual proposals of the commission. These maps strive to reflect the spirit of place of the L'Enfant Plan. The first, #287, extends the street system into the undeveloped hills beyond the original L'Enfant City. Red lines indicated the "Permanent System of Highways, not yet constructed." Buff color denotes "Grounds to which the Public had access but which were not primarily intended as Parks." Green shows "existing Public Open Spaces or Parks." Additional colors on map #289 mark the proposals listed in the report. Comparison of these two maps with the text enables us to synthesize the facets of this remarkable work that changed the course of Washington, and many cities worldwide.

PRESERVING NATURE OR CREATING A FORMAL ALLÉE? TWO SCHEMES

The Senate Park Commission, Rock Creek and Potomac Parkway Plan

TITLE 1: Open Valley Plan, Rock Creek Parkway
TITLE 2: Closed Valley Plan, Rock Creek Parkway
DATE DEPICTED: Title 1: 1901; Title 2: rejected concept
DATE ISSUED: 1901-1902
CARTOGRAPHER: Senate Park Commission
PUBLISHER: Senate Park Commission, 1902
TITLE 1: Ink wash on paper, 9.5 in x 29.5 in
U.S. Commission of Fine Arts

TOWARD IMPLEMENTATION: ENTICING ENVIRONMENTAL SOLUTION

The Langdon Drawing

TITLE: Preliminary Design for Rock Creek & Potomac Parkway
DATE DEPICTED: Future projection
DATE ISSUED: 1916
CARTOGRAPHER: James G. Langdon, landscape architect, Frederick D. Owen, engineer
PUBLISHER: Commission authorized by Congress, March 4, 1913, Directed by Colonel William *W. Harts, U.S.A., Officer in charge of Public Buildings and Grounds, House Doc. No. 114, 64th Congress, 1st Session*

ESSAY BY TIMOTHY DAVIS

Rock Creek and Potomac Parkway played a prominent role in the Senate Park Commission's plan for the improvement of Washington's park system. The parkway was intended to link The Mall and Rock Creek Park in northwest Washington with an attractive tree-lined corridor to provide an appealing escape from city streets while rehabilitating the lower portion of Rock Creek valley, which had degenerated into an eyesore and public health menace. The parkway also was intended to foster urban renewal, transforming the creek's environs from marginal housing and scattered industries into dignified and prosperous residential neighborhoods. This mixture of aesthetic, economic, and environmental motivations epitomized the "City Beautiful" ethos. Rock Creek and Potomac Parkway would join the other parks and parkways outlined in the Commission's report to produce a coordinated system of improvements that would serve as a shining example of the virtues of comprehensive city planning.

The Commission considered two proposals for transforming the degraded valley from an urban wasteland into a civic showpiece. Small sketches of two competing proposals appeared in its report: 1) "culvert" or "closed-valley"; and 2) "open-valley." The first option, known as "culvert" or "closed-valley," entailed enclosing the creek in a large culvert and filling in the valley with soil, from Q Street to Pennsylvania Avenue, to form a level connection between Washington and Georgetown. A dignified formal tree-lined boulevard would have been constructed along the length of the concealed creek, forming the border of an adjacent new residential district. Proponents claimed this alternative would foster economic revitalization of Georgetown while paying for itself through increased property tax revenues. While the closed valley plan might seem radical in retrospect, enclosing urban streams was a common practice across this country at the time and formal boulevards had proven their worth as agents of economic development and civic improvement.

The second option, the "open-valley" scheme, which called for restoring the valley to its original contours and constructing a winding carriage drive at stream-level through a largely man-made "naturalistic" landscape, was in many ways a more radical proposal. Rehabilitating the valley would require extensive excavation efforts, especially between M and P Streets, where enormous banks of unstable debris loomed above the narrow creek. The proposed picturesque landscape was more in keeping with contemporary tastes in parkway development, however, and would provide a more suitable transition from the formality of The Mall to the sylvan seclusion of Rock Creek Park to the north. Placing the carriage drive alongside the creek would help shield parkway users from views of the surrounding city. Since major cross streets would

have to be carried across the valley on impressive bridges, it would also greatly reduce the number of dangerous and disruptive intersections. This was an appealing attribute in the age of horses and buggies that would become increasingly important as the automobile became a dominant factor in twentieth-century American life.

After debating the merits of both proposals at considerable length, the Senate Park Commission ruled in favor of the open valley treatment "on the grounds of economy, convenience, and beauty."

The design of Rock Creek and Potomac Parkway underwent numerous revisions before the project was completed in the 1930s. This 1916 plan (following pages) was the most elaborate of a series of proposals, both in development concept and graphic presentation. Prepared by former Frederick Law Olmsted firm designer James G. Langdon under the auspices of the Office of Public Buildings and Public Grounds, the 1916 plan expanded upon the Senate Park Commission's general recommendations for the creation of an informally landscaped parkway with a creek-side carriage drive.

While the main roadway would wind along the bottom of the valley, formal boulevards would flank the edges of the parkway at street-level. These border roads would accommodate utilitarian traffic, provide a distinctly articulated edge to the parkway reservation, and foster the development of dignified residential districts. The original M Street Bridge would be replaced by a new span between N and O Streets. Numerous roads would provide access from surrounding neighborhoods, most notably below P Street. Here a handsome bridge across Rock Creek led to the main roadway, which remained on the Georgetown side of Rock Creek until it crossed the creek at Oak Hill Cemetery.

The plan's picturesquely located plantings epitomized the reigning Anglo-American tradition of informal landscape design. The ornate planting scheme angered Commission of Fine Arts member Frederick Law Olmsted, Jr., however, who complained that it was an abstract idealization that bore little relation to particularities of the site. Olmsted, who had been the principal author of the landscape design sections of the Senate Park Commission report, sought to save as much of the existing tree-cover as possible, especially in the heavily wooded areas north of Q Street.

While many details of the elaborate 1916 plan were eventually discarded, Rock Creek and Potomac Parkway, now commonly know as Rock Creek Parkway, fulfills many of the goals envisioned by the Senate Park Commission. The informally landscaped parkway serves as an attractive link between major components of Washington's park system. Motorists and recreationalists use the tree-lined parkway to pass from The Mall to Rock Creek Park with scarcely

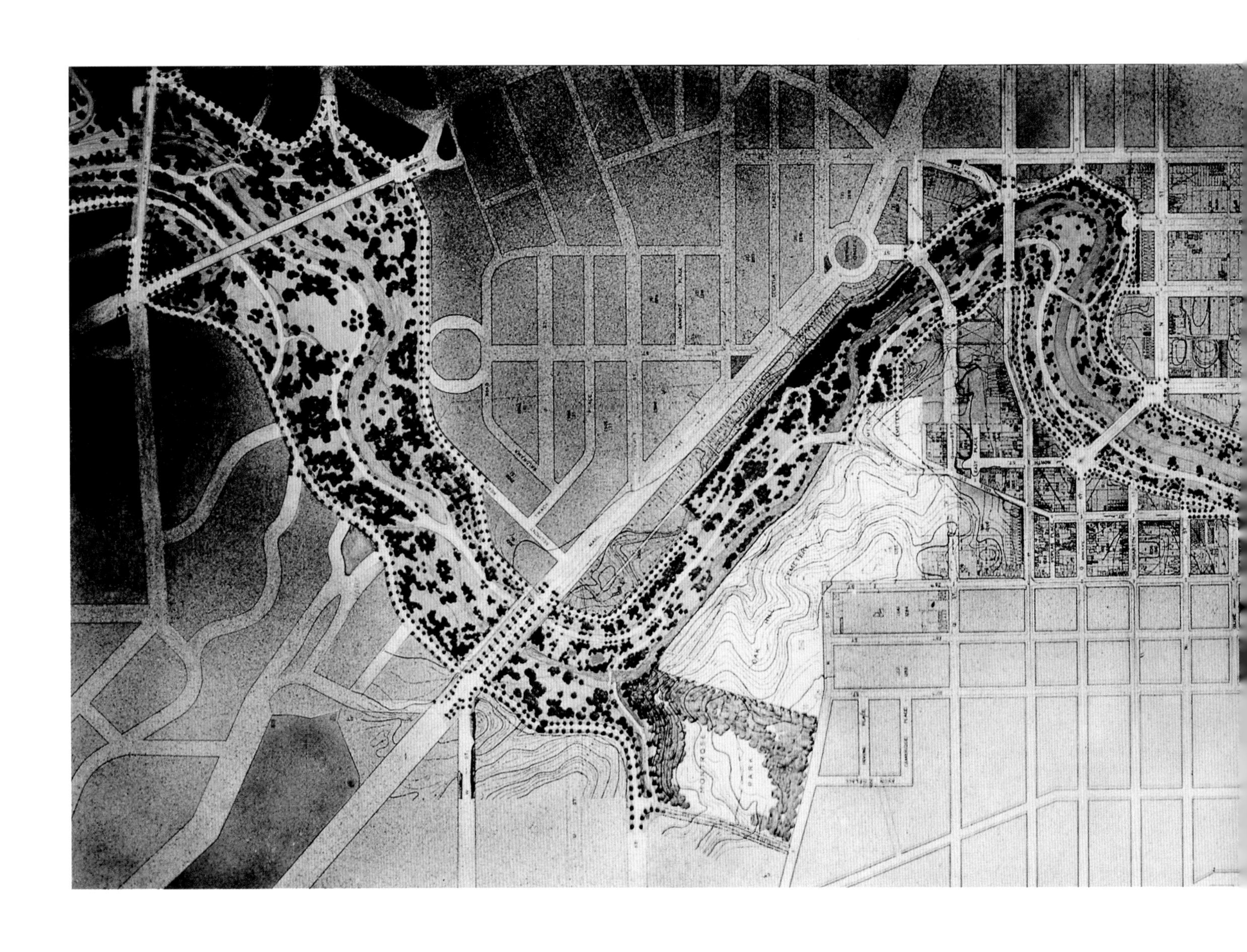

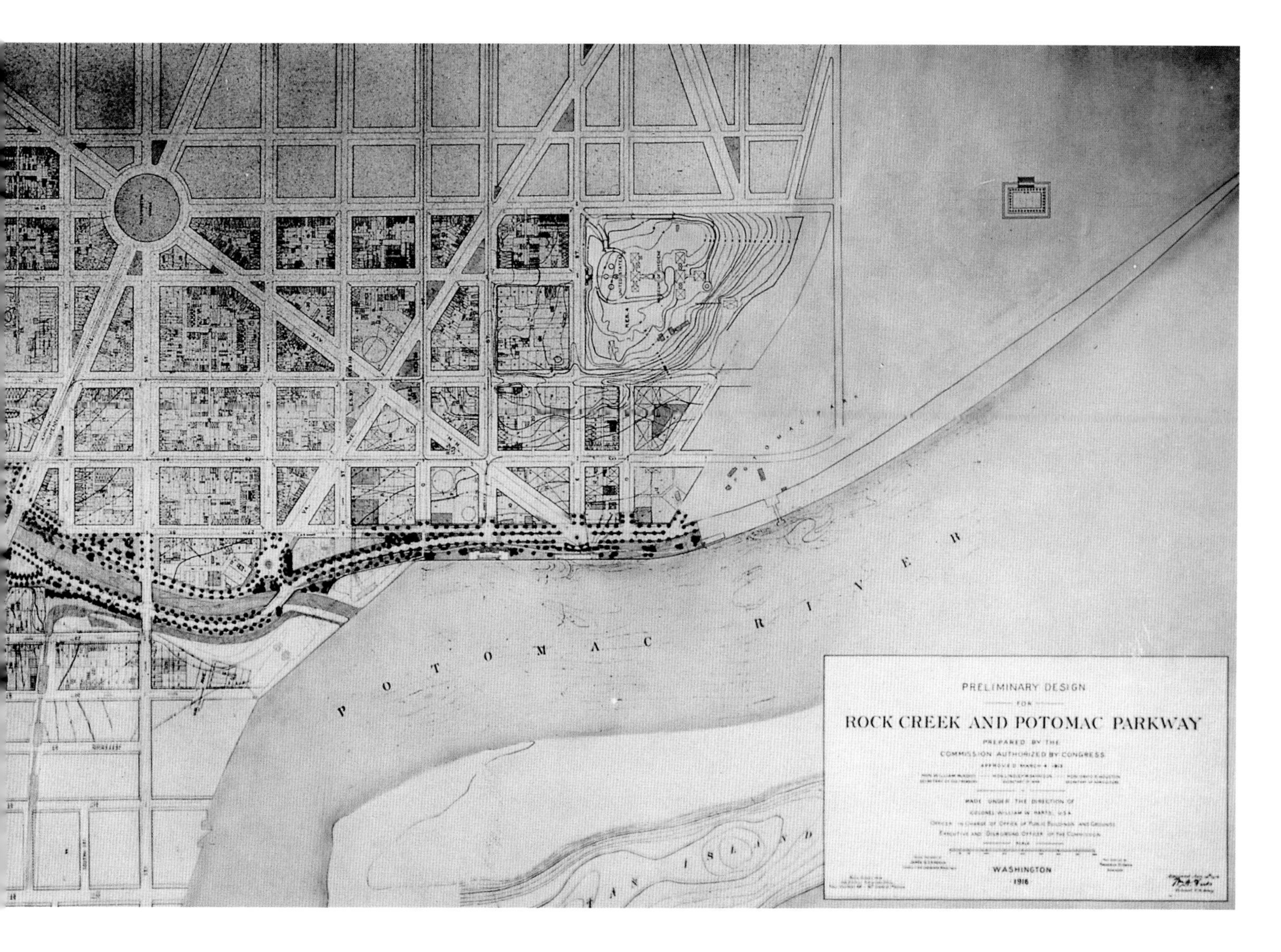

a glimpse of the nearby cityscape. Surrounded by parkland, it provides neighborhoods with a beckoning array of bike paths, pedestrian trails, green lawns, and shady groves.

In the eyes of some, however, the parkway's primary purpose is to serve as a convenient traffic artery, connecting the northwest suburbs with the monumental core and downtown Washington. While the original planners did not fully appreciate the extent to which the parkway might serve a commuting function, its practical value was readily apparent upon its opening in 1936. Numerous proposals aimed at enhancing the parkway's traffic-carrying capacity were floated over the subsequent decades, but with the exception of the Whitehurst Freeway overpasses near Virginia Avenue, the project has survived relatively unscathed, making it one of the best-preserved examples of early twentieth-century parkway development in the country. In recognition of Rock Creek and Potomac Parkway's historical significance, the Historic American Buildings Survey conducted a detailed cultural landscape documentation project in 1991-92. This plan and the accompanying sections are based on aerial photogrammetry augmented by field verification. (These drawings can be found in the Library of Congress, Prints and Photographs Division). The lower portion of the parkway as it appeared in 2000 can be seen in Joseph Passonneau's map (p. 157).

[1] For a detailed account of the planning, development, and subsequent evolution of Rock Creek and Potomac Parkway, see Timothy Davis, "Rock Creek and Potomac Parkway, Washington, D.C.: The Evolution of a Contested Urban Landscape," Studies in the History of Gardens and Designed Landscapes 19 (April-June 1999), 123-237.

[2] U.S. Congress, Senate Committee on the District of Columbia, The Improvement of the Park System of the District of Columbia (Washington, D.C.: Government Printing Office, 1902), 85.

PROFILING AN UNCOMMON SPORT

THE ROCK CREEK PARK ORIENTEERING RECREATIONAL MAP

TITLE: Rock Creek Park, Recreation Map D.C.1, U.S. Orienteering Federation
DATE DEPICTED: 1989
DATE ISSUED: 1989
CARTOGRAPHER: Peggy Dickison, FIELD WORK: Mikell Platt
PUBLISHER: Okay Maps/National Park Service; Base Map: Harvey Map Services Ltd.; Chrismar
Printed colored map, scale [1:15,000], 41 x 17 cm
Library of Congress, Geography and Map Division, G3852.R6E63 1989 .03

The linear ribbon valley of Rock Creek includes four separately designated recreational parks, which provide an ephemeral sense of wilderness passing through the dense urban fabric of northwest Washington. These are Rock Creek Park, National Zoological Park, and the contiguous Rock Creek and Potomac Parkway.

Rock Creek Park begins north of the National Zoo in upper northwest Washington, and continues past the District line into Maryland. This site was proposed for a public park and surveyed by Major Nathaniel Michler at the request of Congress in 1867. Legislation authorizing the park's creation in 1890 aimed to preserve woodland and farm areas as a nature reserve for outer Washington and its suburbs. The park encompasses approximately 1,750 acres. It runs four miles long, varying from nearly one mile wide to less than 300 yards at its narrowest point.

The crooked valley was flanked by steep, thickly wooded hills, with an old growth forest, according to a 1902 Senate Park Commission description. Beech Drive was a paved roadway, extending from the Zoo to Military Road, where it became a long earthen road (today, paved) continuing into Maryland. The Commission expressed concern about future park maintenance and improvements, including new construction and road widening, which might injure the dazzling gorge and beautiful park scenery. In urging caution, Commissioners suggested possibly building another road high above the valley, perhaps a boundary street with homes on the opposite side facing the park. They noted that the acquisition of land at strategic boundaries was essential to ensure future views and prevent structures from obstructing and "intruding into the landscape."

In 1927, Rock Creek Park was among the largest of city parks in the country. A park map of that year, in the Library of Congress (not shown), unveils park contours and major components. The park's diverse range of recreational facilities, which are shown on the 1989 orienteering map, accommodates a host of interests and needs: walking paths, hiking and biking trails, horse stables, nature centers, picnic areas, playing fields, golf, tennis, roads crossing in many directions, bridges fording streams, and institutional facilities such as Carter Barron Amphitheater. From the drawn contour lines, we see great variations in terrain, which bespeak of contrasts in scenery, plants, and water courses.

As an "orienteering" sport map, the comprehensive legend aids in organizing a challenging hiking experience. Using only a map and compass, the participant must follow a designated route from point to point. The sport of "orienteering" offers a variety of challenges to meet one's personal capability. Some examples: a competitive race; an individual or group endeavor on difficult terrain; a game for children and families; or a program for the physically impaired.

One of the other three parks through which Rock Creek flows, the 163-acre National Zoological Park, at the southern tip of Rock Creek Park, was established in 1889 with the support of the Smithsonian Institution. In its 1902 report, the Senate Park Commission noted the importance of the Zoo to the federal capital, and urged continuing preservation of as much land as possible for this purpose.

Rock Creek and Potomac Parkways (pp. 128-131), at the south end of Rock Creek, begin at West Potomac Park terraces, near the riverside at Lincoln Memorial, continuing as a winding four-lane roadway through the narrow creek valley to the Zoo's southern border. Extending 2.5 miles long and a few hundred yards wide, construction of the scenic parkways and recreational area began in the mid-1920s, and was completed in 1936.

Together, the four connected parks provide a tranquil drive through a nature preserve from Maryland to the city's civic core. No longer polluted, the lower Rock Creek valley was reclaimed following a recommendation from the 1902 Senate Park Commission. A multi-use trail, converted from a bridle path in 1970, is used by runners, walkers, roller-bladers and bicyclists. These parks pair a densely forested landscape with extensive open space. Facilities such as picnic areas, dog parks, and active and passive precincts are sufficiently accessible to serve a large urban population. Like an awe-inspiring painting, this picturesque composition of monuments, vistas and nature in sequential "picture-plane" motion is a precious landscape in the nation's capital.

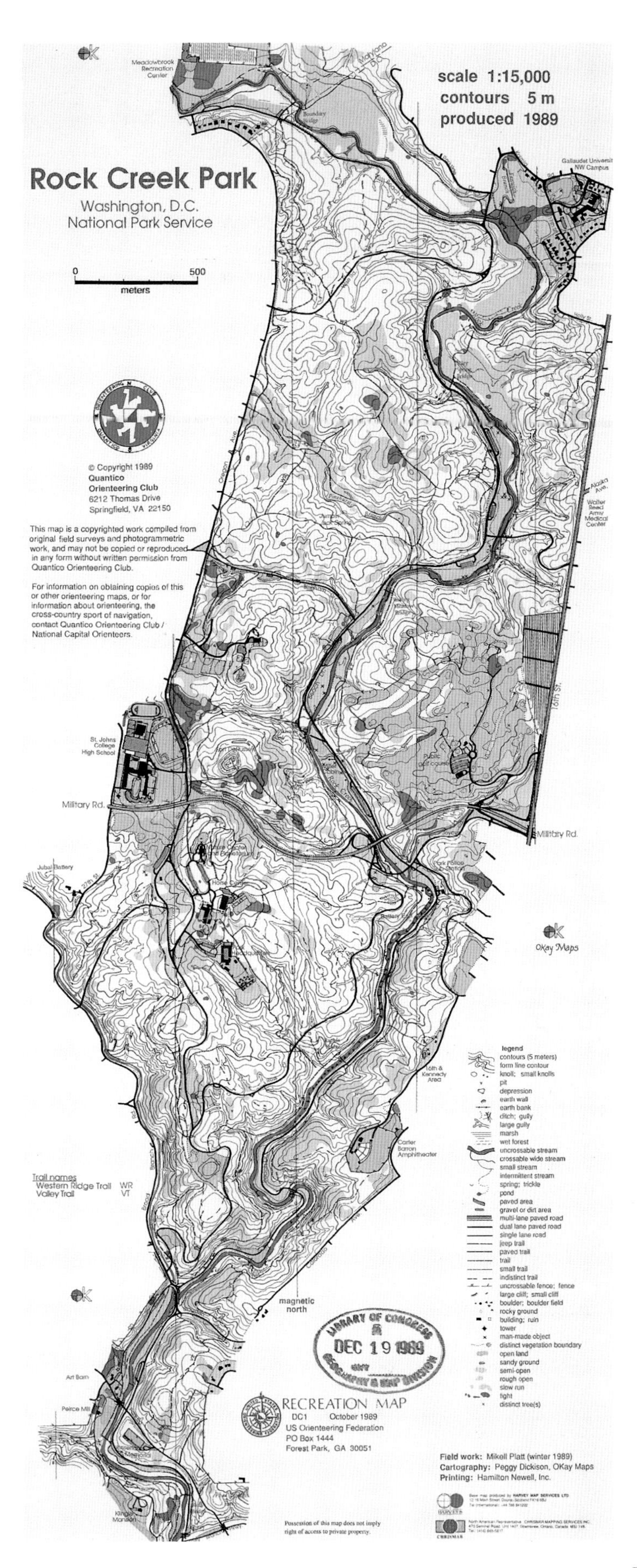
scale 1:15,000
contours 5 m
produced 1989
Rock Creek Park
Washington, D.C.
National Park Service
0 500
meters
Meadowbrook Recreation Center
Gallaudet University NW Campus
© Copyright 1989
Quantico
Orienteering Club
6212 Thomas Drive
Springfield, VA 22150
This map is a copyrighted work compiled from original field surveys and photogrammetric work, and may not be copied or reproduced in any form without written permission from Quantico Orienteering Club.
For information on obtaining copies of this or other orienteering maps, or for information about orienteering, the cross-country sport of navigation, contact Quantico Orienteering Club / National Capital Orienteers.
Walter Reed Army Medical Center
St. Johns College High School
Military Rd.
Military Rd.
16th St.
Jubal Battery
OKay Maps
16th & Kennedy Area
Carter Barron Amphitheater
Trail names
Western Ridge Trail WR
Valley Trail VT
legend
contours (5 meters)
form line contour
knoll; small knolls
pit
depression
earth wall
earth bank
ditch; gully
large gully
marsh
wet forest
uncrossable stream
crossable wide stream
small stream
intermittent stream
spring; trickle
pond
paved area
gravel or dirt area
multi-lane paved road
dual lane paved road
single lane road
jeep trail
paved trail
trail
small trail
indistinct trail
uncrossable fence; fence
large cliff; small cliff
boulder; boulder field
rocky ground
building; ruin
tower
man-made object
distinct vegetation boundary
open land
sandy ground
semi-open
rough open
slow run
fight
distinct tree(s)
magnetic north
LIBRARY OF CONGRESS
DEC 19 1989
GEOGRAPHY & MAP DIVISION
Art Barn
Peirce Mill
RECREATION MAP
DC1 October 1989
US Orienteering Federation
PO Box 1444
Forest Park, GA 30051
Field work: Mikell Platt (winter 1989)
Cartography: Peggy Dickison, OKay Maps
Printing: Hamilton Newell, Inc.
Possession of this map does not imply right of access to private property.

A LANDSCAPE IDEOLOGY: U.S. CAPITOL GROUNDS
The Olmsted Sr. Plan

TITLE: General Plan for the Improvement of the U.S. Capitol Grounds
DATE DEPICTED:
DATE ISSUED: 1874
CARTOGRAPHER: Frederick Law Olmsted, Sr.
PUBLISHER: Architect of the Capitol
Manuscript, ink on paper, watercolor, 43.5 x 33 in
U.S. Capitol, Architect of the Capitol

ESSAY BY HERBERT M. FRANKLIN

In 1874 Frederick Law Olmsted, Sr., the preeminent landscape architect in the United States, was commissioned to lay out the grounds of the United States Capitol. Olmsted was a pioneer in the development of the country's public parks, and many of his designs were influenced by his study of European parks, gardens, and estates.

The Olmsted Plan is notable for the subordination of the grounds to the Capitol. It created a placid green setting for a white jewel, with trees located to provide a shade canopy and unobstructed vistas of the Capitol from various points. The plan omitted flowerbeds or other landscaping that might compete with the gleaming Capitol, and it kept most major trees away from the building. Walls, lanterns, and other features were to be built of stone in earthy or subdued colors. Graceful curvilinear drives and walks contrasted pleasantly with the rectilinear footprint of the Capitol, causing those who approach the building to view it from changing perspectives. Numerous drives and paths, accommodating the many surrounding streets, meant that much of the plan was devoted to circulation, with L'Enfant Plan avenues, such as Maryland and Pennsylvania finding expression as major paths on the west grounds. The eastern approach gave paramount importance to the direct vista of the dome and the central east front portico. (This will be strengthened by a planned Capitol Visitor Center entrance to be positioned on this axis.) The formal symmetry of the plan reflected a nineteenth-century geometrical approach that harmonizes well with the later Senate Park (McMillan) Commission Plan for the nearby National Mall.

Olmsted also addressed a significant optical problem that, for many years, had been neglected: the Capitol was perched on the edge of its hill and from the west appeared to be on the verge of tumbling down. Olmsted's solution was to construct marble terraces on the Capitol's north, west, and south sides, creating a base enabling the building to "gain greatly in the supreme qualities of stability, endurance, and repose." The west terrace and grounds have in recent years become the preferred location for presidential inaugurations.

Olmsted also intended water to play a minor but distinct role in the general plan. A pair of elevated fountains flanking the extension of East Capitol Street as it terminates at the East Plaza was an element of the plan, as are what appear to be two round pools at the corners of the west terrace and a rectangular pool at the base of the west central steps. The corner pools were not developed. A "summer house" of rusticated red brick with a drinking fountain and grotto, designed by Olmsted's associate Thomas Wisedell, was built in the northwest portion of the grounds and remains a somewhat hidden and romantic feature of the landscape nestled in the trees and shrubs. A circular Romanesque fountain was built within the west terrace ensemble.

The drawing is notable also for its outline of the Capitol itself, showing substantial east and west extensions in the central section. It is generally assumed that this was done at the suggestion of the Architect of the Capitol, Edward Clark, to ensure that Olmsted's plan would be compatible with such extensions if they were ever undertaken. The east front extension was intended to correct an optical problem on the east similar to that addressed by Olmsted on the west: the new 1866 dome looked precarious on its small 1826 base.

By 1866, Senate and House wings were complete as well. The extension to the west was probably intended to accommodate the Library of Congress, housed in the Capitol until 1897. A less protruding extension on the east front of the Capitol was ultimately completed in 1958-62; a proposed west front extension was rejected by Congress in 1983, after several decades of extended controversy.

The Olmsted Plan generally has been adhered to, although colorful floral plantings have been added. The trolley lines shown on the plan no longer exist, but the passenger shelters remain as historic features and provide points for excellent oblique views of the Capitol. The grounds of the Capitol have been expanded substantially since 1875, but the segment shown here, sometimes referred to as Capitol Square, remains the heart of the complex. Although there is no statutory prohibition on the erection of memorials on Capitol grounds, the Architect of the Capitol has for many years successfully opposed proposals for such memorials, in order to maintain the integrity of the Olmsted Plan.

B STREET N.
MARYLAND AVE
A STREET N.
E. CAPITOL ST.
A STREET S.
PENNSYLVANIA AVE
B STREET S.
1ST STREET E.
DELAWARE AVE
NEW JERSEY AVE
N. CAPITOL ST.
S. CAPITOL ST.
NEW JERSEY AVE
DELAWARE AVE
1ST STREET W.
B STR. N.
PENNSYLVANIA AVE
MARYLAND AVE
B STREET S.
General Plan
for the
Improvement
OF THE
U. S. CAPITOL GROUNDS.
FRED. LAW OLMSTED
Landscape Architect

ANTICIPATING FUTURE NEEDS

The Architect of the Capitol Master Plan

TITLE: The Master Plan for the United States Capitol
DATE DEPICTED: Future projection
DATE ISSUED: 1981
CARTOGRAPHER:
PUBLISHER: Architect of the Capitol, George M. White
Paper with colored pen, 46 x 86 in
U.S. Capitol, Architect of the Capitol

ESSAY BY HERBERT M. FRANKLIN

The Master Plan for the U.S. Capitol, as shown in this 1981 map, was the result of surveys, studies, and consultations from 1976 to 1981. In 1975, Congress authorized the Architect of the Capitol to develop a master plan for future developments within the existing United States Capitol Grounds, and for the enlargement of such grounds as might be needed to provide for "future expansion, growth and requirements of the legislative branch and such parts of the judiciary branch as deemed appropriate to include in such plan."[1] The legislation directed the Architect to report to Congress upon its completion, and the 1981 map accompanied the final report. Congress did not formally approve the plan. The expansionist assumptions underlying it were subsequently found questionable—in the 1980s, because of federal budgetary deficits; and in the 1990s, for political and technological reasons.

Notwithstanding such assumptions, which were based on projections of continued growth in staff and space needs that had been experienced in the post-war period, the plan is noteworthy in prescribing that future development should respect the integrity of the residential Capitol Hill Historic District, located east of the Capitol Grounds. (This is significant because Congress [which has jurisdiction over the District of Columbia] can override D.C. land use regulations which apply beyond the Capitol Grounds.)

Although the 1981 Master Plan has no legal effect on Congressional decisions, it has served as a guide to the Architect. The Architect oversees 274 acres of grounds and 13 million square feet of buildings in the Capitol complex—including the Capitol, House, and Senate Office Buildings; the U. S. Botanic Garden; the Supreme Court; the Library of Congress; and the Capitol Power Plant. The building marked "G" on the map adjacent to Union Station has since been added t o the complex by the Architect as the Thurgood Marshall Federal Judiciary Building, following urban design guidelines advocated in the plan. The parking contemplated beneath the Capitol's East Plaza ("K") will be supplanted by a planned Capitol Visitor Center, and the proposed Ceremonial Drive ("D") has been precluded by a Congressionally mandated National Garden to be built in its right of way adjacent to the west side of the U.S. Botanic Garden conservatory. Apart from these projects, capital investment in the Capitol Complex has been devoted primarily to upgrading existing buildings for security, telecommunications, system modernization and fire protection purposes.

The Architect has asked, so far unsuccessfully, for funds to update the Master Plan to take account of security, technological and related comprehensive planning, and other considerations. Many urban design principles reflected in the 1981 map—such as the emphasis on the maintenance of presently landscaped areas; the preference for underground rather than surface parking; the creation of a more pedestrian-friendly connection between Union Station and the Capitol; and an improved vista of the Capitol from the south—are likely to remain unchanged in any updating of the plan.

[1] P.L. 94-59 (July 25, 1975); 89 STAT, 288, 289

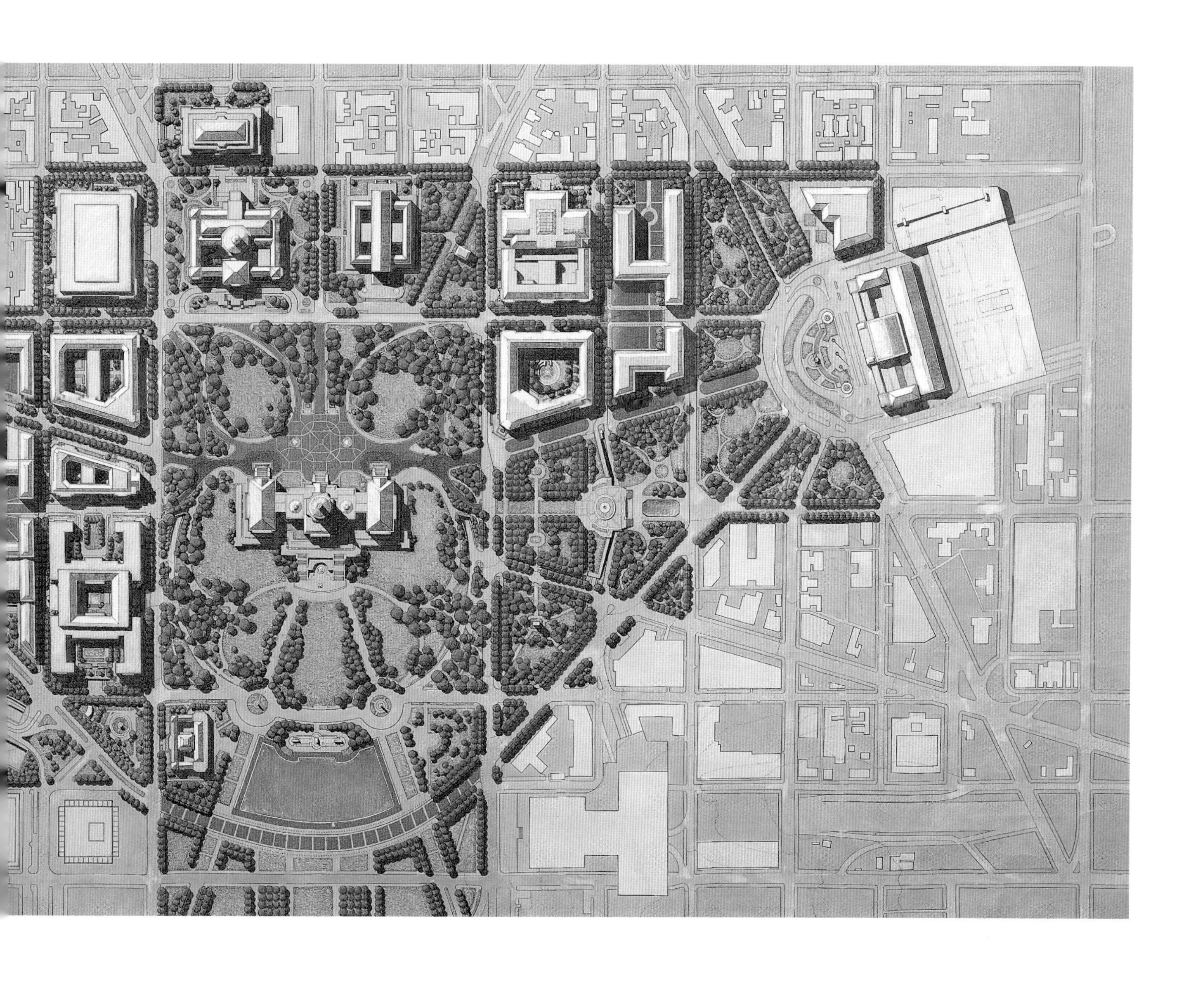

DISTRIBUTION OF PUBLIC RESERVATIONS

The Wilson Map

TITLE: Map of the City of Washington Showing the Public Reservations Under Control of Office of Public Buildings and Grounds
DATE DEPICTED: 1894
DATE ISSUED: 1894
CARTOGRAPHER: John Stewart, surveyor
PUBLISHER: Lieut. Col. John M. Wilson, Corps of Engineers, Washington, D.C.
Printed map, 43 x 58 cm
Library of Congress, Geography and Map Division, G3851.G5 1894 .S7

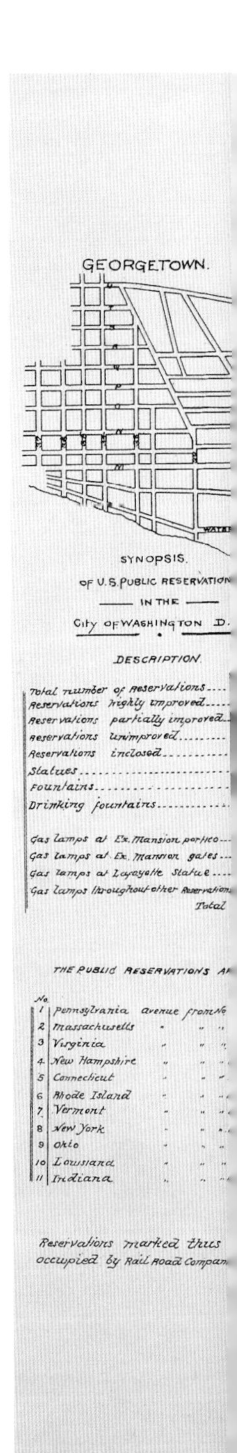

The splendor of Washington—its ingenious gardenesque composition—can be seen in the numerous varied public reservations scattered throughout the city. Today they are often referred to as squares, circles, triangles, civic space and parks. In this adapted "figure-ground" drawing by surveyor John Stewart, C.E., the abundance of open space set aside for public use was delineated with visual clarity. Rendered in solid black, we can imagine these many spaces as green oases shaded by tall trees, modifying the effect of summer heat in the days before air conditioning. A powerful park-like image of the nation's capital was portrayed. This is the model of a "magnificent city" that L'Enfant pictured when he laid out his extraordinary plan.

In the seventy years following the transfer of the seat of government to the District of Columbia, the city resembled a wasteland, barren of trees and amenities befitting a great city. Ultimately, a strong and charismatic leader, D.C. Governor Alexander "Boss" Shepherd, changed that image through a series of public works projects. He planted 60,000 trees, installed gas lamps, and improved other aspects of the city's infrastructure. While his tenure lasted only a few years, from 1871 to 1874, his legacy is still felt today.

The disciplined composition presents a striking hierarchy of public space and linear connections. The grand diagonal avenues, planned by L'Enfant, were greater in width than the grid of streets. Toward the monumental core park grounds were expansive (Presidents-1, Washington-2, Smithsonian-3, Seaton-5/6, Botanical Garden-1, Judiciary-7, and Garfield-17) in contrast to the neighborhood scale of the citywide squares, circles and triangles.

DESCRIPTION:

Total number of Reservations	*301*	*Drinking fountains*	*25*
Reservations highly improved	*92*	*Gas lamps at Ex. Mansion portico*	*6*
Reservations partially improved	*41*	*Gas lamps at Ex. Mansion gates*	*4*
Reservations unimproved	*168*	*Gas lamps at Lafayette Statue*	*2*
Reservations enclosed	*66*	*Gas lamps throughout other Reservations*	*394*
Statues	*14*		
Fountains	*22*		

PUBLIC RESERVATIONS: order of numbers, Avenues:

1 Pennsylvania	*20- 56*	*12 New Jersey*	*190-196*
2 Massachusetts	*57- 93*	*13 Maryland*	*197-213*
3 Virginia	*94-130*	*14 Delaware*	*214-228*
4 New Hampshire	*131-148*	*15 North Carolina*	*229-239*
5 Connecticut	*149-150*	*16 South Carolina*	*240-241*
6 Rhode Island	*151-160*	*17 Georgia*	*242-258*
7 Vermont	*161-170*	*18 Kentucky*	*259-265*
8 New York	*171-185*	*19 Tennessee*	*266-269*
9 Ohio	*86*	*20 Florida*	*270-283*
10 Louisiana	*187*	*21 Canal Street*	*284-293*
11 Indiana	*188-189*	*22 Water Street*	*294-301*

PUBLIC RESERVATIONS: confluence of diagonals and grid:

Parks East of Capitol:		*9*	*Franklin*
14	*Lincoln*	*10*	*Lafayette*
15	*Stanton*	*11*	*Farragut*
17	*Garfield*	*13*	*unnamed*
18	*Marion*	*20*	*Howard*
19	*unnamed*	*26*	*Washington*
Parks West of Capitol:		*60*	*Dupont*
7	*Judiciary*	*63*	*Scott*
8	*Mount Vernon*	*153*	*Iowa*

A statistical table listed the state of improvements and amenities, while another chart recorded the ordered numbering system along the avenues (see above). A hatched triangle pointed out spaces "occupied by the Rail Road Companies." A note in the lower right corner, signed by Colonel John M. Wilson, U.S. Army, stated "To accompany the annual report upon the Improvement and care of public buildings and grounds in the District of Columbia for the fiscal year ending June 30th 1894." For many years, this simply drawn map has been a significant resource of information.

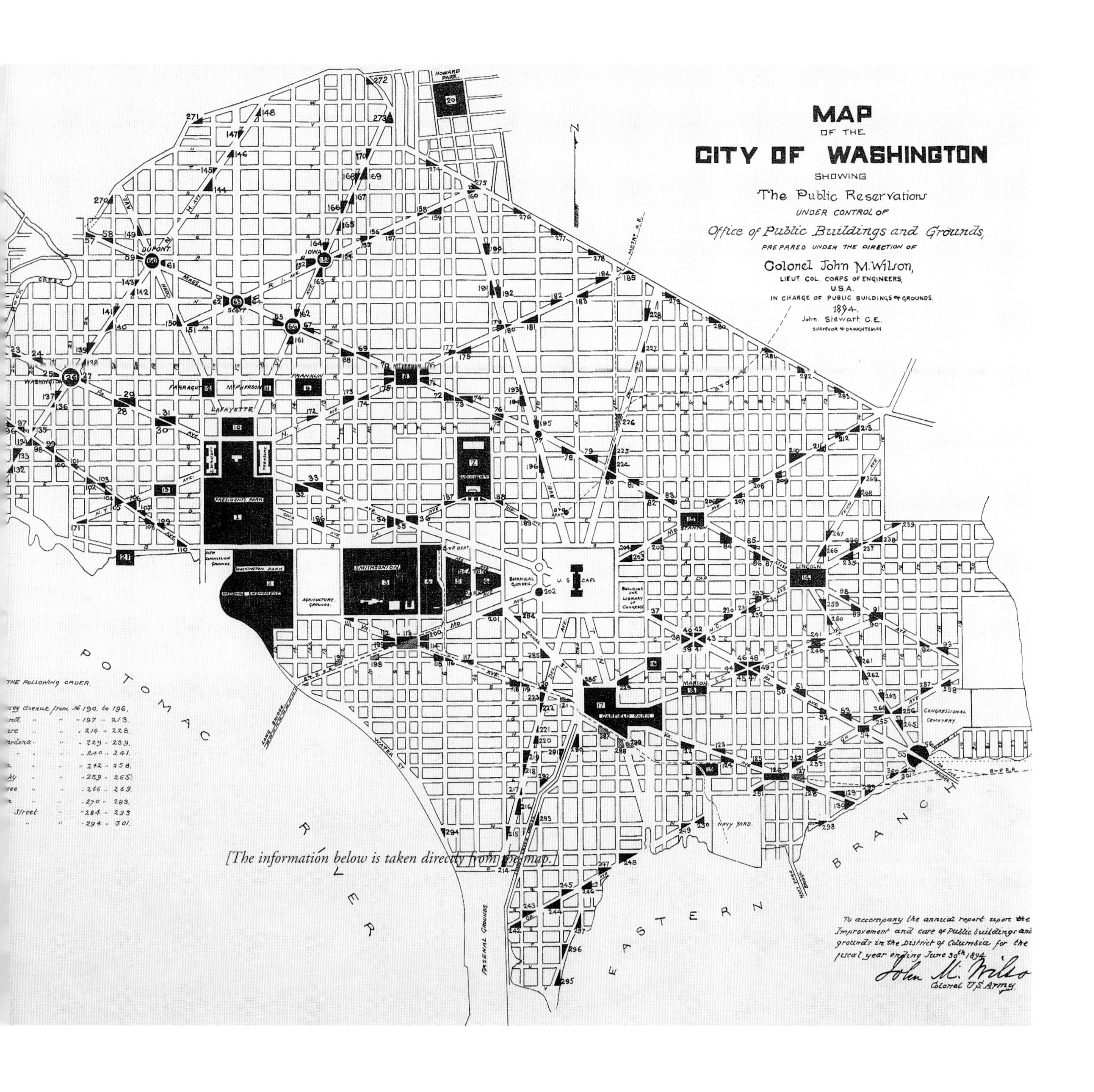
MAP
OF THE
CITY OF WASHINGTON
SHOWING
The Public Reservations
UNDER CONTROL OF
Office of Public Buildings and Grounds,
PREPARED UNDER THE DIRECTION OF
Colonel John M. Wilson,
LIEUT COL. CORPS OF ENGINEERS,
U.S.A.
IN CHARGE OF PUBLIC BUILDINGS & GROUNDS.
1894.
John Stewart C.E.
SURVEYOR & DRAUGHTSMAN
To accompany the annual report upon the Improvement and care of Public buildings and grounds in the District of Columbia for the fiscal year ending June 30th 1894.
John M. Wilson
Colonel U.S. Army
POTOMAC RIVER
EASTERN BRANCH
PRESIDENTS PARK
SMITHSONIAN
U.S. CAPL
GARFIELD PARK
LINCOLN
STANTON
MARION
FRANKLIN
FARRAGUT
LAFAYETTE
DUPONT
IOWA
SCOTT
WASHINGTON
HOWARD PARK
NAVY YARD
CONGRESSIONAL CEMETERY
WASHINGTON PARK
AGRICULTURE GROUNDS
BOTANICAL GARDEN
ARSENAL GROUNDS
WATER ST.
LONG BRIDGE
[The information below is taken directly from the map.]

SPIRIT OF THE TIMES: INFINITE TRANSFORMATIONS
The HABS Comparative Drawings of Reservations

TITLE 1: Parks of the Late Nineteenth Century
TITLE 2: Parks of the Early Twentieth Century
DATE DEPICTED: Nineteenth and Twentieth Centuries
DATE ISSUED: 1993
CARTOGRAPHER: Sandra M.E. Leiva, 1991, Robert R. Argola, 1993, RESEARCH HISTORIAN: Elizabeth Barthold
PUBLISHER: Historic American Buildings Survey, National Park Service, Washington, D.C.
Ink on mylar, 24 x 36 in
Library of Congress, Prints and Photographs Division, DC-668 Sheets 27, 28, and 29 of 32

President George Washington designated seventeen parcels of land for public parks and federal spaces, or "reservations," on March 30, 1791. (The text is attached to the Dermott Map facsimile.) Over 300 reservations were first identified and numbered in 1884. Major parks and traffic circles are typically found at the intersections of broad avenues with major streets. Irregular-shaped parks and triangles are formed where avenues intersect with the grid of numbered and lettered streets. Avenues intersect at almost thirty sites throughout the "Historic City," which is bound by Florida Avenue on the north and the Potomac and Anacostia Rivers on the south and east. This area, developed true to the L'Enfant Plan, was largely preserved by the regulation of street widths and the limitation of building heights, although with a revision in 1910.

The vertical and horizontal streets occasionally shift in longitude or latitude where they intersect with the diagonal avenues, thereby creating rectangular open spaces. Because the reservations fall within road right-of-way, many have been altered for transportation needs, and some have been eliminated altogether. However, the majority of the historic elements of the 200-year-old city plan remain in place.

Major circular or rectangular intersections were landscaped as picturesque parks in the nineteenth century and later redesigned according to the prevailing fashion in the twentieth century. Preservation of the city's reservations was due to continuous federal ownership—the result of the first land transaction with the original proprietors, as overseen by President Washington and Thomas Jefferson in 1791—especially that of the National Park Service. In 1867, jurisdiction of all city parks was transferred from the Department of Interior to the Office of Public Buildings and Grounds, overseen by the US Army Corps of Engineers. From 1871 to 1874, miles of streets were paved and 60,000 trees were planted. In 1933 this responsibility was transferred to the National Park Service. Today, the D.C. Government maintains many reservations, particularly the triangular spaces at the intersections of diagonals and grids, excepting those maintained by private developers who own adjacent parcels.

In the nineteenth century, statuary, fountains, wrought-iron fencing and gates, lighting, and gatehouses became common elements within the parks and reservations. The gatehouse keepers protected the elaborate parks from vandalism. A plethora of equestrian Civil War statuary, typically reflecting for whom each park was dedicated, was criticized by the Senate Park (McMillan) Commission. These sculptures, which have seldom been treated as a part of the design, have been inserted as independent objects valued mainly for their historic or memorial qualities or sometimes for their individual beauty, regardless of the effect on their surroundings.

The City Beautiful Movement, concurrently with the proposed redesign of the park system by the Senate Park Commission in 1901-02, brought new ideas about park design. Parks with curved paths, dense clumps of trees, elaborate flower beds, and exotic plants were replaced with axially symmetrical paths, expansive sodded green panels and sparse plantings of native species, as exemplified by Adolphe Alphand's classical beaux arts redesign of public space in late-nineteenth-century Paris.

Many reservations were re-landscaped in the 1920s and 1930s, through the Depression Era Works Progress Administration.

FRANKLIN SQUARE

Franklin Square is located between I and K, and 13th and 14th Streets, NW. Allocated as Reservation Number 9, it consists of 174,417.5 square feet of open space. This square differs from most of the city's reservations because it is not located within an avenue right-of-way, nor was it part of the original 1791 purchase of property overseen by President Washington and Thomas Jefferson. The rectangular site, originally designated for private development, was low and marshy, containing several natural springs. The property was purchased in 1830 for its freshwater springs; they were to be tapped to provide for the needs of the White House and Treasury Building, and later the State, War, and Navy Building (now the Old Executive Office Building). During the Civil War this water source was most important.

When the wood pipes connecting the spring and White House were repaired in 1871, a central fountain was installed. It was completed in 1873 by engineers who ran pipes from the park to the Potomac River. The fountain's French jets, with ornamental polished Aberdeen granite cop-

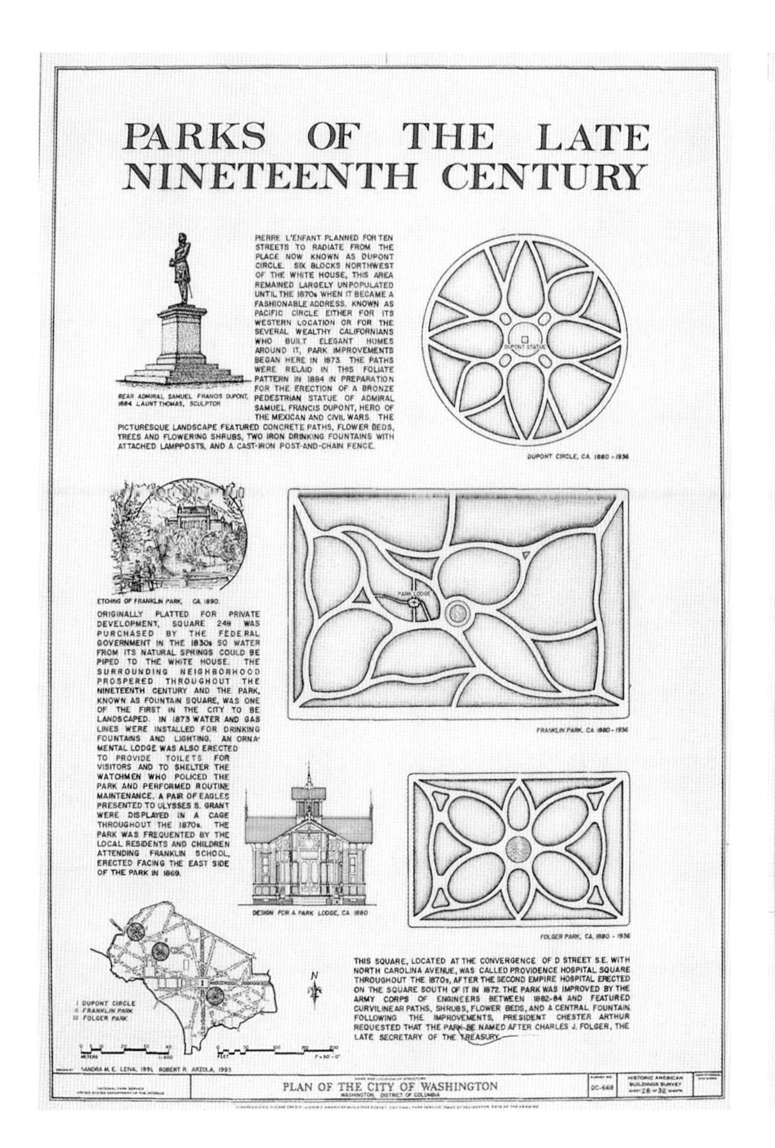

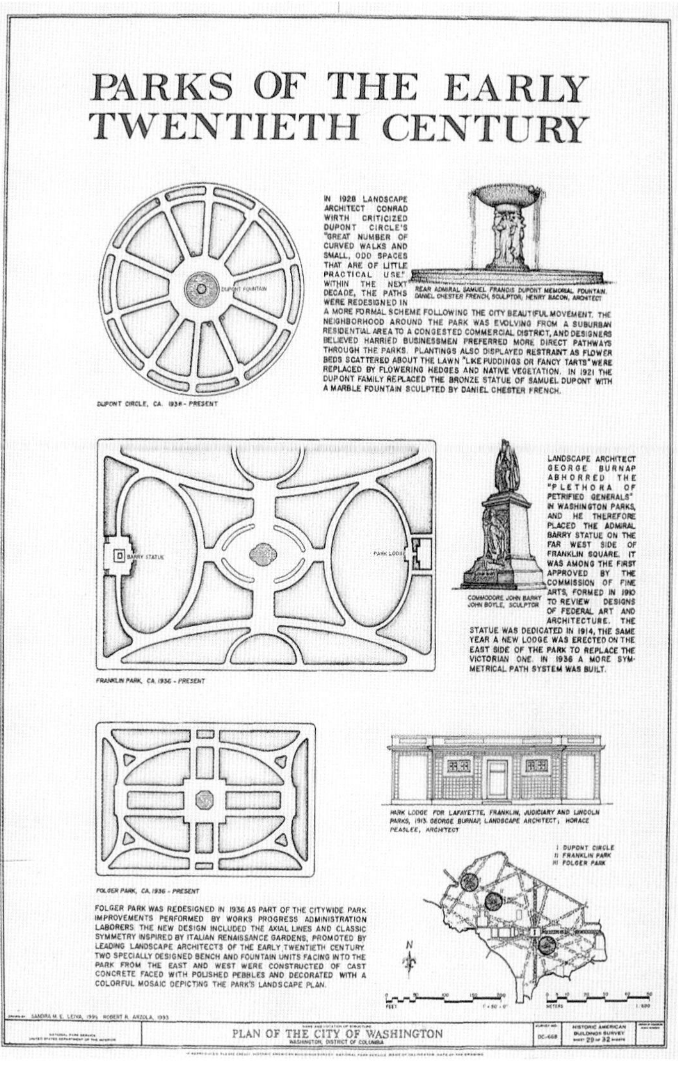

ing, soon became clogged by eels and small fish that swam through the pipes from the river. The problem was corrected by installing an eel trap several yards south of the fountain.

A watchman's lodge was built at the same time about fifty feet west of the fountain. The watchman was responsible for guarding the property during daylight hours. He also fed the American eagles that were kept in a large cage in the park, a tradition that began when one of the birds was sent as a gift to President Ulysses Grant in 1873. The square, which was in the center of a residential neighborhood, was formerly gated with iron fencing similar to squares in London during that period. In the early years, animals could be found grazing within the fenced square.

In 1906, the park was selected as the site for a statue honoring Irish immigrant John Barry, the first American naval officer to capture an enemy vessel in the Revolutionary War. Sculptor John J. Boyle and architect Edward P. Casey designed the eight-foot-tall bronze portrait statue of Com. Barry. The statue, which stands on a marble pedestal with a carved female figure representing Victory, was erected on the west side of Franklin Square in 1914. At the same time, a new lodge identical to those in Lincoln, Judiciary, and Lafayette parks was erected on the east side of the park, and was subsequently removed in 1974.

Franklin Square, like others that had fallen into disrepair, was one of many reservations restored by the Works Progress Administration. In 1936, a new design was installed with symmetrical curved paths and a central terrace of flagstone surrounded by a classical stone balustrade, which was removed in the 1990s. A round central fountain was replaced with one that was quatrefoil shaped with two vertical water jets; it remains to this day. New twelve-foot wide walks were designed to conform to the topography and respect existing specimen trees. This twentieth-century plan, of approximately 4.8 acres, is largely intact.

LINCOLN PARK (NOT SHOWN)

On L'Enfant's plan (p. 34), Lincoln Park is indicated as a rectangular-shaped open area marked with the letter "B," and formed by the convergence of East Capitol Street with

Massachusetts, Kentucky, Tennessee, and North Carolina Avenues between 11th and 13th Streets. Accompanying references describe it as a prominent site for "an historic column—also intended for a Mile or Itinerary Column, from whose station, (a mile from the Federal House) all distances of places throughout the Continent are to be calculated." Set aside as a large rectangle on the Ellicott Plan, Lincoln Park, or Reservation Number 14, has been landscaped as a park since the 1870s and has become significant for its association with black history and the Civil Rights Movement.

Until the Civil War, the surrounding neighborhood was sparsely inhabited. During the War, it became the site of a Union hospital, which soldiers sardonically called Lincoln Hospital in honor of the commander-in-chief. It was officially designated as Lincoln Square by an 1866 act of Congress to memorialize the leader who was assassinated on April 14, 1865.

Improvements to the park were initiated in 1871 and nearly completed in 1875. The organic pattern of curving walks, woodland paths, oval grass panels, and flowerbeds reflected Andrew Jackson Downing's 1851 plan for The Mall (p. 78), and the prevailing romantic ideology about public space. Two ornamental fountains with spray jets were placed at the north and south entrances. A central mound was created for a colossal statue or an historical column. In 1874 a lodge was erected several yards southwest of the mound.

On April 14, 1876, the eleventh anniversary of Lincoln's assassination, a bronze statue of Abraham Lincoln freeing a slave was dedicated. It was placed toward the west-end, facing the Capitol. Called "Emancipation," the statue was designed by sculptor Thomas Ball and was erected by the Western Sanitary Commission, with money contributed solely by freed slaves. Charlotte Scott, a freed slave from Virginia, conceived the idea of honoring the assassinated emancipator with a statue; she donated the first five dollars. The reservation became a popular tourist attraction and came to be known as Freedom Park.

Like many other federal parks and reservations, Lincoln Square was neglected and targeted for rehabilitation by the Works Progress Administration. The concepts of the Senate Park Commission influenced the 1931 design. Old walks were removed, and new linear concrete walkways provided direct access from the surrounding streets to the park.

This plan remained intact until 1973 when it was redesigned to incorporate a bronze statue group honoring educator and social reformer Mary McLeod Bethune (1875-1955). The National Council of Negro Women donated the statue sculpted by Robert Berks. President Lincoln's statue was rotated to face Bethune. Between the two statues a large open green was created. The park has become popular with neighbors for dog walking, jogging and other activities.

DUPONT CIRCLE

On L'Enfant's plan, the area now designated as Dupont Circle, Reservation Number 60, was indicated as a large, amorphous open area formed by the convergence of Connecticut, New Hampshire, and Massachusetts Avenues with 19th and P Streets, NW. Drawn as a circle on Ellicott's plan, it served as a park since its first improvement in 1873. Previously, the park was known as Pacific Circle—said to be named for a group of real estate speculators known as the California Syndicate, who had purchased a large tract of land nearby. In 1884, the reservation was renamed in honor of the naval hero of the Mexican and American Civil Wars, Admiral Samuel Francis Dupont, and a bronze statue of Dupont was dedicated in the center of the park.

Dupont Circle's open space became a gathering place as early as the nineteenth century for homosexuals, intellectuals, and an international community. As more classical homes were built, the park itself was updated along classical lines. Descendants of Admiral Dupont petitioned to move the Victorian bronze statue to Wilmington, Delaware. The statue was replaced in 1921 with a classically inspired marble fountain designed by sculptor Daniel Chester French and architect Henry Bacon. The fountain consisted of a high bowl on a pedestal, centered in a round pool with a wide base that doubled as seating. On sunny days, sparkling water falls from three projecting, gargoyle-like spouts. Niches in the pedestal house a trio of 8.5-foot-tall allegorical figures representing ocean navigation: sea, wind, and stars. The park was redesigned to accommodate the fountain replacing the earlier informal organic, curved paths.

Three concentric walks encircle Reservation Number 60, one abutting the roadway, another about 15-feet inside, and the third surrounding the central fountain creating a plaza. Six straight concrete walkways lead to the fountain, forming a radial spoke pattern on axis with the adjacent streets. Shade trees create a cooling microclimate in summer for benches, which follow the contours of the outer circular walks. Concrete chess tables with stools on the inner walkway are popular gathering spots. Among the circles and squares, Dupont Circle has been one of the most popular gathering places of the twentieth century.

Within Dupont Circle are three federal reservations: one large circle and two smaller quadrilaterals. Reservation Number 60 is an approximately 2.3-acre 80-foot diameter circle; Reservation Number 59 is a quadrilateral of about 5,900 square feet; and Reservation Number 61 is a quadrilateral approximately 73,000 square feet. A 1930 brick lodge with a hipped roof remains in Reservation Number 59.

FOLGER PARK

At the intersection of North Carolina Avenue and Second and Third Streets, SE, Folger Park was part of a large tract of land once owned by Daniel Carroll that reached as far north as K Street, encompassing the Capitol Grounds and extending south to the Anacostia River. Due to the slow development in the city's eastern sector, this open space, otherwise known as Reservation Number 16, remained vacant and unimproved until 1884—though the park had been envisioned almost a century earlier, and was included in Andrew Ellicott's plan of the city.

By 1876, Folger Park was referred to as Providence Hospital Square, after the hospital of the same name erected in 1872. In 1884, President Chester Arthur proposed renaming the park after Charles J. Folger, the late Secretary of the Treasury. An informal design showed two symmetrical, curved paths with a central space for a round fountain.

A new formal design was installed during the Depression in 1936 by the Works Progress Administration, consisting of a cross with flowerbeds inscribed in an oval. A square flagstone patio surrounded an octagonal pool. At the east and west sides, large bench and fountain structures were faced with pebble mosaics depicting the landscape design of the park. This design remains largely intact today.

[Note: A large portion of this information is directly quoted from HABS Survey No. DC-668, DC-669, DC-672, DC-673, DC677]

NEW IMAGE FOR NATION'S MAIN STREET: REAL AND NEAR MISTAKES IN THE AGE OF URBAN RENEWAL AND MONUMENTALISM

The Pennsylvania Avenue Development Corporation Plans

TITLE: Unrealized Plan for Pennsylvania Avenue, Proposal for National Square
DATE DEPICTED: Future projections
DATE ISSUED: 1964
CARTOGRAPHER:
PUBLISHER: Pennsylvania Avenue Development Corporation
Title: Slides, formerly ink on mylar, location of original unknown.
Book: The Pennsylvania Avenue Plan 1974, *Office of Portfolio Management, GSA National Capital Region, Washington, D.C. 1974 (1997)*

ESSAY BY ROBERT L. MILLER

In the 1950s, an apparent epidemic of urban decay—and the then preferred antidote, urban renewal—hit Washington as it did other East Coast cities. Earlier D.C. plans had tackled pockets of slum housing. Nevertheless, the wholesale "blight" of established neighborhoods, downtown and beyond, was pressing.

Many planners favored radical surgery. The only tools available seemed to be "slum clearance"—replaced by public housing and/or private and public redevelopment. In S.W. Washington, perceived as a relative backwater with genuine health hazards, federal and city authorities led by the Redevelopment Land Agency bulldozed over 9000 dwellings on 550 acres, including much of L'Enfant's original street grid. Between 1954 and 1960 some 15,000 poor (both white and black residents), plus businesses, were relocated as part of the single largest land acquisition ever carried out by the federal government.

By 1970, most of Old Southwest's traditional neighborhoods had disappeared forever. Replacing them were modernist super blocks of handsome and livable (but less affordable) low- and mid-rise housing, plus commercial developments, churches and civic facilities arrayed in suburban-style isolation. Ironically by today's planning standards, walkable old row house blocks were leveled while barracks-like housing projects were carefully preserved—a choice that still rankles as Southwest strives to become a safe, diverse, mixed-income community. (See Maps: Olsen 1922, p. 114; Passonneau 1999. p. 156)

Southwest's transformation was not D.C.'s sole example of wholesale clearance and rebuilding, as a glance at Independence Avenue, Foggy Bottom east of the Kennedy Center, or public housing around the District shows. Often spurred by a perceived need to push expressways through the central city, urban renewal planners stamped their technological, anti-congestion aesthetic on the federal city's edges as pervasively as the Senate Park (McMillan) Commission had imposed its "City Beautiful" vision on the core.

A somewhat different approach was clearly called for, however, to curb blight on the "Nation's Main Street," Pennsylvania Avenue, linking the Capitol and White House along a symbolic corridor containing the National Archives, the Treasury and other major institutions and monuments.

The sad state of Pennsylvania Avenue on John F. Kennedy's 1961 inaugural day—by one account only four businesses still functioned along the historic parade route—sparked a motorcade conversation that led to a series of presidential advisory councils and master plans. These in turn inspired a quasi-federal development entity with the mayor and the City Council representatives on its board: The Pennsylvania Avenue Development Corporation (PADC), launched by Congress and President Nixon in 1972. Nathaniel Owings, the first chair, and Senator Daniel Patrick Moynihan were instrumental in the formation of the original Temporary Commission on Pennsylvania Avenue, which launched the beautification effort.

Two pre-1972 advisory commission schemes, one reproduced here, represent ambitious, monumental master plans, typical of over-scaled urbanist mentality of the period; they were designed to attract supporters. In contrast, PADC's 1988 map and the ultimate result reflect a pragmatic, twenty-six year process (ending in 1998 with the opening of the Ronald Reagan International Trade Center and PADC's dissolution) of actually redeveloping individual sites—a retroactive vision, in effect, combining the ideal and the doable. This effort also represented clearance and redevelopment with some rehabilitation. (See completed concept on Passonneau Map, p. 156)

Like their Senate Park Commission predecessors, Pennsylvania Avenue's renewers set forth to reconcile L'Enfant's magnificent intentions with inconvenient reality. The original 160-foot width still gracefully accommodated both parades and rush hours. But other existing conditions presented an odd collage: miscellaneous buildings from a raffish Victorian past, the Federal Triangle's long stolid

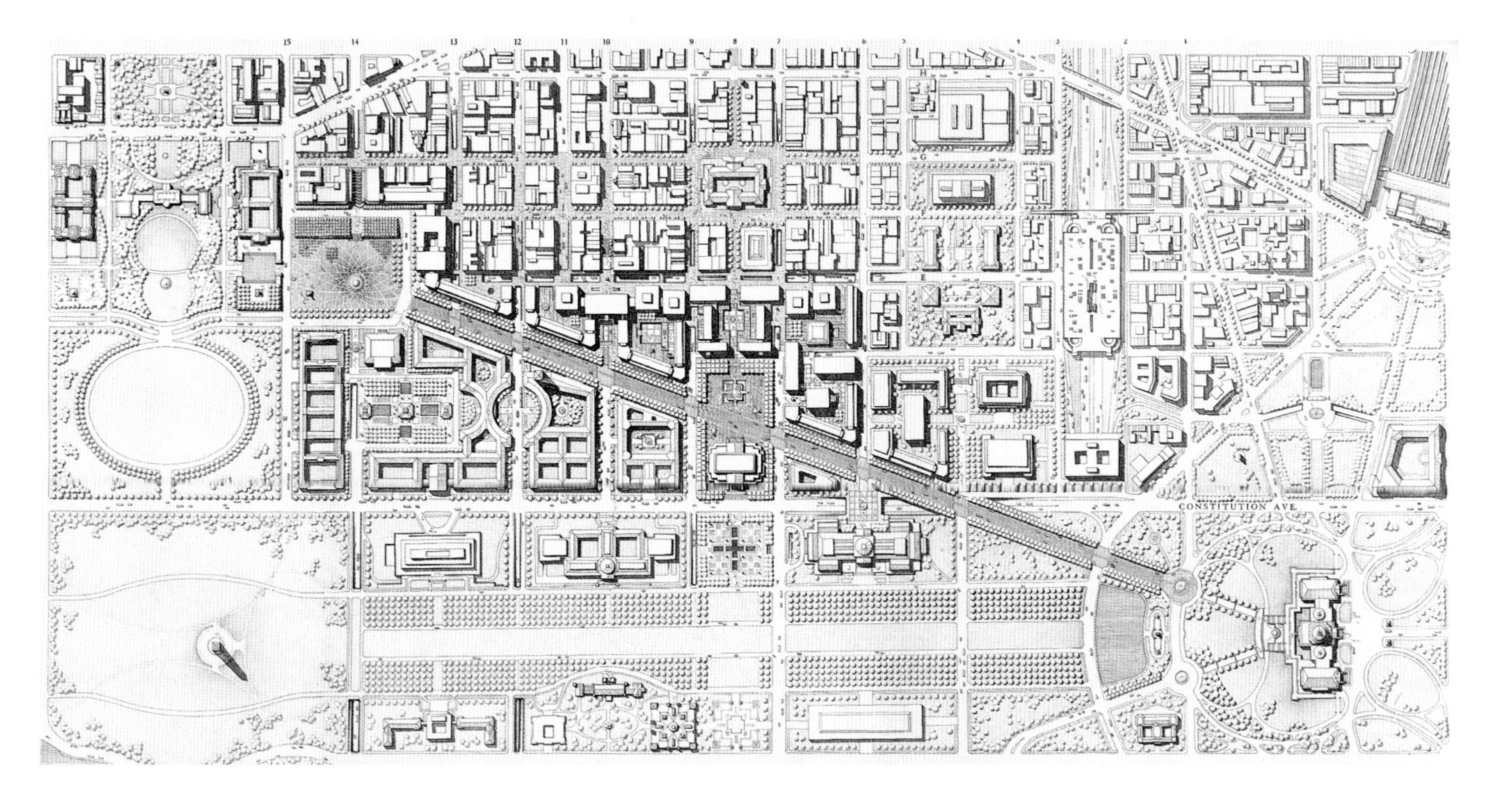

wall, the street plan's ubiquitous little triangles.

The earliest, pre-PADC concepts (1964-1968) were as sweeping as those conceived for Southwest. Fronting the Treasury and rivaling Red Square in Moscow in size, a proposed Great Plaza required (even in later, scaled down versions) demolishing the Willard Hotel and other landmarks. A housing mega-structure, projected for 8th Street, covered four blocks with a hollow square, concealing a kind of hemispherical Italian or pueblo hill town. Today the J. Edgar Hoover FBI Building provides a lone souvenir of the buildings proposed for the Avenue's north side: uniform big-box structure, deeply set back, and devoid of street level activity. Complete clearance of the Avenue with a 50-foot setback began on the north. Then, the 1966 Historic Preservation Act, a national enactment, identified several local buildings to be renovated. This was accomplished and remained in place, thereby changing the emphasis of the plan.

By the time of PADC's arrival in the mid '70s, however, development practicalities, plus the preservation movement and Postmodern architecture's rediscovery of history, all favored a more characteristically-Washington approach of incremental change and compromise.

Campaigns to save the Old Post Office, Willard Hotel, Apex Building, Evening Star Building and others brought preservationists and developers together in creative reuse schemes. Block by block, the results were often splendid—with new paving, street furniture and trees adding visual unity. As a whole, however, the Avenue's collection of setbacks, alignments, styles and materials surely disappointed those who dreamed of a Champs-Elysees. Nor did PADC's acceptance of big new office buildings please dedicated preservationists who admired the lively, quirky commercial street of the old photographs.

Still, PADC often succeeded in urban design as well as development, as seen in several well-used new urban spaces. John Marshall Park, an outdoor open space setting at the eastern end of the Avenue, was created from closing a L'Enfant Street to traffic at 4th Street.

Between 13th and 14th Streets, Freedom Plaza, first called Western Plaza, evolved as a way to smooth traffic patterns, while retaining the idea of a large, open square as a substitute termination for the Avenue (since the Treasury blocks the view to the White House). Originally, the terminus idea was to have been reinforced by two tall, slab-like pylons, designed to frame the Treasury's portico as one came down the Avenue. Remaining from the original Pop Art scheme (which had also included automobile-size models of the Capitol and White House) is the enormous abstraction of the plan of Pennsylvania Avenue inlaid in the granite paving, incised with quotes about the city.

Beyond Freedom Plaza a smaller, contrasting new space, Pershing Park, between 14th and 15th Streets, helps make the transition to the leafy surroundings of the Treasury and White House. Its green perimeter berms open to quiet seating around a pool/ice rink.

Market Square, it has been pointed out, is neither a market nor square. The name has historic roots, however, and the Eighth Street cross-axis it marks has been important in Washington plans since L'Enfant. The plan, however, originally in rectangular form, intended to recreate Market Square in its historic context as a market place for retail, office, and housing. The hemicycle form, embracing both the circular Navy Memorial (another plaza-map, this time of the oceans) and the view of the National Archives across the street, derives from a 1982 American Institute of Architects/Smithsonian Institution Urban Design Charrette—Jean Paul Carlhian and Allan Greenberg, team leaders. Today twin, colonnaded office/apartment buildings reflect the symmetrical and somewhat over-scaled postmodern classicism of the original concept. Market Square, together with the Archives and a vista of the National Portrait Gallery, completes one of the city's few unified, architecturally-framed public outdoor spaces.

FEDERAL AND DISTRICT DOMAIN: COMPREHENSIVE PLANS

The National Capital Planning Commission Plan, Federal Elements, District Elements, and The D.C. Office of Planning Plan

TITLE 1: "Locations for Federal Facilities in the District of Columbia and Immediate Environs, Diagram 2," p. 60, Federal Elements
TITLE 2: "Special Streets and Places, Diagram 1," p. 299, Federal Elements
DATE DEPICTED: Title 1: Future projections, 20 years; Title 2: 1984
DATE ISSUED: Title 1: 1982-1984 Title 2: 1983, Dec. revisions Title 3: 1984, 1985, adopted
CARTOGRAPHER:
PUBLISHER: National Capital Planning Commission
Title 1: NCPC Map File No.61.00/100.00-30126; Title 2: NCPC Map File No. 08.20/100.00-29588
Comprehensive Plan for the National Capital: Federal Elements, *National Capital Planning Commission, Washington, D.C. 1984*

TITLE: District Communities: Place Names
DATE DEPICTED: 1967
DATE ISSUED: 1967
CARTOGRAPHER: Unknown
PUBLISHER: National Capital Planning Commission
The Proposed Comprehensive Plan for the National Capital, *National Capital Planning Commission, U.S. Gov't Printing Office, Washington, D.C., Feb. 1967*

TITLE: District of Columbia Generalized Land Use Map
DATE DEPICTED: May 1995
DATE ISSUED: May 1995
CARTOGRAPHER: District of Columbia Office of Planning
PUBLISHER: ADC of Alexandria, Inc. for Comprehensive Plan Publications
Colored lithograph , scale [1:12,000], 101.2 x 142 cm, 40 x 56 in
Office of Planning, "Comprehensive Plan for the National Capital: District Elements, Title 10," D.C. Municipal Regulations

The L'Enfant Plan of 1791 established the location of the Capitol, White House and the system of avenues, streets and parks in the innermost part of the city. This plan promised to provide a grand setting for the "Federal Establishment." Little was done to improve upon or implement this plan until after the Civil War. Although the 1851 Downing Plan for The Mall and Ellipse focused on the monumental core, the 1902 Senate Park Commission (McMillan) Plan was the first to revisit the original concept of the City of Washington. A product of the "City Beautiful" philosophy, in keeping with the L'Enfant Plan, this plan proposed the completion of The Mall and development of Pennsylvania Avenue between the Capitol and White House for public buildings. In addition, it stressed extending and creating new parks throughout the city.

Other special-purpose plans followed, such as the Zoning Regulations of 1920 and a new park plan in 1924 under the National Capital Park Commission. The National Capital Park and Planning Commission, reconstituted in 1926, however, initiated a comprehensive approach to planning. Its objectives included a public housing program, elimination of alley dwellings, initiation of a system of major highways and parkways, and extension of the park system beyond the boundaries of the District of Columbia. The Capper-Cramton Act of 1930 enabled the commission to purchase land for George Washington Memorial Parkway, for stream valley parks in suburban Maryland, and for parks and playgrounds within the District of Columbia.

Rapid, unanticipated growth of federal employment and regional population during the Depression and World War II soon demonstrated a need for continuous planning for the nation's capital. A new 1950 plan assumed that the region's population would reach 2 million (an increase of 600,000) by 1980. However, growth of the Federal Government had been underestimated. Expansion accelerated to make Washington one of the fastest growing regions in the nation. Population spurted to 2 million in one decade.

The National Capital Planning Commission (NCPC) was further reconstituted in 1952 as the city and federal planning and oversight agency. In that year the comprehensive regional planning function was assigned to the National Capital Regional Planning Council. In 1961, NCPC published "The-Year-2000 Policies" for the City of Washington and its surrounding environs. It addressed housing and other city issues that were not anticipated in the 1950s.

When Home Rule was established by Congress in 1974, the comprehensive planning

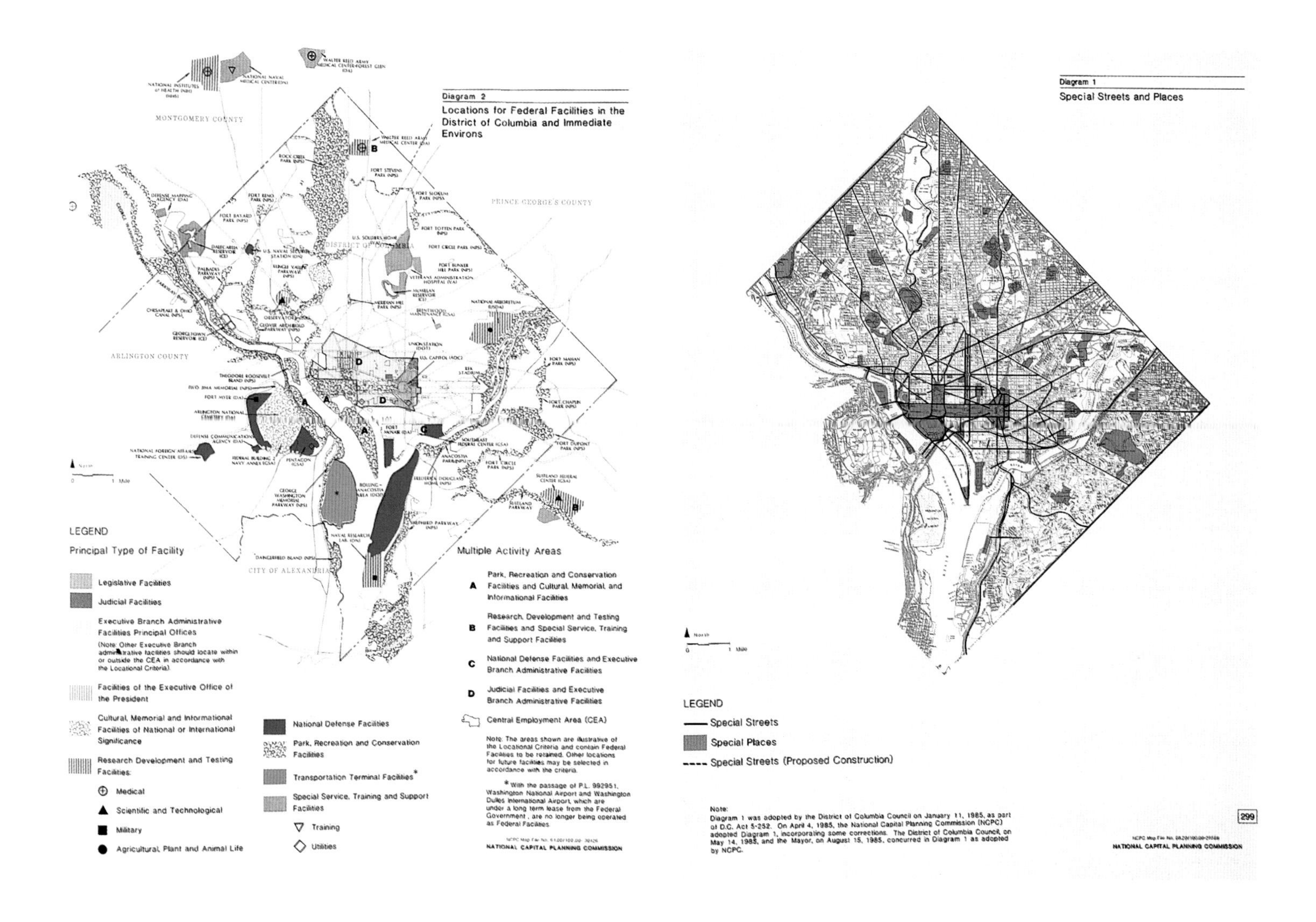

function was divided into Federal Elements and District Elements. The "Federal Elements" portion of the plan addressed the planned growth of federal facilities, international institutions and national interests in the form of federal structures and federally controlled lands, which are documented in its 1995 publication. The "District Elements" concentrated on employment, transportation, land use, economic development, and historic preservation as effective tools for planned growth. The Federal and District Elements together were intended to provide a long term planning mechanism for the city and federal lands in the region.

In the 1960s, NCPC defined three distinct roles for the city and identified problems. First, Washington was viewed as the "Capital City" of a great nation, seat of the federal government, comprising federal office buildings, monuments, tourists, and many national and international organizations dealing with federal agencies. Second, the commission recognized that the city was home to 550,000 people and incorporated provisions for employment, education, and recreation. Third, Washington, as the central city of a great metropolitan region, was the source of business, culture, and other functions for a population much larger than its own. Transportation and accessibility were to play a major role for suburban residents.

Neighborhoods, once known as "in-town suburbs," are differentiated from each other by name, age, architectural vernacular and spacing of buildings and streets, socioeconomic composition or ethnic background of its residents, topographical and landscape settings, and the relationships between the community and other major elements of the city's plan. Since its 1967 publication this NCPC "District Communities" map (p. 148) has been the most comprehensive identification of neighborhoods by name. Agreement as to where one community begins and the other ends has never been established; hence the absence of definitive boundary lines.

The commission found more than 15% of the city's housing was "dilapidated, deteriorating, or sound but lacking some or all plumbing facilities." Nearly half the population lived in sections where schools, parks, and other public facilities generally were substandard. Most buildings in the downtown retail core were showing signs of age, and increased competition from suburban centers inhibited replacement and renewal. Most commercial development outside the downtown area was strung out along miles of former streetcar lines, which had enabled development beyond the original L'Enfant city. At this time, however, mass-transit played a limited role within the metropolitan area.

The social fabric of the city consisted of two extremes: one with a high educational level, largely middle-income; the other, with a standard of living well below the acceptable minimum. The commission decided that the plan must address both populations, providing for their disparate needs while striving to create an environment which would continue to attract and retain more middle-income families.

Federal Elements focussed on the design of buildings, streets, freeways, playgrounds and all other urban elements, taking into account the city's three identities: its natural identity, with distinctive topography, climate and ecology; its symbolic features, which are plentiful in a capital city; and finally, its urban identity, expressed in many ways by a century-and-a-half of building that has made Washington unique.

The commission recognized that Washington possessed a great landscape of parks. The Potomac and Anacostia Rivers served as a significant recreational resource and natural setting. However, the recreational potential is still in part unrealized. The Potomac at Georgetown has been only partially developed, but much of the Anacostia shoreline remains underutilized.

Examples of elements first noted in the 1967 Proposed Comprehensive Plan form the basis for the Plan for the National Capital. The topography of hills, flats and man-made in-fill areas create a "bowl" effect with the federal city in the center. Within this setting are buildings housing the principal national and city governments' functions. As in L'Enfant's concept of the Federal Establishment, the Comprehensive Plan dictates locational criteria for federal facilities. The generalized Land Use Map indicates various facilities by region. Legislative, Judicial, and Executive Facilities are marked with shades of gray and are primarily situated on either side of The Mall extending from the Capitol along Pennsylvania Avenue, past the White House on the west. Major memorials and monuments are concentrated here. A web of parks, recreation areas, and street greenery unify the urban setting. Central to the overall composition is The Mall, the city's greatest open space. "Parks, Recreation and Conservation Facilities" are noted with blue, while gray indicates "Cultural, Memorial and Informational Facilities of National or International Significance."

Criteria exist for international and cultural facilities—such as Chanceries and the offices of Foreign Missions (privately owned and used for diplomatic, consular or other governmental activities) found throughout many of Washington's neighborhoods. "Permitted Chancery Facility Locations" (not shown), the "International Cultural and Trade Center" also known as the Ronald Reagan Building, and the "International Center," which includes a chancery compound.

Streets and avenues provide the framework for buildings and circulation. In an architectural sense, the rhythmic variations of building mass and frequency of intersecting streets are corridors of space that can serve social needs and provide visual ties to other elements of the cityscape. The Federal Elements of NCPC set forth policies to ensure the mainstay of "Special Streets and Places." Protecting, maintaining, and enhancing existing "special places" set in place by the L'Enfant Plan is the primary focus of this aspect of the Federal Elements.

Extending the Legacy, referred to as the "Legacy Framework," published by NCPC in

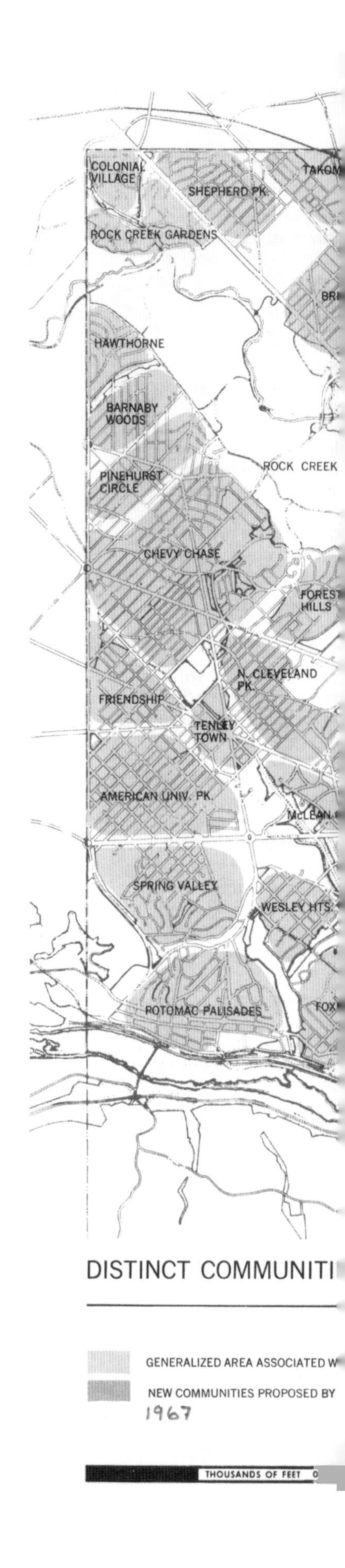

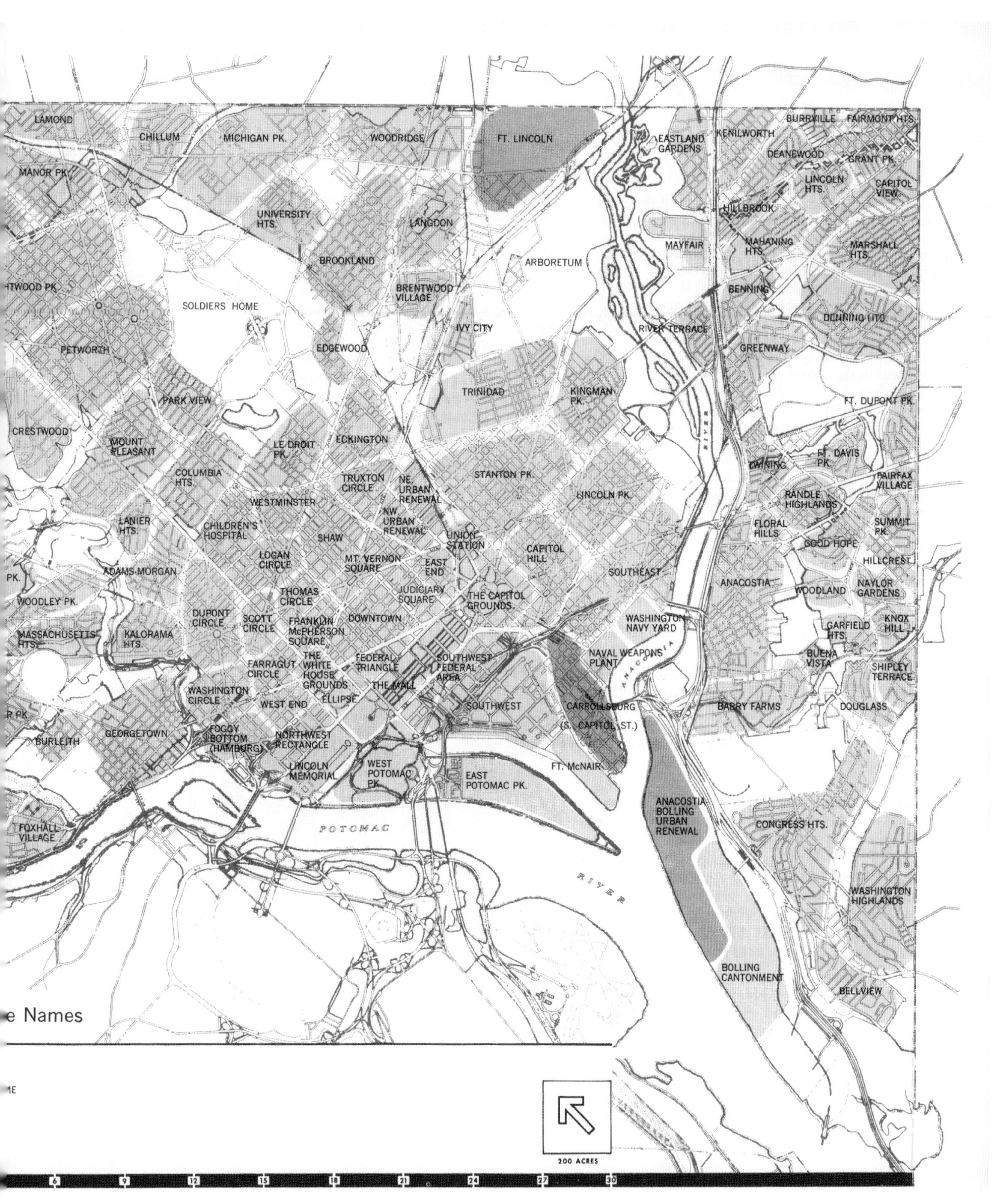
LAMOND
CHILLUM
MICHIGAN PK.
WOODRIDGE
FT. LINCOLN
EASTLAND GARDENS
KENILWORTH
BURRVILLE
FAIRMONT HTS.
DEANEWOOD
GRANT PK.
MANOR PK.
LINCOLN HTS.
CAPITOL VIEW
UNIVERSITY HTS.
LANGDON
HILLBROOK
MAYFAIR
MAHANING HTS.
MARSHALL HTS.
BROOKLAND
ARBORETUM
BRENTWOOD VILLAGE
BENNING
SOLDIERS HOME
IVY CITY
RIVER TERRACE
PETWORTH
EDGEWOOD
GREENWAY
TRINIDAD
KINGMAN PK.
PARK VIEW
FT. DUPONT PK.
CRESTWOOD
MOUNT PLEASANT
LE DROIT PK.
ECKINGTON
RIVER
FT. DAVIS PK.
COLUMBIA HTS.
TRUXTON CIRCLE
NE. URBAN RENEWAL
STANTON PK.
FAIRFAX VILLAGE
WESTMINSTER
LINCOLN PK.
RANDLE HIGHLANDS
NW. URBAN RENEWAL
LANIER HTS.
CHILDREN'S HOSPITAL
SHAW
UNION STATION
FLORAL HILLS
SUMMIT PK.
GOOD HOPE
LOGAN CIRCLE
MT. VERNON SQUARE
EAST END
CAPITOL HILL
HILLCREST
ADAMS-MORGAN
SOUTHEAST
ANACOSTIA
NAYLOR GARDENS
WOODLEY PK.
THOMAS CIRCLE
JUDICIARY SQUARE
THE CAPITOL GROUNDS
WOODLAND
DUPONT CIRCLE
SCOTT CIRCLE
FRANKLIN McPHERSON SQUARE
DOWNTOWN
WASHINGTON NAVY YARD
GARFIELD HTS.
KNOX HILL
MASSACHUSETTS HTS.
KALORAMA HTS.
FARRAGUT CIRCLE
THE WHITE HOUSE GROUNDS
FEDERAL TRIANGLE
SOUTHWEST FEDERAL AREA
NAVAL WEAPONS PLANT
BUENA VISTA
SHIPLEY TERRACE
WASHINGTON CIRCLE
WEST END
ELLIPSE
THE MALL
SOUTHWEST
CARROLLSBURG (S. CAPITOL ST.)
BARRY FARMS
DOUGLASS
BURLEITH
GEORGETOWN
FOGGY BOTTOM (HAMBURG)
NORTHWEST RECTANGLE
LINCOLN MEMORIAL
WEST POTOMAC PK.
EAST POTOMAC PK.
FT. McNAIR
ANACOSTIA BOLLING URBAN RENEWAL
FOXHALL VILLAGE
POTOMAC
CONGRESS HTS.
RIVER
WASHINGTON HIGHLANDS
BOLLING CANTONMENT
BELLVIEW
200 ACRES
6
9
12
15
18
21
24
27
30

1997, "gamely, even bravely tried to re-imagine the whole of Washington as it might develop in the twenty-first century," according to Robert E. Miller. He continues, "The document and elegant graphics reaffirm the L'Enfant and Senate Park Commission plans and principles for a modern, monumental, and also living, prosperous, pollution-free city, with renewed focus centered upon the Capitol, and higlighting the value of planning and urban design." Miller observes: "Imagine Daniel Burnham's famous 'Make no little plans' quote rewritten as 'Make no little frameworks,' and the limits of NCPC's power to stir men's souls become clear."

COMPREHENSIVE PLAN FOR THE NATIONAL CAPITAL: DISTRICT ELEMENTS

Pursuant to the Home Rule Act of 1973, the District of Columbia became responsible for city planning, building onto the L'Enfant Plan legacy of 1791. In 1984-1985 the mayor proposed, and the City Council of the District adopted, the District's Comprehensive Plan (after NCPC reviewed for any adverse impact to the federal interest) and began implementing the local land use elements. The elements have been reviewed, amended and implemented on a periodic basis as recently as 1998 with formal reports submitted to the City Council every two years.

Twelve District Elements compose the 1989 Comprehensive Plan for the National Capital: General Provisions, Economic Development, Housing, Environmental Protection, Transportation, Public Facilities, Urban Design, Preservation and Historic Features, Downtown, Human Services, Land Use, and Ward Plans. Each element incorporates the concepts of the General Land Use Map to provide general guidance for the overall objectives of the Comprehensive Plan. Shown here is the most recent General Land Use Map. Updating has begun for the next edition.

The local elements of the plan entertain several themes, including neighborhood stabilization through maintenance and enhancement of existing housing stock, and enhanced local economic growth by an increased share of regional employment and economic growth. A specific focus on developing and maintaining a living Downtown provides for a mix of land uses, especially a critical mass of residential dwellings to attract and serve a variety of users. The local elements recognize the District's cultural and natural amenities, and seek to preserve and promote these advantages. Additional attention is paid to respecting and improving the physical character of the District and ensuring community input into the implementation process. A consistent emphasis on preserving the historic character of the District can be found throughout the local elements of the Comprehensive Plan (corresponding to the Federal Element on Historic Preservation).

The Home Rule Act established a statutory requirement that zoning "shall not be inconsistent with" the Comprehensive Plan. The Land Use Map and other related illustrative maps are employed to guide in such determinations. Building height limits also contribute to a cohesive urban setting. The Capitol dome, the Washington Monument, National Cathedral's bell tower, dome and spires of the Shrine of the Immaculate Conception, Post Office tower, and Healy Hall's tower at Georgetown University punctuate the horizontal skyline. Each landmark belongs to a distinct neighborhood.

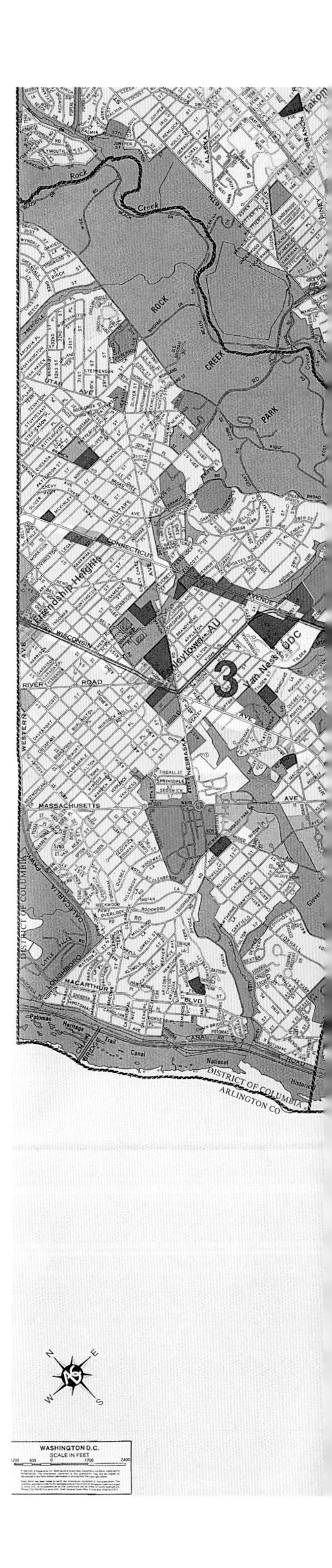

4
5
1
6
7
2
8
POTOMAC
RIVER
McMillan Reservoir
Rhode Island Avenue
Minnesota Avenue
Benning Road
Federal Triangle
The Mall
White House
The Ellipse
Smithsonian
Waterfront
Tidal Basin
Washington Monument
Roosevelt Island
Arlington Memorial Bridge
Washington Channel
East Potomac Park
West Potomac Park
Anacostia
Massachusetts
Wisconsin
South Dakota Ave
Fort Dupont Park
District of Columbia
Virginia

GRAPHIC DESIGN TELLS TRANSPORTATION STORY

The WMATA Metro Plan

TITLE: Metro System Map
DATE DEPICTED: ca. 1990
DATE ISSUED: ca. 1990
CARTOGRAPHER: Unknown
PUBLISHER: Washington Metropolitan Area Transit Authority, Washington, D.C.
Colored map, 84 x 85 cm, on sheet 124 x 85 cm
Library of Congress, Geography and Map Division, C3851.P33 1990 .W35 Oversize

"Quality of life" is enhanced by transit-oriented, livable communities. Recognizing the many advantages of metro transit, legislators first mandated a mass-transit system for Washington, D.C. in 1952 when they passed the National Capital Planning Act. The intent was to create a model unified Metrorail system with a sequence of stations—safe and secure, reliable and clean—with all of the latest modern materials and design.

Stations generally were to be "lookalikes." Platforms of rust-toned terra cotta tile would project warm elegance under foot against unpainted finely detailed repetitive pre-cast concrete coffers of continuous wall-to-ceiling barrel-vaults. The image is a dignified modern transformation of Neoclassical design. Today, works by local artists embellish many stations, through competitions by Art in Public Places sponsored by D.C. Commission on the Arts and Humanities.

Of paramount importance was a legible contemporary graphics and signage system—particularly necessary for visually impaired and non-English speakers. Helvetica typeface, white on black ground was selected. Freestanding rectangular tubes were fabricated for platforms to supplement wall signs. An abstract graphic map (shown here), in primary colors, depicts rail routes and stations—existing and proposed. Schematic maps, produced by WMATA, for walls, brochures, and all metro literature clearly convey metro data.

Construction commenced at Judiciary Square subway station at the end of 1969 and "Metrorail" opened in 1976. More than 51,000 people turned out to test-ride the 4.2-mile, u-shaped Red Line when it opened on March 27. Subway cars carried them through five stations—from Rhode Island Avenue, in N.E. Washington, to Farragut North station, downtown. A year later the Blue Line went into service. It stretched from Washington's Stadium-Armory station across the Potomac River to National Airport in Virginia, now linking to Maryland as well. The Orange Line, which debuted in 1978, links D.C. and Virginia—as does the Yellow Line, which opened in 1983, adding more direct service from Downtown to Reagan National Airport.

The most recent newcomer to the system, the Green Line, opened in 1991. Designed to ford the Anacostia River, this line serves less affluent residents of S.E. Washington—the last to benefit from the transit system. Meanwhile, in prosperous Georgetown, the community successfully lobbied to keep the subway away from its exclusive terrain.

Metrorail's innovative award-winning subway and aboveground railway system radiates out from center city, covering 103 miles, and could be expanded, as in the past. Currently Metrorail has 83 stations, 48 of which are underground. WMATA is a cross-jurisdictional entity serving Maryland, Virginia, and the District of Columbia. It was created in 1966 to oversee the Metrorail and Metrobus transit system. Metrorail stations linked to commuter trains are Virginia's Railway Express and Maryland's MARC Commuter Rail Services.

The Federal Highway Act of 1973 required Metrorail stations to be accessible for disabled riders. Some stations were forced to delay opening until elevators were installed to comply with the Act. Synthesized voice guides and rippled walking and edge surface guides were installed for hearing and sight-impaired.

As ridership increased, hours of operation have likewise been extended. However, city planners and commuters have complained that city-centered Metrorail does not fit suburb-to-suburb commuting needs. Smart-growth advocates have encouraged extension of the system. Studies are underway to develop a new "cross-town" line that might parallel parts of the Beltway ring road. A proposal was released in January 2001 to extend Metrorail (a purple line) with four stations in route to Dulles International Airport in Virginia by 2010-2015. The Metrorail success—of aesthetic and functional consequence—reflects the pulse of modern times. Parallel to the ambitious public works projects of Alexander Shepherd of the 1870s, Metrorail is an expression of the inherent relationship between form and place.

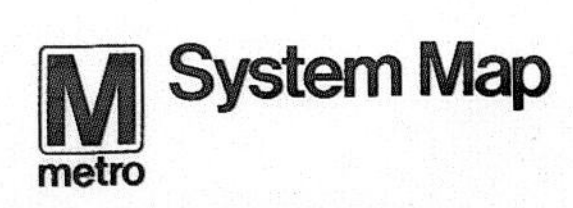

Legend

- Red Line • Glenmont/Shady Grove
- Orange Line • New Carrollton/Vienna
- Blue Line • Addison Road/Franconia-Springfield
- Green Line • Branch Avenue/Greenbelt
- Yellow Line • Huntington/U Street-Cardozo

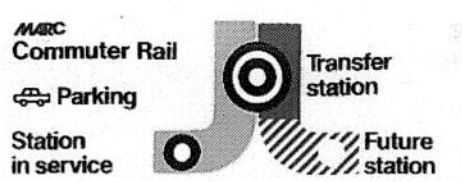

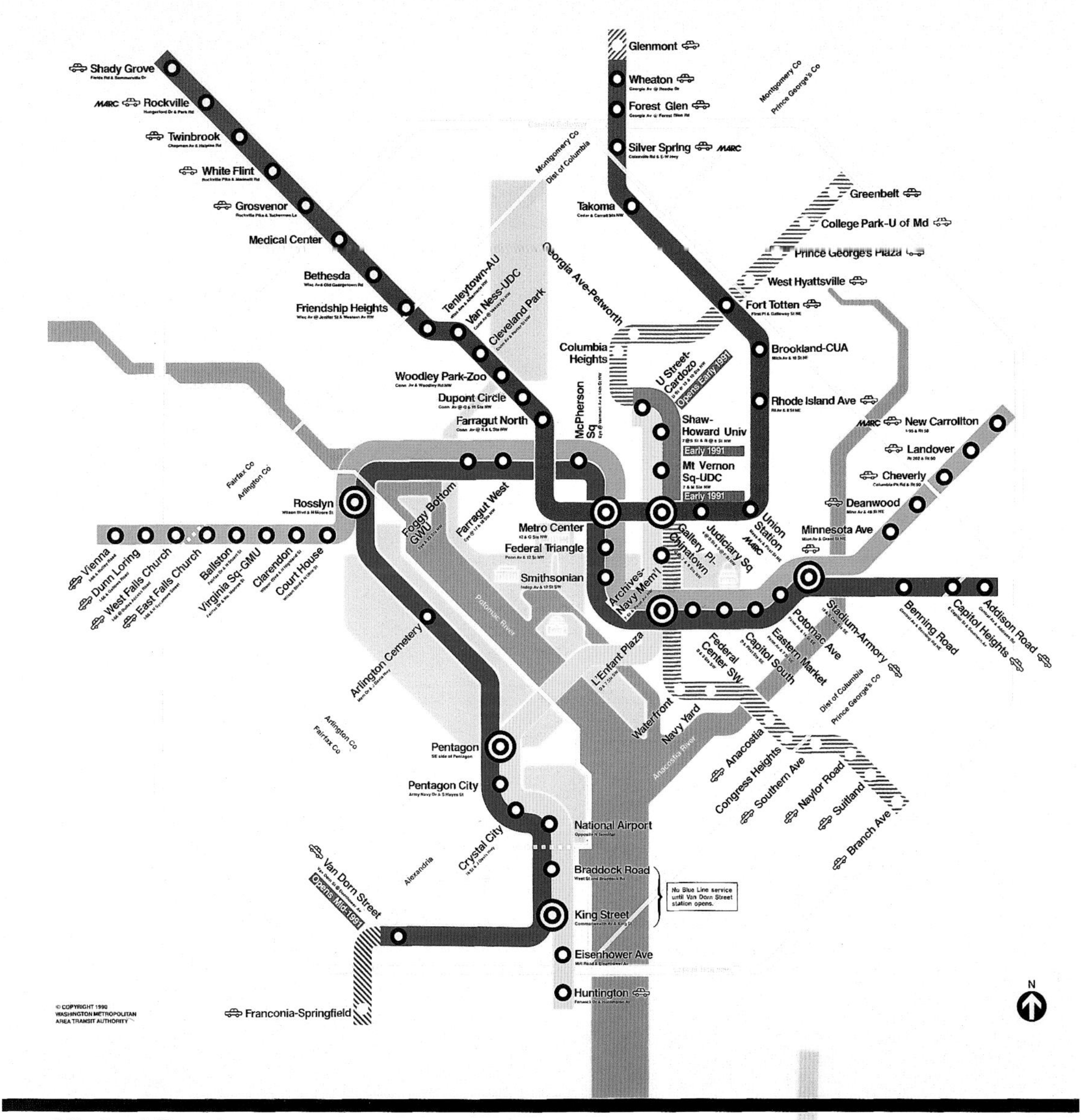

Minutes & Fares to:

Station & Minutes to:	Line	Fares Non-Rush	Fares Rush
Addison Road			
Anacostia			
Archives-Navy Mem'l			
Arlington Cemetery			
Ballston			
Benning Road			
Bethesda			
Braddock Road			
Branch Ave			
Brookland-CUA			
Capitol Heights			
Capitol South			
Cheverly			
Clarendon			
Cleveland Park			
College Park-U of Md			
Columbia Heights			
Congress Heights			
Court House			
Crystal City			
Deanwood			
Dunn Loring			
Dupont Circle			
Eastern Market			
East Falls Church			
Eisenhower Ave			
Farragut North			
Farragut West			
Federal Center SW			
Federal Triangle			
Foggy Bottom-GWU			
Forest Glen			
Fort Totten			
Franconia-Springfield			
Friendship Heights			
Gallery Pl-Chinatown			
Georgia Ave-Petworth			
Glenmont			
Greenbelt			
Grosvenor			
Huntington			
Judiciary Square			
King Street			
Landover			
L'Enfant Plaza			
McPherson Square			
Medical Center			
Metro Center			
Minnesota Ave			
Mt Vernon Sq-UDC			
National Airport			
Navy Yard			
Naylor Road			
New Carrollton			
Pentagon			
Pentagon City			
Potomac Ave			
Prince George's Plaza			
Rhode Island Ave			
Rockville			
Rosslyn			
Shady Grove			
Shaw-Howard Univ			
Silver Spring			
Smithsonian			
Southern Ave			
Stadium-Armory			
Suitland			
Takoma			
Tenleytown-AU			
Twinbrook			
U Street-Cardozo			
Union Station			
Van Dorn Street			
Van Ness-UDC			
Vienna			
Virginia Square-GMU			
Waterfront			
West Falls Church			
West Hyattsville			
Wheaton			
White Flint			
Woodley Park-Zoo			

Rush Hours 5:30 to 9:30 AM & 3:00 to 7:00 PM, Monday to Friday

All other times are Non-Rush

Station Mezzanine System Map

CITY PORTRAYED IN BLACK AND WHITE
The Thadani Figure-Ground Plan

TITLE: The Monumental Core of Washington, DC
DATE DEPICTED: 1991 ©
DATE ISSUED: 1982-1997 ©, 1995 CADD
CARTOGRAPHER: Dhiru Thadani
PUBLISHER: Thadani Hetzel Partnership, Architects ©
Pen and ink on mylar, scanned and auto-vectorized into cadd format, from 18 panels at 15 in x 20 in
Thadani Hetzel Partnership Collection

Figure-ground is a clear visual representation of the relationship between buildings and public space. Against white ground, solid black represents objects—whole blocks or significant buildings. A pattern, or cluster, depicts density of settlement, urban design typology, and may also suggest building scale. Compare the southwest sector of south of The Mall with the 1922 Olsen Map (p. 114) to note the change in building scale resulting from the 1970s urban renewal programs.

Washington's Baroque plan, of an imprecise street grid bisected by diagonal avenues linked to a repetitive sequence of public reservations, is explicit. Unbuilt, undefined land areas and parklands are contrasted against the figural shape of monumental civic space and waterways. This map type, a favorite among urban designers and architecture professors, exposes the designer's intention.

Why are some blocks large while others seem tightly squeezed? Why are the streets not evenly spaced? To the schooled map-reader, two answers stand out. First, the plan derives from the European-Baroque influence of the seventeenth and eighteenth centuries and its French designer, Pierre Charles L'Enfant. It expresses hierarchies in space, use, and symbolism. Second, the plan responds to the undulating topography and other natural features. To walk the streets with map in hand, the subtle level changes are sensed along with the simple grandeur. Thus, the artistry of L'Enfant's whole composition can be found. However, resulting from alterations to the original L'Enfant Plan, roadway connections at intersections of the grid and diagonals are more precise, more "rational." Lost was the inspiring European irregularity of L'Enfant's painterly eye.

This figure-ground map exposes deviations in the urban fabric. A variety of edge-conditions can be seen, such as at Rock Creek, the old Florida Avenue northern boundary, and the Potomac River. The interface between two major transportation networks—the railroad and Interstate 395—is evident.

Prompted by an interest in urbanism and city form, in 1982 the map's authors initiated this project for use with their students and The Catholic University of America. Also, it was to be a resource for the design community. The original idea, to draw a "Nolli" Plan of Washington like that of the revered 1748 Plan of Rome by Jean Baptiste Nolli, drew a host of dedicated weekend volunteers and consultants, who participated by doing research, analyses, verification of existing conditions, and drawing. The first hand-drawn base maps of ink on mylar were prepared at a scale of 1"=200' while the plans of approximately 250 public buildings were drawn at 1'=80', reduced and inserted into the base map. An additional 150 buildings were partially completed.

Seen together, all at the same scale, ideas embodied in the floor plans exposed unanticipated revelations that became the focus of obsessive study. The documentation covered approximately seven miles within the 1791 L'Enfant Plan.

The base data, collected over thirteen years, was scanned and converted in 1995 to a computer-based image. Data was reconfigured as a vector-based computer drawing. The product was a CD-ROM data file that is organized in layers so that the data can be usefully manipulated. It can be seen on the computer screen in color to designate printing output line weight, while the printout image varies in tones of gray and black. Building type, landscape, and text are provided on ten layers each.

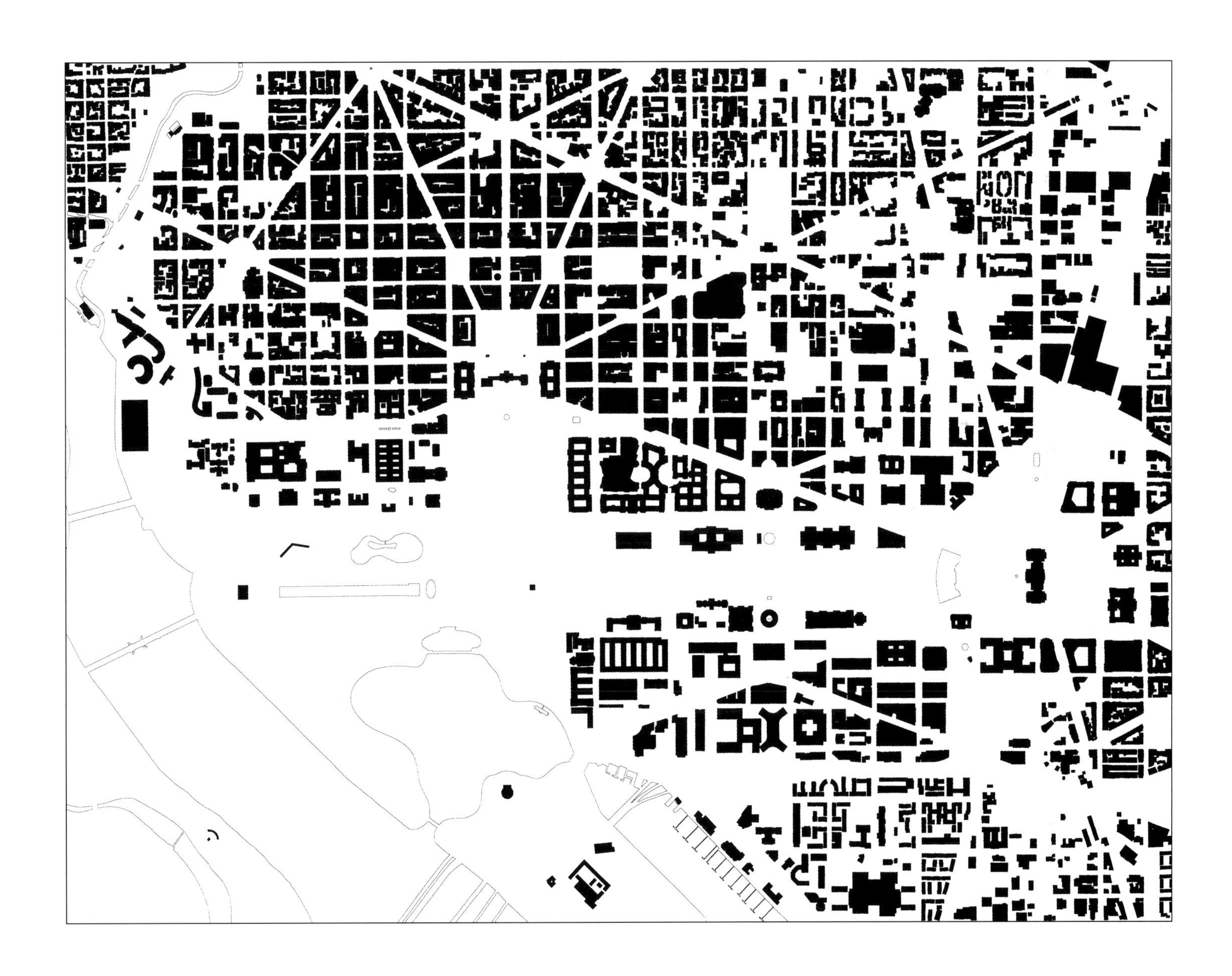

LOGIC OF MIXED-USE AND AXONOMETRIC PROJECTION PLANS

The Passonneau Maps

TITLE 1: Monumental Center of Washington, DC
TITLE 2: The Year 2000 Monumental and Commercial Center of the National Capital and Surrounding Residential Neighborhoods
DATE DEPICTED: 2000
DATE ISSUED: 1999
CARTOGRAPHER: Title 1: Joseph Passonneau, drawn by Tamara King, Elizabeth Blount, Azlina Abu Harsan , and Vitaly Givorkian;
Title 2: Joseph Passonneau, drawn by Vitaly Givorkian and David B. Akopian
PUBLISHER: Joseph Passonneau & Partners
Colored lithograph from ink on mylar, 33.75 x 42.75 in
Joseph Passonneau Collection

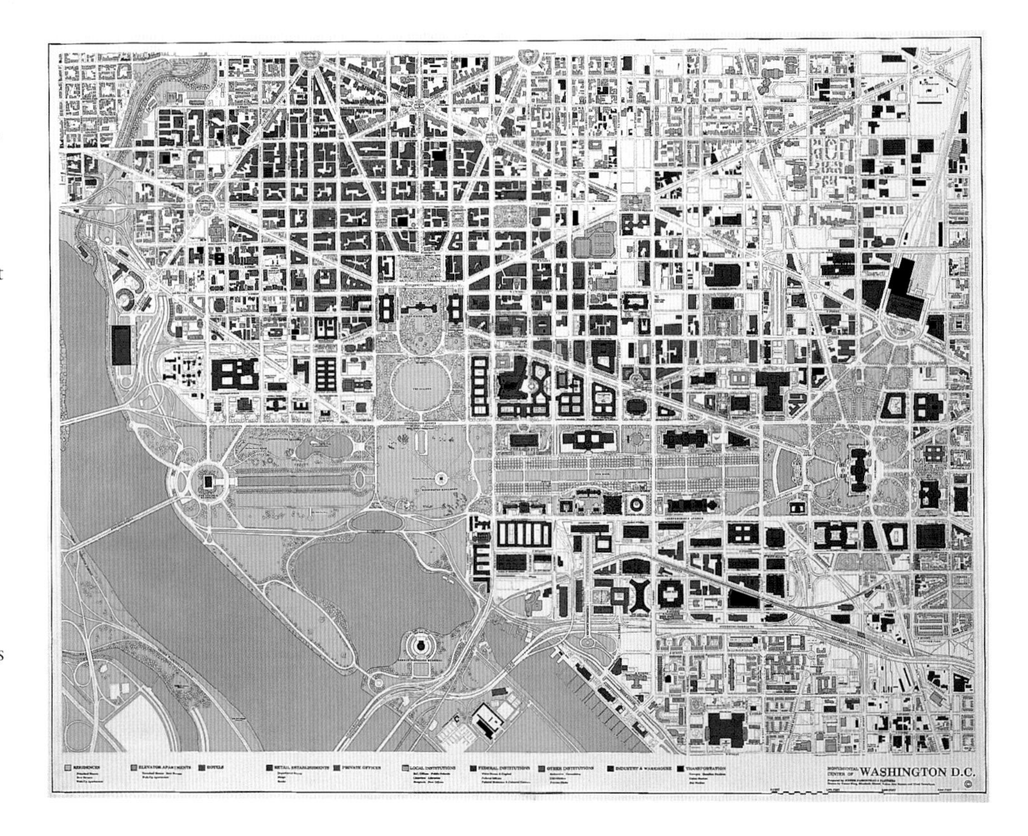

Joseph Passonneau, a dedicated urbanist and prolific mapmaker, has made a significant contribution to our understanding of the dynamics of urban change and the role of maps in weaving history, land uses, and the language of city-building over time. He has depicted the city in its many stages of development, including street closings, traffic studies, building and population density. His name is synonymous with maps in Washington.

Passonneau's 1999 Land-Use Plan is a fascinating graphic composition. Bright primary and secondary colors reveal at a glance urban implications of various clustered uses. This document showcases the impact of change. Poor government decision-making, new technology, economic development and life-style cycles, and suburban sprawl drove out remnants of residential uses from much of the central city, making way for construction of large-scale commercial buildings. In turn, there followed a loss of small-scale retail, individual entrepreneurs, and general decline of the downtown. Recent economic improvements and the perception of inner city vitality have once again created a potential for a living urbanism befitting our national capital as we begin the twenty-first century.

The monumental core's decorous federal presence is contrasted with an overwhelming percentage of office space encircled by a narrow ensemble of ground-level retail concentrated in the downtown. Scattered institutions fill many quarters. Spacious tree-lined streets and avenues, fanning out to facilitate communication and movement, belie the reality of gridlock, especially severe since the closing of the streets around the White House. Paramount, the large expanse of open park space presents a majestic picture of the public realm.

Text and images on the verso with a sequence of dates set the stage to tell a story, mixing commentary and historical fact. The story of a small village in 1800 thrust into the spotlight, building its streets, its public spaces, its neighborhoods and civic structures unfolds along with the drama of human nature. Infrastructure is only partially visible, while architectonic aspects of buildings and monuments impart the quintessential tension

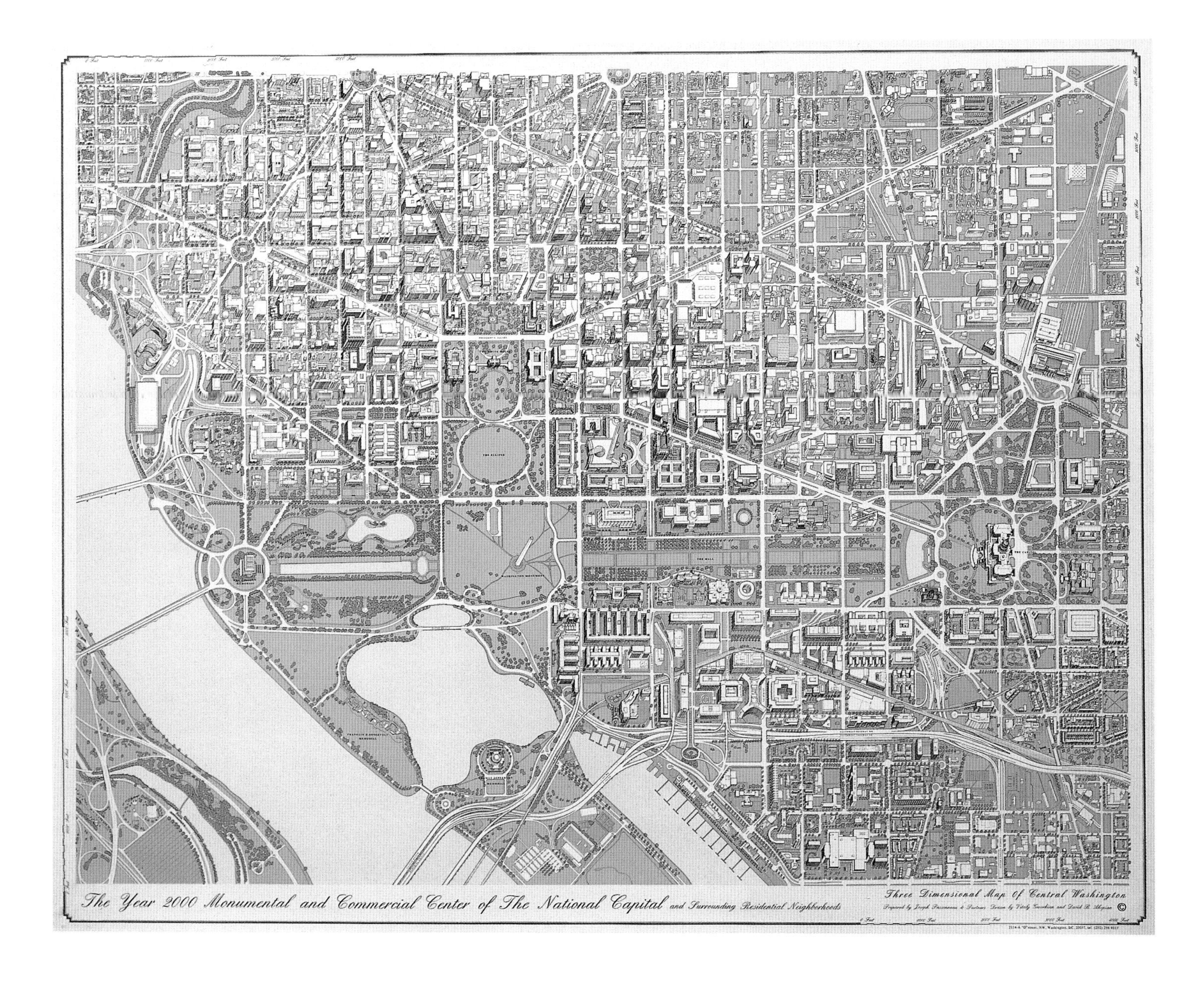

between formal federal imagery and idiosyncratic facets of the neighborhoods.

The reader is engaged in a dialogue about population growth, traffic and transportation, tourism, and transformations of street imagery. In text on his map, Passonneau extols three aspects of the city legacy: Metrorail as an "impressive public works project of this century"; Pennsylvania Avenue, a grand boulevard representing the national "Main Street"; and the Mall, a "wonderfully useful and active public open space."

The spirit of Washington is further captured in another of Passonneau's maps—an axonometric projection (rendered in three dimensions). This map reveals the fundamental gardensque character of Washington, and the balance between public space and buildings. The verso, uncolored, contains names of many buildings. Axonometric maps are first laid out in plan—streets and building footprints. Structures are drawn at an angle to scale, accurately portrayed in size and bulk. These maps attain a quality of reality, somewhat like a birds-eye view. When viewed as a series (comprising six prior periods beginning with 1800), they disclose changes over time in urban form and style. They are dispassionate, allowing us to analyze what is, what was, and what might have been.

URBAN FORM IMBUED WITH VALUE

The Miller Vision

TITLE: Visions of Washington, Composite Plan of Urban Interventions
DATE DEPICTED: Future projections based upon Charrettes, 1980s
DATE ISSUED: 1991
CARTOGRAPHER: Iris Miller, with Carlos Posada, Mariela Corrochano, Chinh Doan; **BASE MAP:** Joseph Passoneau
PUBLISHER: Iris Miller, PRINTER: Williams & Heintz Co., Washington, D.C.; **CALLIGRAPHY:** Sheila Waters
Ink on mylar, assembled photographically 107 x 214 cm, 96 x42 in; Photo lithograph on archival paper, Part A: Plan; Part B: Text, 58 x 102 cm, Library of Congress, Geography and Map Division, C.3851.G45 1991 .M4 Vault Oversize; C.3851.G45.1991 .M5
Also, in Iris Miller Collection and National Capitol Planning Commission
1985 Study, Judiciary Square, Jean Paul Carlhian, Iris Miller Collection
1986 Study, New York Avenue/White House, Roger Lewis, Library of Congress, Geography and Map Division

"The Charrettes," as Urban Design Seminar/Charrettes soon were known, were developed in 1982 for professional continuing education. They were envisioned as a living laboratory to explore strategies for urban enhancement and revitalization. Spearheaded by the Washington, D.C. Chapter/American Institute of Architects, with co-sponsors such as the Smithsonian Institution Resident Associates, and other professional organizations and universities, the Charrettes established a continuous 8-year conceptual record, offering a narrative and framework in which to question and formulate appropriate values of urbanism. Comprising processes for community and government participation, a multi-disciplinary group of citizens and designers garnered ideas for design solutions grounded in social and economic issues. The program sensitized participants toward a humane understanding of local traditions, communal memory, and formal urban relationships that create a "quality of place."

The Charrettes were both pedagogical and practical, assimilating and applying urban theory to actual situations—an innovative vehicle for "brainstorming" in an age of increasingly complex urban parameters imposed upon the design process. In 1983, Cleveland, St. Louis, and Dallas held adapted versions of the Charrettes. The well-publicized programs, with nationally recognized team leaders, became a flexible, transferable model for new and older communities, private offices, university programs, and even for new urbanism.

The design goals of the Washington Charrettes are delineated as a visionary composite map, accompanied by a "gazette" page of explanatory text, reminiscent of Colton's 1855 atlas (p. 82). The gazette portrays the French eighteenth-century milieu—precedents and inspiration—for the L'Enfant's plan. "Washington, D.C., the Place" is described, offering geographic and historical data. The gazette also provides a full description of the charrette process, its objectives, and an accounting of the urban design issues explored for each site during its first eight years.

The text, with philosphical basis, of "Visions of Washington," written in fine calligraphy, is quoted below, and team leaders of the Washington Charrettes are listed by sites and program years.

OBSERVATIONS: I. URBAN SITES in need of refinement were selected because of their prominent civic locations and their present day use. Urban form and aesthetics were juxtaposed with [the reality of] citizen needs and aspirations. THESE "CHARRETTES" were created in Washington from 1982-1989 as intensive educational programs with lectures, town meetings, [site visits], & design teams. Noted professional designers, assisted by area architectural educators, led [multidisciplinary] teams of designers, students, and lay public to generate design concepts. Study sites were often extended to emphasize linkages [connections within or to nearby areas] and continuity.

WITH DRAWINGS as the medium to expound ideas, these action-based seminars offered an opportunity for professionals of varying experience levels to work together [with community members] in a democratic forum to develop urban design strategies & problem-solving methodologies for shaping the Federal City.

II. TO ACHIEVE THIS COMPOSITE PLAN the AUTHOR has selected, elaborated upon, and often recomposed the visions, and occasionally initiated a new design while holding true to the general intentions of "charrette teams." (The term "en charrette" [cart] in architecture originated from the cart on which student design work was collected] at l'Ecole des Beaux Arts in France, referring to intense work associated with completion deadlines for student and professional design projects.) URBAN INTERVENTIONS define public space by providing enclosure and clear edge conditions, linear connections and extensions, points of terminus and vista, [historical allusions, context and "place-making"]. Devices include special elements such as fountains, columns and statuary, allées of trees and other landscaping, paving and infill buildings [drawn in heavy black lines. Relevant social, economic, political, and infrastructure concerns are accounted for herein.]

III. THIS PLAN, representing a substantial consensus among many professionals [and enlightened public] about the essence of urban fabric, is

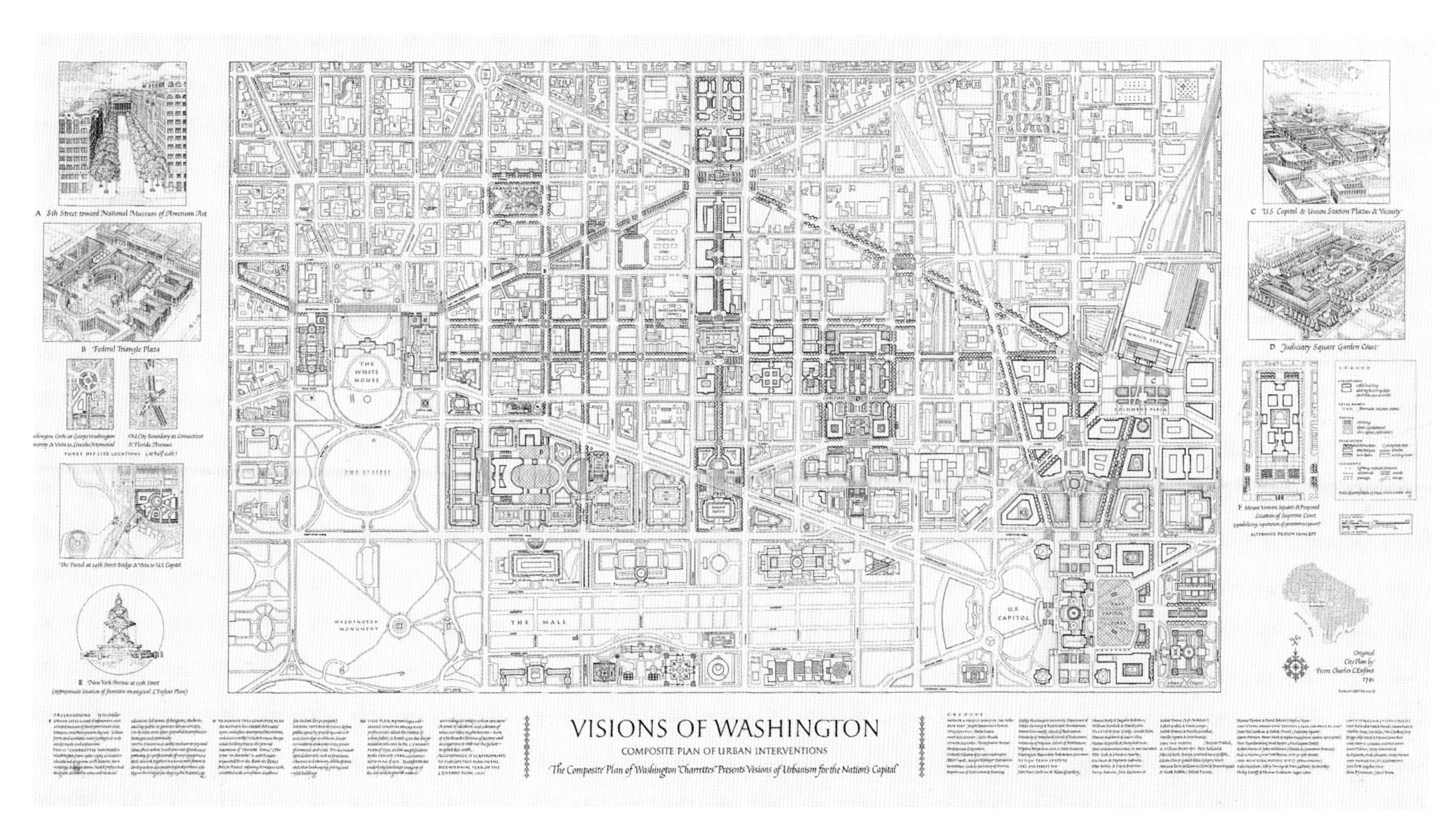

A 8th Street toward National Museum of American Art
B Federal Triangle Plaza
THE WHITE HOUSE
WASHINGTON MONUMENT
THE MALL
U.S. CAPITOL
UNION STATION
COLUMBUS PLAZA
C U.S. Capitol & Union Station Plazas & Vicinity
D Judiciary Square Garden Court
E New York Avenue at 15th Street
VISIONS OF WASHINGTON
COMPOSITE PLAN OF URBAN INTERVENTIONS
The Composite Plan of Washington "Charrettes" Presents Visions of Urbanism for the Nation's Capital

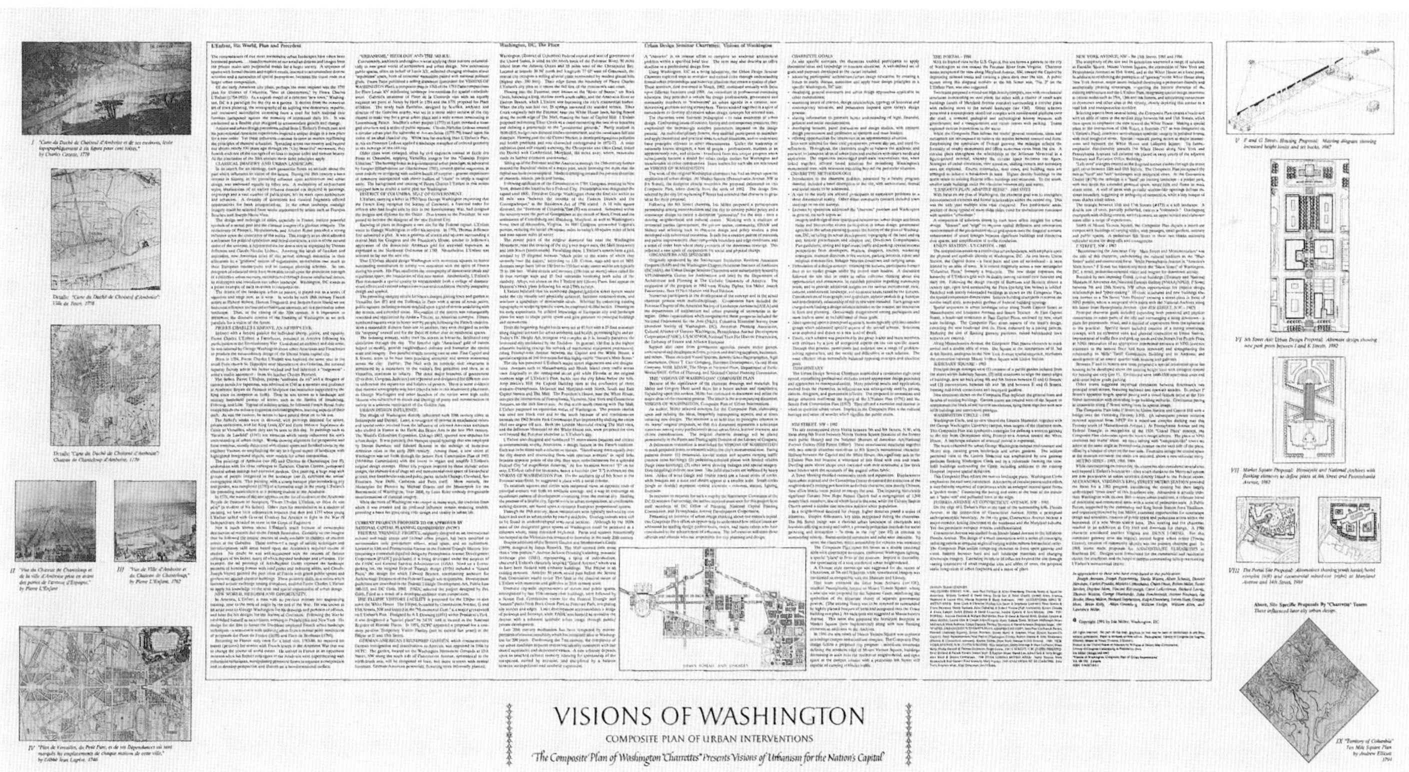

VISIONS OF WASHINGTON
COMPOSITE PLAN OF URBAN INTERVENTIONS
The Composite Plan of Washington "Charrettes" Presents Visions of Urbanism for the Nation's Capital

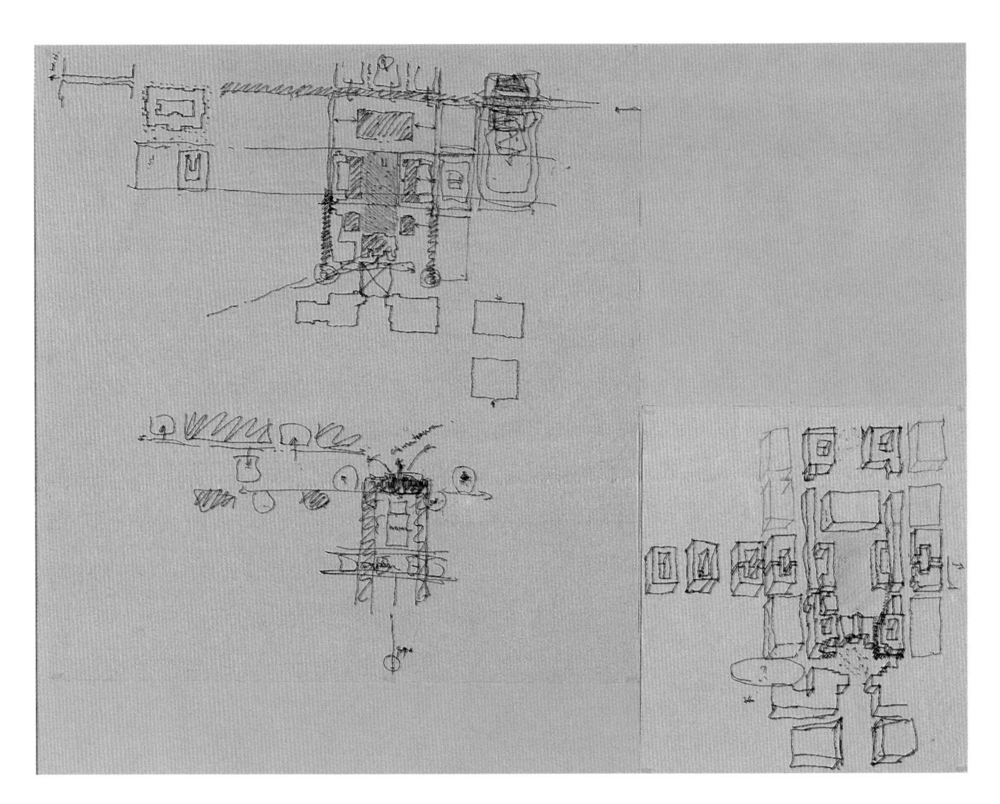

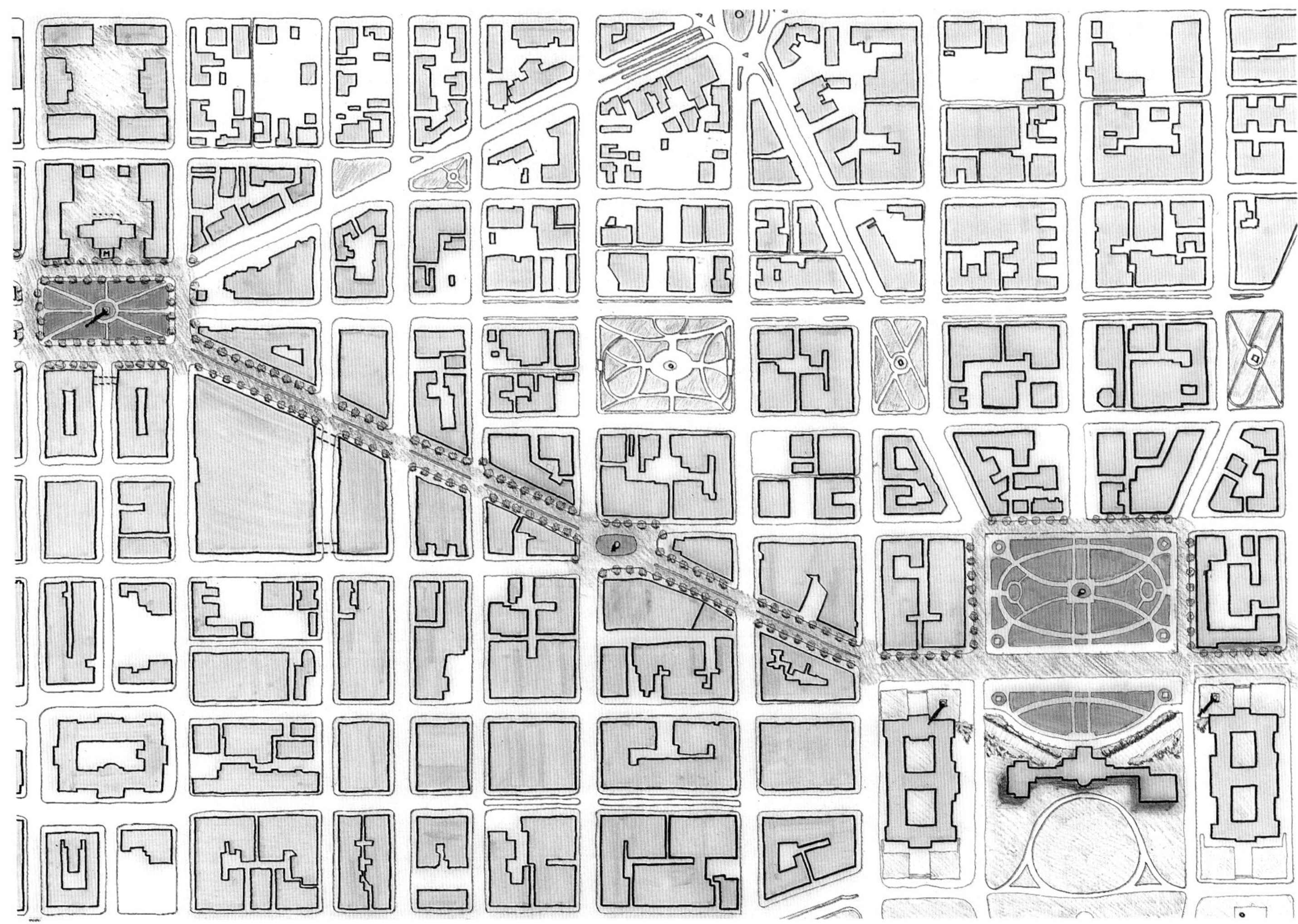

based upon the design tradition inherent in the L'ENFANT PLAN of 1791, and its amplification by the SENATE PARK COMMISSION PLAN of 1901. It confronts underlying landscape imagery of the city and its growth without unraveling its unique urban structure. A sense of idealism and a dream of what our cities might become—born of a faith in the lessons of history and an eagerness to embrace the future—inspired this work. ACCORDINGLY, IT IS APPROPRIATE TO PUBLISH THIS PLAN IN THE BICENTENNIAL YEAR OF THE L'ENFANT PLAN, 1991.

DESIGN TEAM LEADERS: 1982 8TH St,NW—Jean Paul Carlhian, Allan Greenberg; Thomas Beeby, Jaquelin Robertson; William Turnbull, David Lewis; David Lee, Peter Gisolfi; Gerald Allen, Frances Halsband, Laurie Olin; Marius Reynolds, Harry Robinson. 1983 Alexandria King St Metro—Peter Cook, Christine Hawley; Lou Sauer, Stephanie Ledowitz; Peter Bohlin, Frank Bauman; Danny Samuels, John Casbarian, Robert Timme; Robert Geddes, Frank Lawyer; Judith Dimaio, David Levanthal; Neville Epstein, Ann Munley. 1984 The Portal—Antoine Predock, William Bechhoefer; Peter Rolland, Harold Roth; Donlyn Lyndon, Mary Griffith; Laurie Olin, Gerald Allen; Sussana Torre, William McDonald; Ulrich Franzen, Thomas Thorton, David Acheson; 1985 "5" Sites, in Urban Core—Steven Peterson, Steven Hurtt, Stephen Muse (Union Station/U.S. Capitol); Jean Paul Carlhian, Patrick Pinnell (Judiciary Square); Peter Papademetriou/Neal Payton (Washington Circle); Robert Hanna, John Weibenson (Florida-Connecticut Avenues); Anders Nierim (New York Avenue). 1986 New York Avenue—Roger Lewis; Milo Meachum; Sibley Jennings, Marc La Pierre; Phillip Esocoff, Thomas Eichbaum. 1987 F Street—Peter Rolland, Patrick Pinnell; Steven Hurtt, Stephen Muse; David Lee, John Clark, Jerry Briggs; Alex Ward, Dennis Carmichael. 1988 Dunn Loring Metro—Samir Younis; Mary Konsoulis, Rod Garrett; Fred Schwartz; Mary Warner. 1989 Anacostia/St. Elizabeths—John Torti; Stephen Muse; Alan Dynerman; Jan O'Brien.

Below, four representative projects are described; vignette drawings are shown only for Judiciary Square and New York Avenue. The Library of Congress, Geography and Map Division, website will include this map along with the nearly 1,000 Charrette drawings.

1) 1982—8th Street, N.W., Jean Paul Carlhian and Allan Greenberg: This historic 8th Street axis, midway between the Capitol and White House, interpenetrated the monumental core and a residential neighborhood. Merging mixed-uses of housing, civic structures and pedestrian/street-oriented commerce in a serene "room in the city," the scheme produced spontaneity in an ordered urban setting.

2) 1985—Union Station/U.S. Capitol, Steven Peterson, Steven Hurtt, Stephen Muse: Emphasis upon Washington's physical and symbolic identity envisioned as a continuous urban landscape with spatial sequences marking arrival, gateway, procession, and reciprocal vistas. The Capitol dome, icon of nationhood, was the focal point framed by an allee of trees. Spatial composition was defined by grand public plazas, semi-public garden courtyards, and tree-lined streets, all with a sense of enclosure and prescribed edges, representing the duality of national civic function and the daily life of a town.

3) 1985—Judiciary Square, (opposite, top) *Jean Paul Carlhian and Patrick Pinnell: A grand public courtyard isolated from the street, such as Palais Royal in Paris, was created within Judiciary Square. The surrounding buildings, set back from the street, were offset by the addition of slim new corresponding structures to complete the "street wall." Nearby office buildings were retrofitted with a pattern of inviting mid-block connections to smaller courtyards and passages.*

4) 1986—New York Avenue, N.W., (opposite, bottom) *Roger K. Lewis: Reinforcing the notion of "gateway," this prominent avenue leading to the White House, recaptured its historic character. Revisiting the L'Enfant Plan, its stately allee of trees framed a majestic fountain at the 13th Street intersection. Emphasis was upon articulation of critical nodes, directionality and reciprocal views, shifting centers of focus, distinctive hard and softscape, and strong linkages to downtown. Pennsylvania Avenue, in the blocks connecting Lafayette Square and the White House, received special treatment, as a symbolic gesture to honor the president's residence and office.*

POWER OF AN ENLIGHTENED STRATEGY

THE KRIER VISION

TITLE: The Completion of Washington, DC-Master Plan for the Bicentennial Year 2000
DATE DEPICTED: Future projection for the Year 2000
DATE ISSUED: 1985
CARTOGRAPHER: Leon Krier
PUBLISHER: Museum of Modern Art, New York
Manuscripts: 1. Plans, ink and crayon on vellum, 96.8 x 74.9 cm, 38.5 x 29.5 in;
2. Plan, ink, colored pencil and crayon on vellum, 57.2 x 107.3 cm, 22.5 x 42.25 in;
3. Aerial perspective, ink, colored pencil and crayon on vellum 76.2 x 106.7 cm, 30 x 42 in
Museum of Modern Art of New York, 1. 397.85; 2. 396.95; 3. 395.95

An ideal city masterplan, Leon Krier states, "is both of a topographical and moral nature ... satisfying all material and spiritual needs within walking distance ... independently of its architecture, people and vegetation .... Conversely, not even the worst intent or taste will be able to erase the beneficial order of a good masterplan."[1]

According to Krier: "Washington's monumental core is but an outline sketch of a great city to be, a grand skeleton with noble limbs but little flesh. In those hundreds of empty acres I see but an unfinished canvas, an incomplete portrait which craves for completion. The seed of this inspiring national project was placed by its founders and past builders; we must not leave it to rest before it has borne fruit."[2]

Leon Krier, architect and philosopher, was commissioned by the Museum of Modern Art, New York City, to create a visionary plan for the bicentennial of Washington, D.C. Krier's idealistic plan, with text manuscript and drawings, is an indictment against twentieth-century planners who have "mutilated L'Enfant's symbolic, topographical and social vision."[3]

Replete with symbolism and metaphor associated with the District of Columbia, the plan reflects the density of European "urbanisme" of its author. Few United States cities share the communal polis proposed here. Throughout—central business district, monumental core, even The Mall—Krier's imposed building infill creates a critical mass to define neighborhoods. Contrasting with Washington's existing mono-centric configuration which consumes exhaustive hours in commuting, Krier envisions sixteen cities within the city—pedestrian-oriented neighborhoods with "poly-functional-centers" surrounding "The Federal City."

Four independent towns, each the size of Georgetown, will comprise the federal city—a city of 80,000 residents. "All the amenities, comforts and pleasures of a national and international Capital [will be provided] ... tree-lined streets, squares and gardens, and within walking distance of [people's] professional activities."[4] The National Gallery's cornice line will set the height limit for all non-public buildings. A federal development agency will enforce guidelines and ensure the sale of non-public, newly divided blocks—using revenue earned to enhance and adorn public architecture, gardens and spaces.

Modeled after Venice, a gradually narrowing Grand Canal shall traverse The Mall from Washington Monument to an immense "Constitution Square" below the Capitol. Arcades, plantings and varied facades will result in richly articulated elevations along The Mall. A network of public passages, streets and squares will penetrate the Federal Triangle. "Pyramid Lake," newly created (named for a proposed monument), seizes much of the public recreational parkland, enlarging the Tidal Basin to link with the Reflecting Pool fronting the Lincoln Memorial.

Adding a screen of buildings to surround the White House (north, east, west) the breadth of the Ellipse and President's Grounds are thus diminished to focus upon the symbolic prominence of the executive domain. Yet for Americans, the existing setting is distinctive. Its prominence is consistent with the stature of this residence in this gardenesque urban capital.

Chiding L'Enfant for merely mentioning, in passing, a Supreme Court site, Krier laments

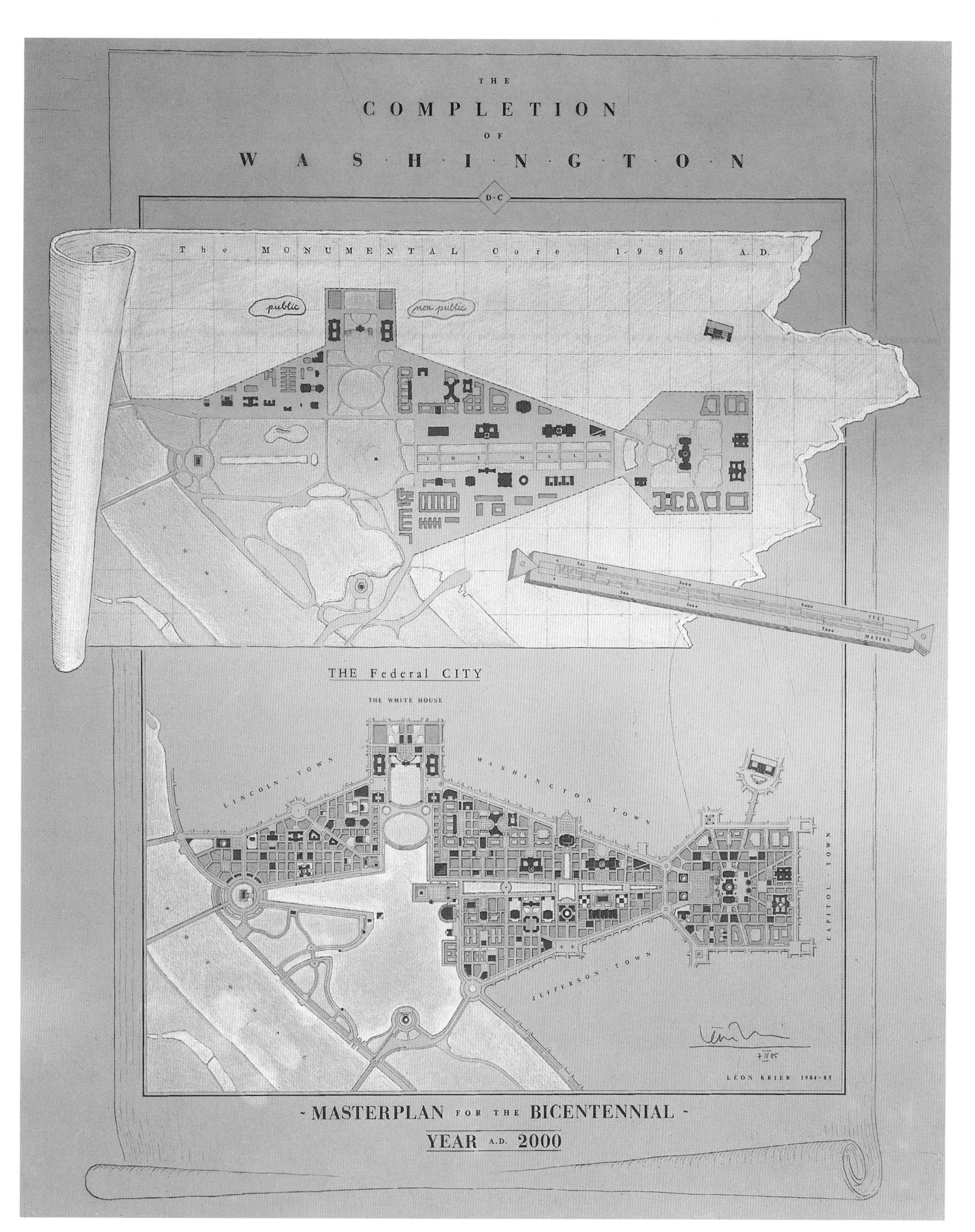
THE
COMPLETION
OF
W A S · H · I · N · G · T · O · N
D·C
The MONUMENTAL Core 1-985 A.D.
public
non public
THE MALL
FEET
METERS
THE Federal CITY
THE WHITE HOUSE
LINCOLN · TOWN
WASHINGTON TOWN
JEFFERSON · TOWN
CAPITOL · TOWN
LÉON KRIER 1984-85
~ MASTERPLAN FOR THE BICENTENNIAL ~
YEAR A.D. 2000

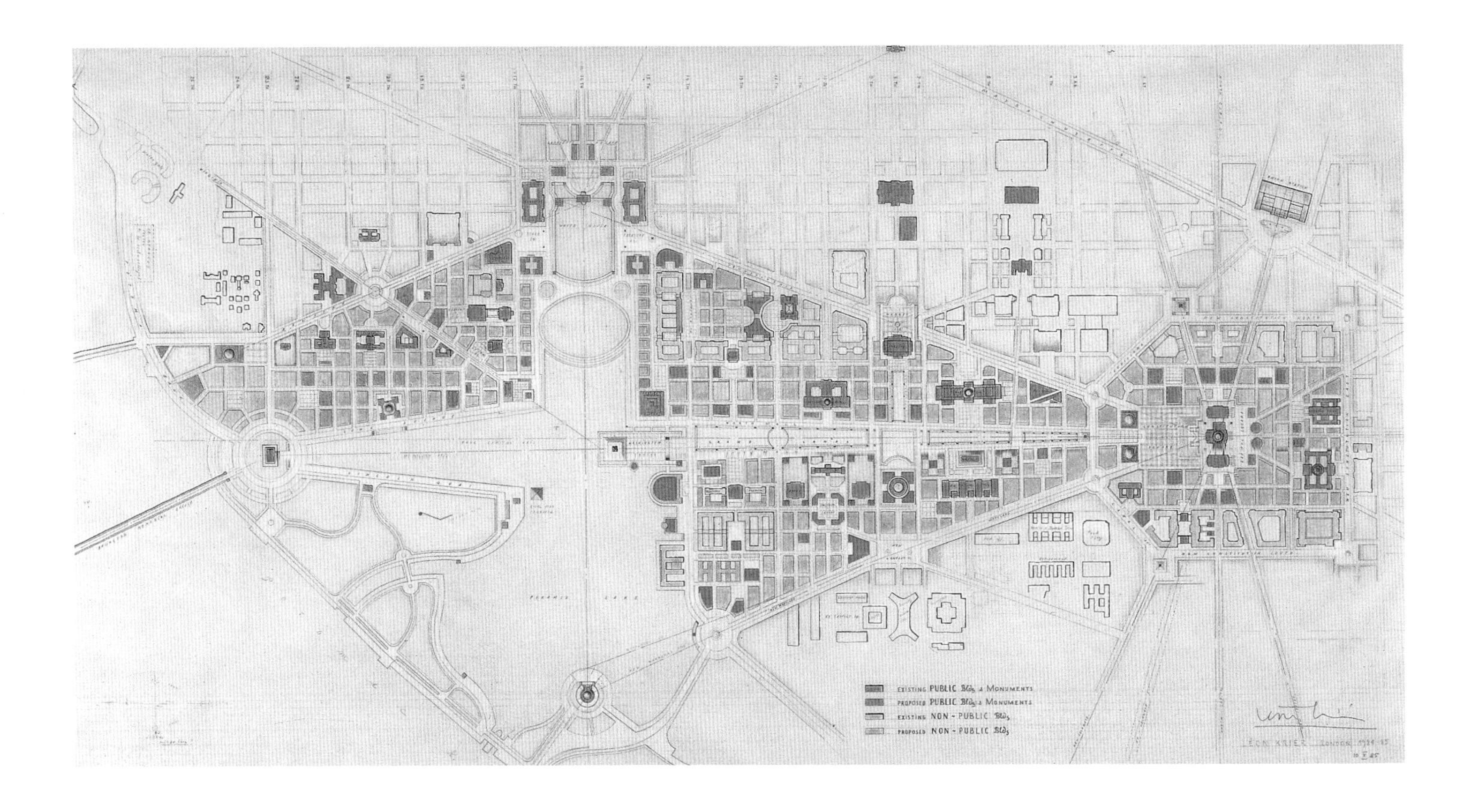

the missed opportunity of a symbolic location for the co-equal third branch of government. He suggests per chance a lost opportunity in the sites now occupied by the Jefferson and Lincoln Memorials.

In his quest to signify esteem for this City of Washington, its plan, and the symbolic presence of democratic values, Krier's design offers a strong point of departure for debate and instruction about the public realm. At the beginning of the twenty-first century, emphasis upon neighborhoods has become the highest priority of city government.

Notwithstanding its obvious merits, Krier's plan has its shortcomings. Who would want to fill The Mall and adjacent parkland with buildings and water? Surely not anybody who has been privileged to be among the crowd on Independence Day, or has played Frisbee or ball within the back drop of the Capitol, or has taken a turn on the Carousel. Surely, not the many whose sports teams compete nightly; nor families who come to picnic or stroll; nor those who attend grand festivals or rallies to address national grievances and concerns.

Washington "is destined to become the criterion for the rebirth of urban life ... for our

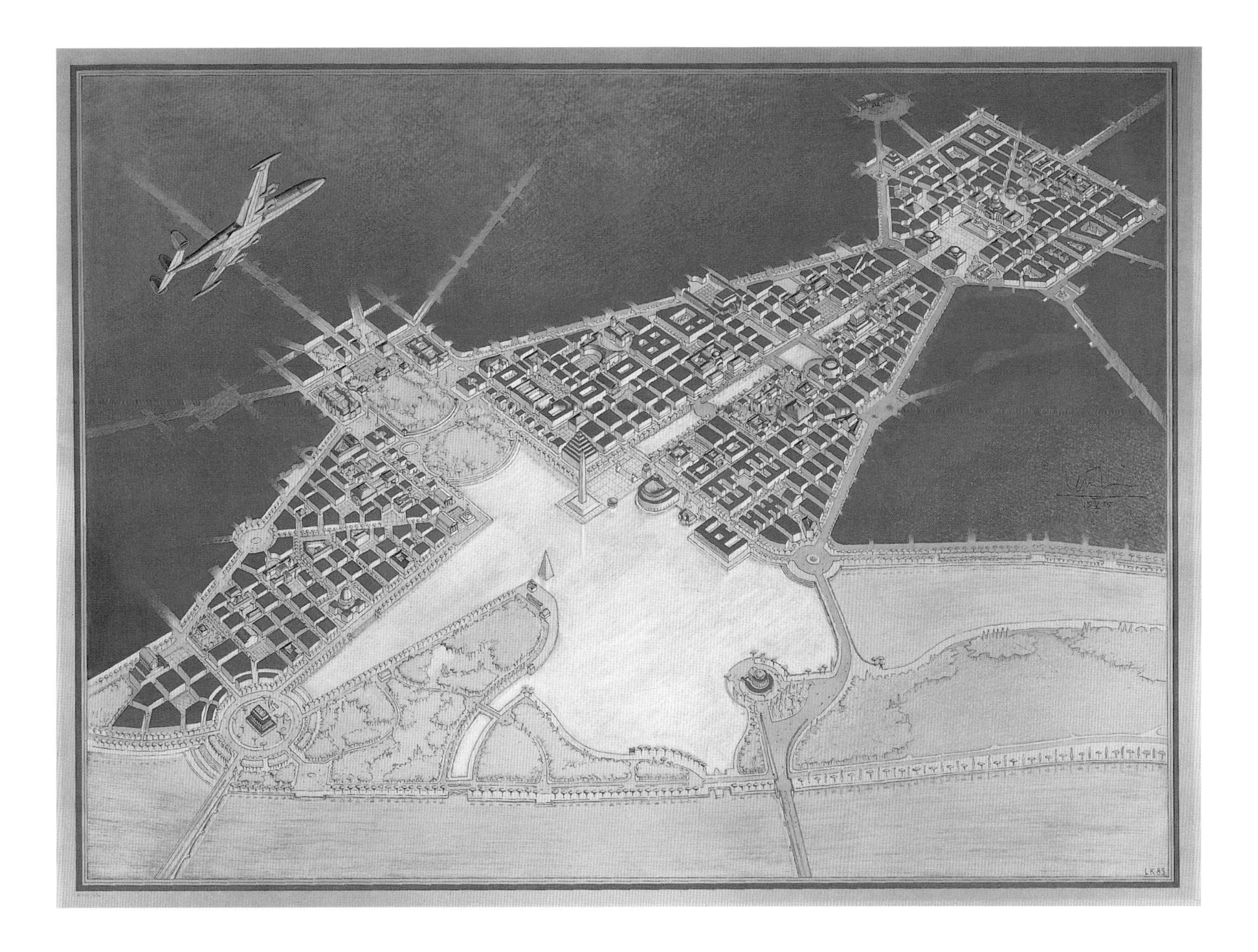

children, the ultimate urban paradigm."[5] Leon Krier affirms that a homeland is its people and their history, all that "our senses will embrace," pride in our hearts, memories of splendid landscapes and cities imparted to coming generations. "My generation is faced with a colossal and almost inhuman task of global ecological reconstruction ... of disasters [greater] than it is able to solve ... [We must] cut environmental problems down to manageable size [through better] town planning and architecture ... and use of resources and time."[6]

Notes: 1-6 in Archives d'Architecture Modern, MCMLXXXVI, N30 1986, Brussels: 1, p.18; 2, p.17; 3, p.19; 4, p.20; 5, p.26; 6, p.13.

SATELLITE TECHNOLOGY: A NEW ORDER

THE SPOT METROVIEW OF WASHINGTON

TITLE: SPOT Satellite Image of Washington, DC
DATE DEPICTED: Image 1: October 15, 1999; Image 2: 2000 Composite
DATE ISSUED: 1999; 2000
CARTOGRAPHER: SPOT Image Corporation ©
PUBLISHER: CNES 2000
Satellite Photograph K,J:623, 272, GeoTIFF@360dpi, SPOT Image Corporation
Courtesy SPOT Image Corporation ©

People dream of seeing the earth from above, of mastering the elusive skill of birds. This has probably always been part of the human psyche. For centuries, artists drew often charming, but imprecise aerial perspectives. With the invention of the camera, 3-dimensional models were built from which to photograph a scene. Kites, and later balloons, were sent airborne with cameras attached. Satellite photography brought a new level of technology to aerial information-gathering capability.

Reconnaissance needs in World War I spurred rapid advancement in the development of aerial photography equipment, and thousands of images were taken daily. In World War II, technological advancement allowed for the taking of millions of images. Advancement continued during and in the wake of the Korean War. Today, the U.S. National Archives contains the largest public aerial photography collection in the world.

When used for military purposes, aerial photography can measure water depths, locate underwater objects, determine building heights, identify types and quantities of vegetation. It can also serve civilian purposes in archeology, geology, meteorology, urban and road planning, traffic analysis, and cultural and forestry surveys. Aerial photography (at one-meter resolution) is especially effective in showing details of a small area, and in charting new areas to update old maps. Among recent Washington aerial photographs taken for planning and development purposes were the 1981 and 1991 flyovers as a cooperative venture by National Capital Planning Commission and D.C. Department of Public Works.

In the 1960s a new kind of photography was developed to complement aerial photography—remote sensing, or satellite photography. Satellite photography (at ten-meter resolution) covers a broader range than more conventional technology, and can detect imperceptible environmental conditions not otherwise observable. Use of satellite photography rather than aerial photography thus depends upon project objectives. Formerly protected for military use, satellite images are now available through commercial companies.

Rizzoli's *Manhattan in Maps*, includes an aerial survey map of Manhattan Island from 1921 which appears to be "seamless," but is in fact compiled of one hundred photographs taken at 1000 feet altitude by a continuously operating camera, and later pieced together to create a single image. Only the center of each is truly vertical. In contrast, fewer satellite images are required to create one photograph. For example, a recent Spot satellite image of the Washington area, from Dulles Airport on the west to Annapolis on the east, required only three shots to create one seamless mosaic; its seam size was 37x37 miles.

Photo-maps from space were first taken in the 1960s by NASA satellites. In 1986 SPOT (Systeme Probatiore d'Observation de la Terre) satellite Earth Observation System, designed by CNES (Centre National d'Etudes Spatiales) in France with participation by Belgium and Sweden, launched its first of four Spot satellites. Three of these are currently in operation. Spot has twenty-six ground tracking stations around the globe. Spot gained attention for its detailed views of Chernobyl nuclear reactor accident that same year.

SPOT features high resolution, stereo imaging, and revisit capability enabling data collection for various applications such as cartography, agriculture, environment, land use, land cover, and ozone. "SPOT 4" maintains a polar axis, circular orbit on a 26-day cycle, with a nominal orbital period of 101.4 minutes to complete an orbit. Its pixel size is 20x20 meters (multi-spectral) and 10x10 meters (panchromatic). Both spectral modes can operate simultaneously or individually. Its swath width for vertical viewing is 60 kilometers (37miles); for oblique viewing the angle is adjustable through +/-27 degrees relative to vertical and can be used to increase the frequency, which varies with latitude.

Unique to SPOT is its location grid reference system, which located this Washington image at k,j:623-272 (k refers to longitude 623 and j refers to latitude 272). At the equator, SPOT's orbital altitude is 822 kilometers (511

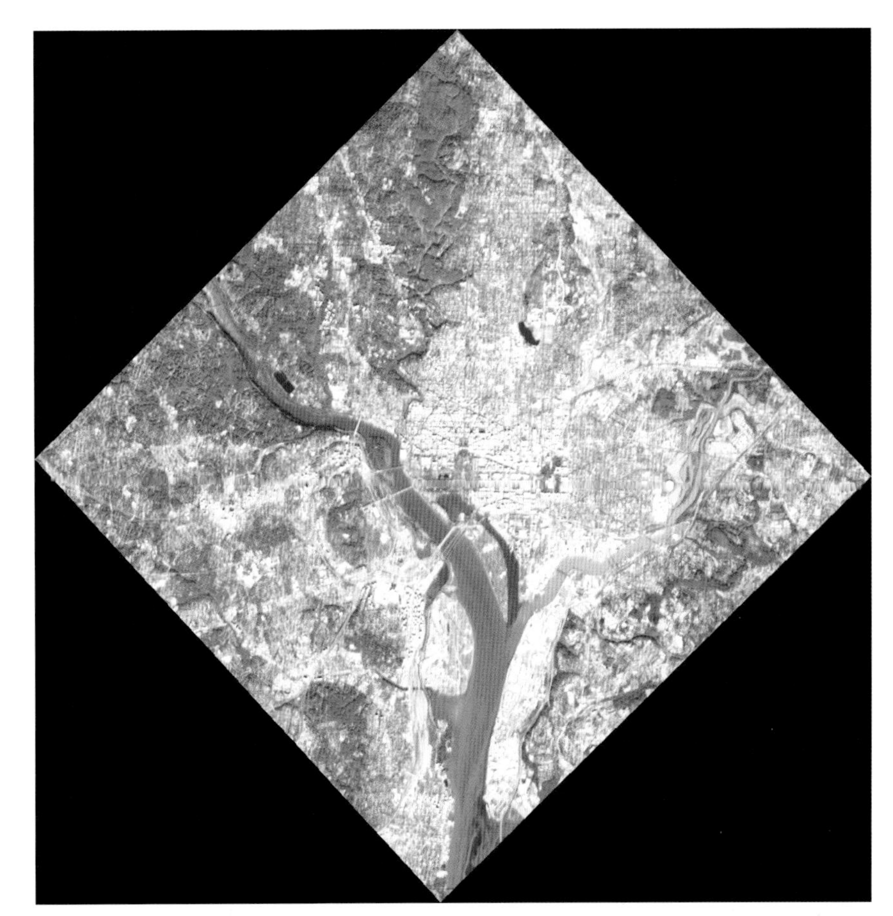

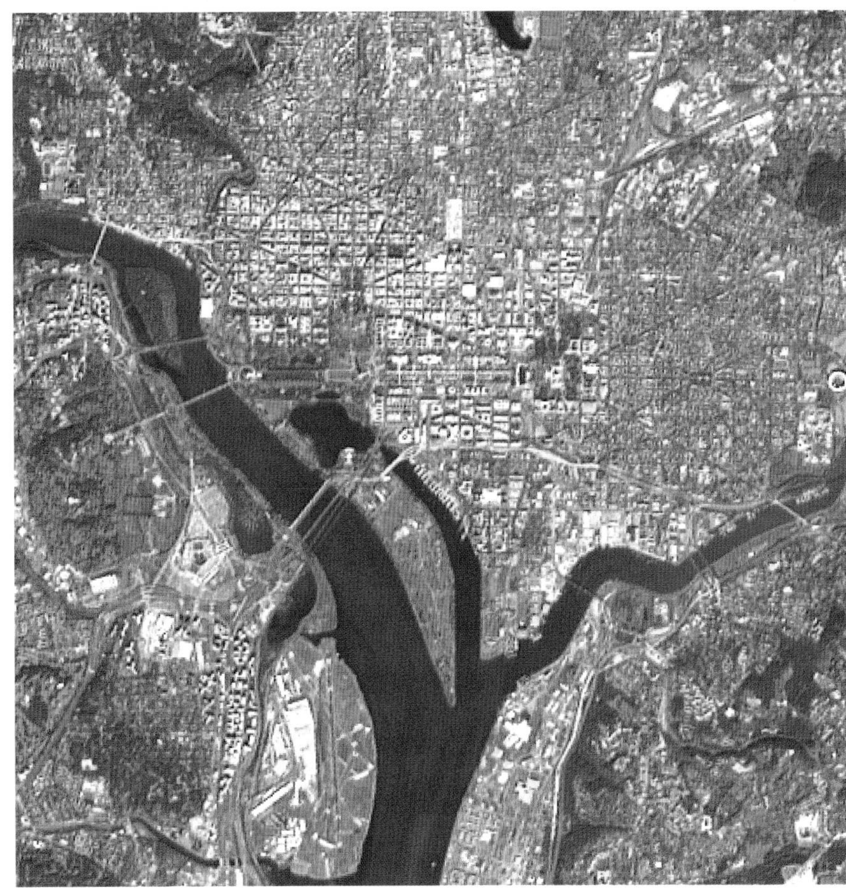

miles); at Washington, D.C. it is 828 kilometers (515 miles). The spectral bands of the file are displayed as 421 in RGB, referring to red (4), green (2), blue (1). Colors are based upon reflectance value. For example, water absorbs most of the near-infrared; its reflectance value is high, therefore, it appears as black. Areas seen as red are actually green. This is called "false color" because colors are not true to ground colors. Color distortion results from images passed through color filters. Each wave band carries different information. Near-infrared band is not visible to the eye.

What can be seen in this image? SPOT's high-resolution cameras can recognize details, thus enabling us to see equally the density of urban fabric, patterns of streets and open space, and individual buildings. The L'Enfant Plan is here clearly articulated—a gardenesque urban landscape plan with a hierarchy of open spaces, a repetitive pattern of squares, and a relaxed grid of streets, bisected by diagonal avenues.

It is thrilling to see the parks and significant buildings from above. The Mall, President's Park, and parkland surrounding the Capitol as well as other parks, basins, bridges, and notable buildings (the Capitol, the White House, the Kennedy Center, Union Station, various sports facilities, the Pentagon, the National Airport, etc.) are held in the arms of a tree-like shape of the two rivers—the Potomac and the Anacostia—proudly heralding our nation's capital.

SELECTED BIBLIOGRAPHY

Adams, William Howard. *The Eye of Thomas Jefferson.* Exhibition catalogue. Charlottesville: University of Virginia Press, 1981.

Alphand, Adolphe. *Les Promenades de Paris.* Paris: J. Rothchild. New York: Princeton Architectural Press, reprint 1984.

Applewhite, E.J. *Washington Itself: An Informal Guide to the Capital of the United States.* New York: Alfred A. Knopf, 1983.

Augustyn, Robert T., and Paul E. Cohen. *Manhattan in Maps: 1527- 1995.* New York: Rizzoli, 1997.

Bennett, Edward H., and Daniel H. Burnham. *Plan of Chicago.* ed. Charles Moore. Commercial Club of Chicago, 1908. New York: Princeton Architectural Press, reprint 1993.

Beveridge, Charles E., and Paul Rocheleau. *Frederick Law Olmsted: Designing the American Landscape.* New York: Rizzoli 1998.

Bowling, Kenneth R., *The Creation of Washington, D.C.: the Idea and Location of the American Capital.* Fairfax, Virginia: George Mason University Press, 1991.

Boyd, Julian P., ed. *"Fixing the Seat of Government." The Papers of Thomas Jefferson.* 20: 3-72. Princeton: Princeton University Press, 1982.

Bramucci, Nancy M., Robert J.H. Janson-La Palme, Russell Morrison, and Edward Papenfuse. *On the Map: Maryland and the Chesapeake Bay.* Chestertown: Washington College, 1983.

Brown, Glenn, comp. *Papers Relating to the Improvement of the City of Washington, District of Columbia.* Washington: 56th Congress, 2nd Session, Senate Document 94, 1901.

Bryan, Wilhelmus Bogart. *A History of the National Capital from its Foundation Through the Period of the Adoption of the Organic Act.* 2 vols. New York: Macmillan Co., 1914-16.

Caemmerer, Paul H.The *Life of Pierre Charles L'Enfant, Planner of the City Beautiful: The City of Washington.* Washington: National Republic Publishing Co., 1950.

Caemmerer, Paul H. *Washington: The National Capital.* Senate Document No. 332, 71st Congress, 3rd Session. Washington: U.S. Government Printing Office, 1932.

Crew, H. W. *Centennial History of the City of Washington, D.C.* Dayton, Ohio: W. J. Shuey, 1892.

Cox, William V., comp. *Celebration of the One Hundredth Anniversary of the Establishment of the Seat of Government in the District of Columbia.* Washington, D.C.: Government Printing Office, 1901.

Davis, Timothy. "Rock Creek and Potomac Parkway, Washington, DC: the Evolution of a Contested Urban Landscape." *Studies in the History of Gardens and Designed Landscapes.* Volume 19:123-257, Number 2, Summer. Washington and London: Taylor and Francis, 1999.

Dennis, Michael. *Court and Garden.* Cambridge, Mass: MIT Press, 1986.

Downing, Andrew Jackson. "Explanatory Notes to Accompany the Plan for Improving the Public Grounds At Washington, DC." Records of the Commission of Public Buildings, Letters Received, RG42LR, vol. 32. Washington: National Archives, March 3, 1851.

Downing, Andrew Jackson. *A Treatise on the Theory and Practice of Landscape Gardening Adapted to North America.* New York, 1850.

Ehrenberg, Ralph E. "Mapping the Nation's Capital: The Surveyor's Office, 1791- 1818." *The Quarterly Journal of the Library of Congress.* ed. Frederick Mohr. Volume 36, Number 3, Summer. Washington: Library of Congress, 1979. PG.279

Ehrenberg, Ralph E. *Scholars' Guide to Washington, D.C. for Cartography and Remote Sensing Imagery.* ed. Zdenek V. David. Washington: Smithsonian Institution Press, 1987.

Fitzpatrick, John C., ed. *The Diaries of George Washington, 1748-1799.* 4 volumes. Boston: Houghton Mifflin Co., 1925.

Friis, Herman R. *Guidebook: Geographical Reconnaissance of the Potomac River Tidewater Fringe of Virginia from Arlington Memorial Bridge to Mount Vernon.* Washington: Association of American Geographers, 1968.

Fronceck, Thomas, ed. *The City of Washington: An Illustrated History.* comp. The Junior League of Washington. New York: Alfred A. Knopf, 1977.

Goode, George Brown, ed. *The Smithsonian Institution 1846-1896, The History of the First Half of the Century.* Washington: Smithsonian Institution, 1897.

Goode, James M. *Capital Losses: A Cultural History of Washington's Destroyed Buildings.* Washington: Smithsonian Institution Press, 1979.

Goode, James M. *The Outdoor Sculpture of Washington: A Comprehensive Historic Guide.* Washington: Smithsonian Institution Press, 1974.

Green, Constance M. *Washington: Capital City, 1879-1850.* Princeton: Princeton University Press, 1963.

Greenberg, Allan. *George Washington: Architect.* London: Andreas Papadakis, 1999.

Gutheim, Frederick, and Wilcomb E. Washburn. *The Federal City: Plans and Realities.* Washington: Smithsonian Institution Press with National Capital Park and Planning Commission 1976.

Gutheim, Frederick. *Worthy of a Nation: The History of Planning for the National Capital.* Washington: Smithsonian Institution Press, 1977.

Hegemann, Werner, and Elbert Peets. *The American Vitruvius: An Architects' Handbook of Civic Art.* New York: The Architectural Book Publishing Co., 1922. New York: Princeton Architectural Press, reprint 1988.

Hodges, Allan A., and Carol A. Hodges. *Washington on Foot.* Washington: American Institute of Planners, 1976.

Jennings Jr., J. L. Sibley. "Artistry as Design: L'Enfant's Extraordinary City." *The Quarterly Journal of the Library of Congress.* ed. Frederick Mohr. Volume 36, Number 3, Summer. Washington: Library of Congress, 1979.

Kite, Elizabeth S. *L'Enfant and Washington, 1791-1792.* Baltimore: Johns Hopkins University Press, 1929.

Konvitz, Joseph W. *Cartography in France, 1660-1848: Science, Engineering, and Statecraft.* Chicago: University of Chicago Press, 1987.

Kostov, Spiro. *The City Shaped: Urban Patterns and Meanings Through History.* Boston: Little, Brown & Co., 1991.

Krier, Leon. "Completion Achevement de Washington, D.C., Bicentennial Masterplan for the Year 2000 Plan Directeur General Pour le Bicentenaire." No. 30. *Brussells: Aux Archives d'Architecture Moderne,* 1986.

Laine, Christian K., and Iris Miller. *The Plan of Washington, D.C., Leon Krier and the City Beautiful Movement.* Exhibition catalogue. Washington: American Architectural Foundation, 1987.

Laugier, Marc-Antoine. *Essaie sur l'architecture, and Observations sur l'architecture.* trans. Hermann, Wolfgang and Annis. Los Angeles: Hennessey and Ingalls, 1997.

Lindall, William. *Origin and Government of the District of Columbia.* Washington, D.C.: Government Printing Office, 1908.

Longstreth, Richard, ed. *The Mall in Washington, 1791-1991.* CASVA Symposium Papers XIV, Studies in the History of Art, No.30. Washington: National Gallery of Art, 1991.

Maddox, Diane. *Historic Buildings of Washington, D.C.* Pittsburgh: Ober Park Associates, 1973.

Miller, Iris. *L'Enfant's Plan: Visions of Washington.* Exhibition Catalogue. Washington: Charles Sumner School Museum and Archives, 1992.

Moore, Charles ed. *The Improvement of the Park System of the District of Columbia.* Washington: 57th Congress, 1st Session, Senate Report 166, 1902.

Moore, Charles. *Daniel H. Burnham: Architect, Planner of Cities.* Volumes I & II. Boston: Houghton Mifflin Co., 1921.

Nichols, Frederick D. and Ralph E. Griswold. *Thomas Jefferson, Landscape Architect.* Charlottesville: University of Virginia Press, 1978.

Padover, Saul K., ed. *Thomas Jefferson and the National Capital.* Washington: Government Printing Office, 1946.

Partridge, William T. *L'Enfant's Methods and Features of His Plan for the Federal City.* Washington: National Capital Parks and Planning Commission, 1930.

Patte, Pierre. *Plan of Selected Competition Projects for Place Louis XV, Paris 1767.* Musee du Carnavalet, Paris.

Reed, Robert. *Old Washington, D.C. in Early Photographs: 1846- 1932.* New York: Dover Publications, 1980.

Reps, John W. *The Making of Urban America: A History of City Planning in the United States.* Princeton: Princeton University Press, 1965.

Reps, John W. *Monumental Washington: The Planning and Development of the Capital Center.* Princeton: Princeton University Press, 1967.

Reps, John W. *Tidewater Towns: City Planning in Colonial Virginia and Maryland.* Williamsburg: Colonial Williamsburg Foundation, 1972.

Reps, John W. *Washington on View: The Nation's Capital Since 1790.* Chapel Hill: University of North Carolina Press, 1991.

Rice, Howard C. *Thomas Jefferson's Paris.* Princeton: Princeton University Press, 1976.

Rice, Howard C and Anne S. K. Brown, ed. and trans. *The American Campaigns of Rochambeau's Army 1780, 1781, 1782, 1783.* Princeton: Princeton University Press, 1972.

Ristow, Walter W. "Aborted American Atlases." *The Quarterly Journal of the Library of Congress.* ed. Frederick Mohr. Volume 36, Number 3, Summer. Washington: Library of Congress 1979. PG320

Rose Jr., Cornelia Bruere. "The Alexandria Canal." *Arlington County Virginia: A History.* pp. 78-79. Arlington Historical Society, 1976.

Slawson, Allan B. *A History of the City of Washington, Its Men and Institutions.* pp.1-148. Washington, D.C.: Washington Post Co., 1903.

Spreiregen, Paul D., ed. *On the Art of Designing Cities: Selected Essays of Elbert Peets.* Cambridge, Mass.: MIT Press, 1968.

Stephenson, Richard W. *"A Plan Whol(l)y New": Pierre Charles L' Enfant's Plan of the City of Washington.* Washington: Library of Congress, 1993.

Stephenson, Richard W. "The Delineation of a Grand Plan." *The Quarterly Journal of the Library of Congress.* ed. Frederick Mohr. Volume 36, Number 3, Summer. Washington: Washington Library of Congress, 1979.

Taft, William Howard. "Washington: Its Beginning, Its Growth, and Its Future." *National Geographic Magazine.* Volume XXVII, Number 3. Washington: National Geographic Society, March 1915.

Walton, Thomas, dissert. *The 1901 McMillan Commission: Beaux Arts Plan for the Nation's Capital.* Washington: The Catholic University of America, 1980.

White, George, FAIA. *The Master Plan for the United States Capitol.* Washington: The Architect of the Capitol, 1981.

Wiebenson, Dora. *Picturesque Garden in France.* Princeton: Princeton University Press,1978.

Washington History, Magazine of the Historical Society of Washington, as of 1989. Previously *Records of the Columbia Historical Society,* Vol. 1-12

AUTHOR

IRIS MILLER

Iris Miller maintains a practice in landscape architecture and urban design in Washington, D.C., where she is Director of Landscape Studies in the School of Architecture and Planning at The Catholic University of America. She specializes in park and city design, and consults and lectures on urban landscape and planning for cities in the United States, France, Japan, and China. She has served as curator for Washington map exhibitions and has written widely on urbanism. She has served on the National Fulbright Scholarship Selection Jury and was founder of numerous innovative programs.

In 1995 she received a grant from the Government of France to research Pierre Charles L'Enfant and the inspiration for the Washington plan. She has also received four grants-in-aid to support the writing of this book from The Catholic University of America, the U.S. Department of Interior Historic Preservation Fund/D.C. State Historic Preservation, the University of Maryland, and Washington Map Society.

CONTRIBUTING AUTHORS

TIMOTHY DAVIS

Timothy Davis is an historian with the National Park Service's Historic American Engineering Record. He received an A.B. from Harvard College's Visual and Environmental Studies Program and a Ph.D. in American Studies from the University of Texas at Austin. A recipient of fellowships from the National Gallery of Art, the Smithsonian, and Dumbarton Oaks, he has written on the evolution of parks, parkways, and other cultural landscapes.

HERBERT M. FRANKLIN

Herbert M. Franklin is former Administrative Assistant to the Architect of the Capitol and also represented the Architect on the District of Columbia Zoning Commission and the National Capital Memorial Commission. A graduate of Harvard College and Harvard Law School, he is a founding trustee of the National Building Museum in Washington, D. C.

LUCINDA PROUT JANKE

Lucinda Prout Janke has a B.A. in Political Science from Wellesley College and an M.A. in Museum Studies from George Washington University. She is Curator, The Kiplinger Washington Collection. She is a board member of the Historical Society of Washington, D.C., and a member of its City Museum Task Force. She was a former board member of the Congressional Cemetery Association. Janke is co-author with Ruth Ann Overbeck of "William Prout—Capitol Hill's Community Builder," in *Washington History,* Spring/Summer 2000.

CHARLENE DREW JARVIS

Charlene Drew Jarvis, a native Washingtonian and a former research scientist, holds a doctorate in neuropsychology from the University of Maryland. She was appointed as Southeastern University President in July 1996. Dr. Jarvis served for 21 years on the City Council of the District of Columbia, in Washington, D.C., and was Chair of the Committee on Economic Development and President Pro Tempore. Dr. Jarvis is President of the District of Columbia Chamber of Commerce and on the Board of the Greater Washington Board of Trade, chairing its Community Business Partnership Committee. She is the daughter of Dr. Charles R. Drew, noted blood bank pioneer.

AUSTIN H. KIPLINGER

Austin H. Kiplinger, chairman of the Kiplinger Washington Editors, Inc., is former editor of the *Kiplinger Washington Letter* and *Kiplinger's* monthly magazine. A Washingtonian by birth, he has maintained a lifelong interest in the history of the capital city, where his company maintains the Kiplinger Collection of Washingtoniana, which includes many early maps, drawings and engravings of the site selected by George Washington for the original District of Columbia. Mr. Kiplinger serves as co-chairman of the Leadership Committee of the City Museum of Washington, which will open in 2003 in the former Carnegie Library Building at Mt. Vernon Square in the nation's capital.

ROBERT L. MILLER

Washington, D.C. architect, public relations executive, writer, and artist, Robert L. Miller has degrees from Yale University in architecture and architectural history. He worked for public relations firm Hill and Knowlton for six years before starting his own corporate and marketing communications consulting firm, Robert L. Miller Associates, in 1985. His articles on Washington architecture, preservation, and urban design have appeared in *Historic Preservation, The Washingtonian, Progressive Architecture, Architectural Record,* and *The Washington Post.*

WHAYNE S. QUIN

Whayne S. Quin is a partner with the Law Firm of Holland & Knight LLP, specializing in zoning, urban planning, building and housing codes, and historic preservation. In 1980, after ten years of litigation, Mr. Quin's research and legal theory culminated in the landmark D.C. Court of Appeals decision that "original" alleys in the District of Columbia, which had been established around 1791 pursuant to original Deeds of Trust with the City's founders, were titled in the abutting property owners and not in the United States.

INDEX

	Prior to 1775	1800	1825	1850	
HISTORICAL EVENTS	1608 Capt. John Smith sails the Potomac to future site of Washington; creates first area map 1662 First land patent for future site of District of Columbia (DC) granted 1751 Georgetown established 1775 American Revolution begins	1776 Signing of Declaration of Independence 1783 Treaty of Paris Revolutionary War ends Independence of United States of America 1785 Land Grant Ordinance enacted 1788 Virginia and Maryland offer land for development of Federal District 1789 US Constitution ratified; Georgetown University established 1790 Act to establish DC by Congress 1791 Pierre Charles L' Enfant creates Plan of Washington 1800 US Government moves from Philadelphia to Washington Adams first president to occupy White House Stamp Office established *Population 3,000*	1801 Jeferson first president to be inaugurated in Washington Congress designates DC as Federal Territory. 1802 City of Washington charted 1808 Construction begins on Washington Canal. 1812 War of 1812 begins 1814 Great Britan burns White House and Capitol 1815 War of 1812 ends with signing of Treaty of Ghent Washington Canal finished 1820 Washington residents gain right to vote for mayor and city council 1821 George Washington University established	1829 James Smithson of Great Britain funds creation of Smithsonian Institution 1835 B&O Railroad sited on The Mall; National Theatre opens 1846 Retrocession of land from DC to Virginia beyond Potomac River Congress establishes Smithsonian Institution Mexican War begins 1848 Mexican War ends 1850 C&O Canal completed Congress abolishes slave trade in DC *Population 40,000*	1861 Civil War begins 1862 Congress frees all slaves in DC 1865 Lincoln assassinated Civil War ends 1863 Emancipation Proclamation issued 1867 DC granted suffrage Howard University chartered by Congress 1870 Washington Canal closed, moved underground 1871 Georgetown incorporated into DC New Territory form of Government for DC
PRESIDENTS		George Washington, 1789-1797 John Adams, 1797-1801 Thomas Jefferson, 1801-1809 James Madison, 1809-1817 James Monroe, 1817-1825 John Quincy Adams, 1825-1829	Andrew Jackson, 1829-1837 Martin Van Buren, 1837-1841 William Henry Harrison, 1841 John Tyler, 1841-1845 James Knox Polk, 1845-1849 Zachary Taylor, 1849-1850	Millard Fillmore, 1850-1853 Franklin Pierce, 1853-1857 James Buchanan, 1857-1861 Abraham Lincoln, 1861-1865 Andrew Johnson, 1865-1869	Rutherford Birchard Ha James Abram Gar Cheste Ulysses Simpson Grant, 18
PUBLIC SPACE & MONUMENTS		White House (1792-1907) US Capitol (1793-1802) Library of Congress(1800)		Washington Monument (1848-1884)	
PRINCIPAL PLANS/ HERITAGE PLANNING		1791 L'Enfant Plan 1792 Ellicott Plan 1797 Dermott "Tin Case" Map (first official signed map by Pres. Washington)		1841 Mills Plan (The Mall)	1851 Downing Plan (The Mall and Ellipes) 1874 Toner Map: Washington in Embryo (original land owners)